SMP interact

for GCSE mathematics **Higher**

CAMBRIDGE
UNIVERSITY PRESS

PUBLISHED BY THE PRESS SYNDICATE OF THE UNIVERSITY OF CAMBRIDGE
The Pitt Building, Trumpington Street, Cambridge, United Kingdom

CAMBRIDGE UNIVERSITY PRESS
The Edinburgh Building, Cambridge CB2 2RU, UK
40 West 20th Street, New York, NY 10011-4211, USA
477 Williamstown Road, Port Melbourne, VIC 3207, Australia
Ruiz de Alarcón 13, 28014 Madrid, Spain
Dock House, The Waterfront, Cape Town 8001, South Africa

http://www.cambridge.org/

© The School Mathematics Project 2003
First published 2003

Printed in Italy by G. Canale & C. S.p.A., Borgaro T.se, (Turin)

Typeface Minion   *System* QuarkXPress®

A catalogue record for this book is available from the British Library

ISBN 0 521 89022 5 paperback

Typesetting by The School Mathematics Project
Technical illustrations by The School Mathematics Project and Jeff Edwards
Illustration on page 115 by David Parkins; page 283 (top) by Matthew Soley; page 330 (top) by Chris Evans
Other illustrations by Robert Calow and Steve Lach at Eikon Illustration
Cover image © Getty Images/Nick Koudis
Cover design by Angela Ashton

The publishers thank the following for supplying photographs:
page 120 © 1995–2000 by the Association of Universities for Research in Astronomy, Inc. All Rights Reserved; page 124 Astrid and
Hanns-Frieder Michler/Science photo library (amoeba), Manfred Kage/SPL (diatom, radiolarium), Custom Medical Stock Photo/SPL
(red blood cells), Biophoto Associate/SPL (chromosomes), CNRI/SPL (basophil), Barry Dowsett/SPL (ebola virus), EM Unit,
VLA/SPL (scrapie virus), Dr Kari Lounatmaa/SPL (influenza virus); page 207 Tiffany Passmore; page 208 Royalty-free/CORBIS;
pages 245, 254, 330 and 331 Paul Scruton; page 255 Graham Portlock

The authors and publishers are grateful to the following Examination Boards
for permission to reproduce questions from past examination papers:

AQA(NEAB)	Assessment and Qualifications Alliance
AQA(SEG)	Assessment and Qualifications Alliance
Edexcel	Edexcel Foundation
OCR	Oxford, Cambridge and RSA Examinations
WJEC	Welsh Joint Education Committee

Data on page 234 from *Transport statistics Great Britain* (Department of Transport, 1996); page 237, Home Office; pages 238–239,
Statistical yearbook (United Nations) and *World almanc*, 1998; page 240, *Pocket Britain in figures* (The Economist, 1997); page 241
from www.amstat.org/publications/jse/datasets/titanic.txt and www.amstat.org/publications/jse/datasets/titanic.dat; Thames barrier
data on page 248 courtesy of the consulting engineers High-Point Rendel and the Environment Agency; Illustration on page 332,
which originally appeared in *Time*, 9 April 1979, redrawn from *The Visual Display of Quantitative Information*, Edward R. Tufte
(Graphics Press, 1983), p. 62; extracts on page 332 (bottom) from *On Being the Right Size and Other Essays*, J. B. S. Haldane (Oxford
University Press, 1985), pp. 1–2

Contents

1 Surveys and experiments

This work will help you plan and carry out a project in data handling.
You will need to be able to make frequency tables, draw bar charts or
pie charts and use scatter diagrams.

You will learn how to
- ◆ write an effective questionnaire
- ◆ carry out experiments to get data

A *The data handling cycle*

Specifying the question	The starting point is a **question** or an area of interest, for example: • If the local library can be open for only ten hours a week, at what times would it be best to open? • Do people remember words better than numbers?
Collecting data	To answer the question, we need to decide what information or data we need. We have to plan how to **collect** it and how we will use it to help answer our question. If we have to collect the data ourselves, for example by asking people questions or by counting or measuring something, then the data is called **primary data**. If the data has already been collected by someone else, it is called **secondary data**.
Processing and representing the data	To help answer the question, the data has to be **processed** (for example, by working out percentages, finding frequencies, calculating means, and so on). It is often helpful to **represent** the data in pictorial form (for example, frequency chart, scatter diagram, pie chart).
Interpreting the data to answer the question	Processing and representing the data allows us to **interpret** it to help answer the question we started with. The result may suggest that some more data needs to be collected. It might also suggest other questions which need answering. So we may go back to an earlier stage of the cycle and repeat.

Primary data and secondary data

Primary data is data which you collect yourself.
For example, you are collecting primary data when you give
people questionnaires to fill in. You are also collecting
primary data when you make measurements in an experiment.

Height	152 cm	172 cm
Weight	57 kg	78 kg
Pulse rate before exercise	89 b.p.m.	75 b.p.m.
Pulse rate after exercise	127 b.p.m.	133 b.p.m.

Secondary data is data which someone else has collected and organised.
For example, data about crime which is published by the government is
secondary data.

Vandalism, per 10 000 households		
Year 1981	1993	1995
Cases 1481	1638	1614

Sometimes data does not fit easily into either type.
For example, suppose you collect information about the prices of
secondhand cars from newspaper adverts. Is this primary or secondary data?
It feels more like primary data because although it's written by someone else,
it isn't organised in any way.

Ford Fiesta 1.4, 1992, 65000 miles. Blue vgc.
MOT. £1250 ono.
Ford Sierra 2.0LX, J reg, 1992, 34000 miles. Red.
One owner. FSH. MOT, taxed to Aug. £2600.
Ford Sierra 1.8 estate, 1994, Grey. No rust. Recent
service. £2900 ono.

B Surveys

School uniform
Report by Chris and Melanie

The school council discussed changing the school uniform. Some people didn't like the colour and some wanted sweatshirts instead of blazers. A lot of pupils thought that there shouldn't be a uniform at all.

We decided to find out what other students felt about the uniform. We thought that boys and girls might feel differently and so might different year groups. We wrote a questionnaire and we decided to give it to some students in every year group. (There are about 180 in each year.)

Here is our questionnaire.

1 What year group are you in? (Please tick.)   Y7   Y8   Y9   Y10   Y11

2 Are you male or female?   Male   Female

3 Do you think there should be a school uniform?   Yes   No

4 If there has to be a uniform, would you prefer   blazer   sweatshirt

5 What colours would you like the uniform to be?

6 'Students should be allowed to wear jewellery.' What do you think?

    Strongly agree   Agree   Not sure   Disagree   Strongly disagree

---

Questions for discussion

- What do you think of the questions? Are they easy to answer?
  Are they clear – will they mean the same to everyone who answers them?
  Will the responses be easy to analyse?

- Who would you give the questionnaire to?
  How would you collect their responses?
  How many people would you give it to?

I'll ask everyone in the school.

I'll ask all my friends.

I'll ask 5 people in each year.

I'll ask everyone in the choir.

The report continues like this.

> Our teacher told us it was a good idea to pilot a questionnaire. This means giving it to a few people to see if there are any 'bugs' (problems).
>
> We gave it to 10 people. Some of them thought there should be a question about ties. Two people said that 'jewellery' was too vague: ordinary rings could be allowed but not nose rings.
>
> We also found that everybody had written different colours that they liked, sometimes three or four colours, e.g. dark blue, red, yellow. It would be difficult to analyse the answers to this question.

- Look back at the questionnaire.
  How could you improve it to avoid these problems?

In their report, Chris and Melanie made tables of the replies they got to the questions in their questionnaire.

> This table shows the replies we got to the question 'Would you prefer blazer or sweatshirt?'
>
> | Year | | 7 | 8 | 9 | 10 | 11 |
> |---|---|---|---|---|---|---|
> | Boys | Blazer | 7 | 7 | 4 | 3 | 3 |
> | | Sweatshirt | 8 | 11 | 12 | 11 | 13 |
> | Girls | Blazer | 7 | 7 | 6 | 6 | 7 |
> | | Sweatshirt | 5 | 8 | 9 | 10 | 9 |

B1 Draw a chart, or charts, to illustrate this data.
Explain why you chose your type of chart.

B2 What conclusions would you draw about
the preferences?

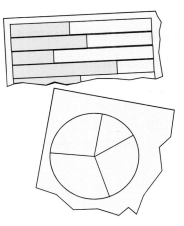

Question types

Here are some types of question you could use in a questionnaire.
Questions which ask for boxes to be ticked (or letters to be ringed)
make it easier to collect all the data together afterwards.

Yes/no questions

| A | Have you passed the driving test? | YES ☐ | NO ☐ |

The question must have a clear 'yes' or 'no' answer.
(If you think someone might not know the answer,
then you could include **DON'T KNOW** ☐).

Multiple-choice questions

B	Which age group are you in?
	0–19 ☐ 20–39 ☐ 40–59 ☐ 60–79 ☐ 80 or over ☐

C	Which of these statements best describes how you plan what you will watch on TV?
	A I plan days ahead.
	B I decide on the day.
	C I just flick around to see what's on.
	Please ring A B C

In examples B and C above, the person chooses one response.
In example D below, they can choose more than one.

D	Which of these languages do you study for GCSE?
	French ☐ German ☐ Spanish ☐ Gujerati ☐ Latin ☐

Questions which give a scale of responses

E	Which statement best describes how you feel about maths?
	A I like it a lot.
	B I quite like it.
	C It's all right.
	D I don't like it very much.
	E I hate it.
	Please ring A B C D E

Questions which ask for a number

| F | How many subjects are you taking at GCSE? | Number .... |

If you don't need to know the number exactly,
then it is better to give groups (as in example B).

Questions which ask for an order of preference

| G | What kind of music do you prefer?
Put in order of priority (1 for your favourite, 5 for your least).
Hard rock ☐   Pop ☐   Jazz ☐   Easy listening ☐   Classical ☐ |

Open questions

| H | What do you think about school lunches? |

This kind of question is good for finding out people's own ideas,
but it is hard to summarise the answers.

Things to avoid!

- Don't ask questions which could be embarrassing. ('How old are you?')
- Don't ask questions which try to lead people to answer in one way.
  ('Would you like to see the safety of our children improved by banning traffic
  from the road in front of the school?') These are called **leading questions**.
- Don't ask questions which are difficult to answer precisely.
  ('How many hours of TV do you usually watch each week?')

B3   Criticise these questions and try to improve them.

| How much do you earn?   £................ |
| How many are there in your family?   .... |
| Where do you shop? Please tick.   Asda ☐   Sainsbury's ☐   Safeway ☐   Tesco ☐ |
| How much do you spend a week on food?   £................ |
| How do you think supermarket fruit and vegetables compare with
the real fruit and vegetables you buy direct from a farm? |

Carrying out a survey

1. Be clear about the purpose of your survey.

2. Write a draft questionnaire.

3. Pilot your draft questionnaire with a small number of people.

4. Improve the questions if necessary.

5. Decide who to give the questionnaire to, and how many people to ask.

6. Decide whether you will see people and ask the questions, or give them the questionnaire to fill in.

7. Collect all the responses together, analyse them and write a **report**.

In your report

- State the purpose of your survey. Describe how you carried it out, any difficulties you had to overcome and any changes of plan.

- Include your final questionnaire.

- Say how many people responded.

- Summarise the responses to each question. Use tables and charts where appropriate.

  If you are comparing the responses of different groups (e.g. boys and girls), summarise them **separately**. You could use a table something like this.

Hours of TV	0-9	10-19	20-29	30+
Girls	17	12	15	9
Boys	12	10	19	10

- Write a conclusion.

Points for discussion

Music charts

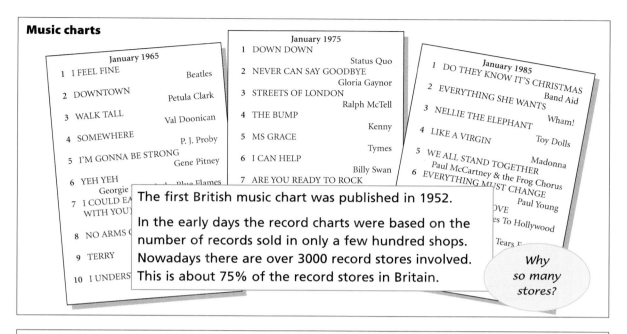

January 1965

1 I FEEL FINE   Beatles

2 DOWNTOWN   Petula Clark

3 WALK TALL   Val Doonican

4 SOMEWHERE   P. J. Proby

5 I'M GONNA BE STRONG   Gene Pitney

6 YEH YEH   Georgie ...

7 I COULD EA... WITH YOU...

8 NO ARMS ...

9 TERRY

10 I UNDERS...

January 1975

1 DOWN DOWN   Status Quo

2 NEVER CAN SAY GOODBYE   Gloria Gaynor

3 STREETS OF LONDON   Ralph McTell

4 THE BUMP   Kenny

5 MS GRACE   Tymes

6 I CAN HELP   Billy Swan

7 ARE YOU READY TO ROCK

January 1985

1 DO THEY KNOW IT'S CHRISTMAS   Band Aid

2 EVERYTHING SHE WANTS   Wham!

3 NELLIE THE ELEPHANT   Toy Dolls

4 LIKE A VIRGIN   Madonna

5 WE ALL STAND TOGETHER   Paul McCartney & the Frog Chorus

6 EVERYTHING MUST CHANGE   Paul Young

...OVE   ...es To Hollywood

Tears ...

The first British music chart was published in 1952.

In the early days the record charts were based on the number of records sold in only a few hundred shops. Nowadays there are over 3000 record stores involved. This is about 75% of the record stores in Britain.

Why so many stores?

US elections

Surveys got a bad name in 1936. In that year the US Presidential elections were held. There were two candidates, Landon (who represented the better off) and Roosevelt (for the less well off).

A magazine did a postal survey on who people would vote for. They obtained the names and addresses from telephone directories and car registrations.

Over 2 million of the 10 million sent questionnaires replied. These predicted a massive victory for Landon.

In fact Roosevelt won by a massive majority!

Why do you think the result was so different?

Honest!

About 60 years ago, an American survey contained the question

What do you think of of the new metallic Metals Law?

The option boxes included 'I don't know' as an option, but fewer than 25% ticked it. Everyone else ticked an opinion.

In fact the 'new metallic Metals Law' was completely fictitious!

Why did over 75% of people express an opinion?

C Experiments

Priya and Ben decided to investigate how good people are at remembering words, numbers and pictures.
They wrote a report on their findings.

Remembering words, pictures and numbers
by Priya and Ben

We wanted to see if there was any difference between how good young people are at remembering words, pictures and numbers.

We both thought it would be easiest to remember pictures.

We decided to test years 10 to 13, who are mostly between 14 and 18 years old.

How we got our results
We made up some experiments.

We chose:
- 10 words – we tried to make sure there were no links between them (like 'pencil' and 'paper')
- 10 pictures
- 10 numbers between 1 and 100

We showed our class the 10 words for 30 seconds and gave them 60 seconds to write down as many as they could remember. The order didn't matter.

We did the same with the pictures and the numbers.

Each correct word, picture or number scored 1 point.

Each student had three scores out of 10 and wrote them on a slip of paper. Our class is in year 10 and we wanted results from years 10 to 13. We couldn't use year 11 because they were on exam leave so we asked our teacher Mr Cassell to do the same experiment on his year 12 and 13 mathematics groups.

Our results
We collected all the slips of paper and chose 10 at random from each year so that we had the results for 30 students.

school
heather
lamp
sky
hate
spoon
necklace
birthday
hair
leaf

Here are some questions for discussion.
Explain each of your answers as fully as you can.

C1 Did you find Priya's and Ben's description of their memory experiments easy to follow?

C2 They made up a list of 10 words for one experiment.
Why do you think they tried to have no links between their words?

C3 Why do you think they used the same number of words, numbers
and pictures in their experiments?

C4 Was it a good idea for Mr Cassell to collect the data from his
year 12 and 13 mathematics groups?

C5 Why do you think they chose 10 students at random from each year?
Do you think this was a good idea?

We made a table for each year but we analysed all the results together.
W stands for Words; P stands for Pictures; N stands for Numbers.

Year 10		
W	P	N
8	7	5
5	6	6
5	3	7
7	7	5
8	7	7
10	10	8
9	10	8
10	10	9
10	9	10
9	8	6

Year 12		
W	P	N
8	9	6
9	10	9
10	10	8
9	9	9
7	9	8
10	10	7
7	8	4
7	8	10
8	7	8
9	10	7

Year 13		
W	P	N
10	**10**	**9**
9	9	6
9	10	10
7	9	4
8	10	7
8	9	5
8	10	4
8	8	4
8	8	6
10	10	6

Each row shows the scores for one student.
For example, the first row in the year 13 table shows that a student correctly remembered 10 words, 10 pictures and 9 numbers.

Analysing our results

We drew bar charts for our results.

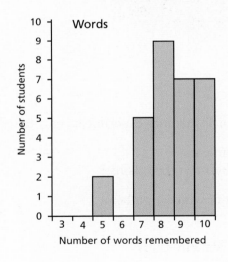

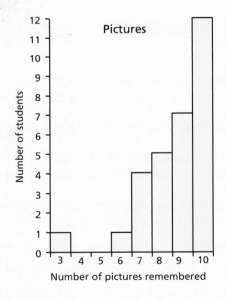

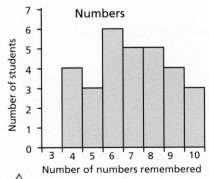

Conclusion

From the shape of our charts, we think that young people are best at remembering pictures, then words and then numbers. We expected that people would be best at remembering pictures.

For this sample:

C6 One student remembered only 3 pictures.
How many words and numbers did this person remember?

C7 How many students in year 13 remembered all 10 pictures?

C8 How many students from all three years remembered fewer than 7 pictures?

C9 How many students from all three years remembered more than 8 words?

C10 Do you think that the bar charts show that the students are best at
remembering pictures, then words and then numbers ?

We then decided to calculate the mean, median and range for each set of results.

	Words	Pictures	Numbers
Mean	250 ÷ 30 ≈ **8.3** words	260 ÷ 30 ≈ **8.7** pictures	208 ÷ 30 ≈ **6.9** numbers
Median	**8** words	**9** pictures	**7** numbers
Range	10 − 5 = **5** words	10 − 3 = **7** pictures	10 − 4 = **6** numbers

The means and medians show that our conclusion is correct.
Young people are best at remembering pictures, then words and then numbers.
The ranges show that the results for the pictures are more spread out.

C11 Why do they think that the values for the means and medians show
that their conclusion is correct ?

C12 Do you think they have enough evidence to say that young people are
best at remembering pictures, then words and then numbers ?

C13 Why do you think they did not compare students from years 10, 12 and 13 ?

Extension
We then investigated if there was a link between
our memory for words and for numbers.

We drew a scatter diagram for our results.

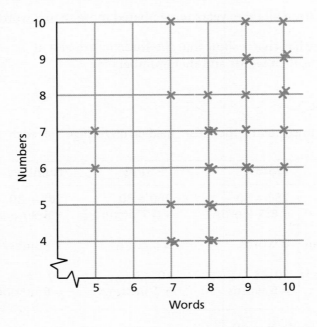

Conclusion
The crosses are quite spread out.
We don't think this shows a link between our memory for words and for numbers.

Possible further work
If we had more time we would have tried to find if there was a link between
our memory for words and for pictures and also for numbers and pictures.

We also think that we would like to do the experiments again with more than
10 words, pictures and numbers. A lot of people scored 10 in at least one
experiment.

Next week we are going to ask our class how many of the words, pictures and
numbers they can remember (but we haven't told them this). We want to see
how good they are at remembering these things after a week has passed.

C14 Do you agree with Priya and Ben that the scatter diagram shows
there is no link between our memory for words and for numbers?

C15 For the students in Priya's and Ben's investigation, draw a scatter diagram
to decide if there is a link between their memory for words and for pictures.

C16 Why do you think Priya and Ben want to repeat their experiments
with more than 10 numbers, words and pictures?

C17 Investigate to see if you come to the same conclusions
as Priya and Ben for students in your school.
(There are 10 pictures on sheet G12 that you can use.)

D Ideas for primary data projects

These projects are described in more detail on sheets G124 to G132.

Remember, remember …

Investigate aspects of memory, such as whether age affects memory or whether background music makes it easier to remember.

Food for thought

Investigate aspects of healthy and unhealthy eating.

First names

Investigate aspects of people's first names, such as popularity or length.

Lunchtime menu

Carry out a survey of people's eating habits in order to decide what to include in the lunchtime menu of a café.

Groovers

Investigate people's preferences for different styles of music, for example to programme the output of a radio station.

Computer games

Investigate opinions about computer games and their popularity.

Wine gums

Compare different makes of wine gum: cost, taste, and so on.

Town and country houses

Use the information given in estate agents' adverts to investigate aspects of houses, such as how their value varies from place to place.

Helicopter seeds

Some trees have seeds with wings that rotate in the wind as the seed falls. Investigate how these seeds fly by making simple paper models.

The tangent function

Before starting this work, you need to know that the ratio $\frac{b}{a}$ is the multiplier from a to b. You will be doing your first work in trigonometry, which is about the relationship between lengths and angles.

A Finding an 'opposite' side

Draw this right-angled triangle accurately.

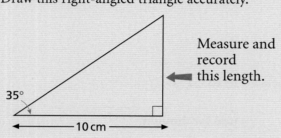

Measure and record ← this length.

If you draw another vertical line 5 cm from the 35° vertex, how long should it be? Draw and measure to check.

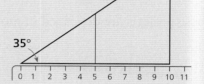

Add vertical lines, 1 cm, 2 cm, 3 cm, ... from the 35° vertex. Work out how long each of them should be. Then check by measuring.

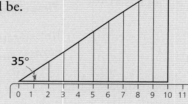

A1 Use your results to work out the missing lengths here.

(a)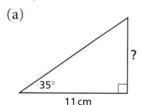

(b)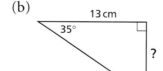

(c)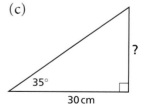

(d)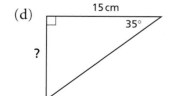

(e)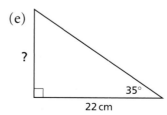

(f)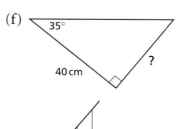

A2 Do a drawing experiment as above but using an angle of 50°.

A3 Use what you discovered in A2 to find the missing lengths here.

(a)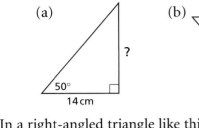

(b)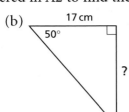

(c)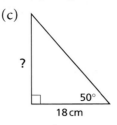

(d)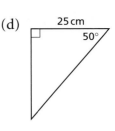

In a right-angled triangle like this,

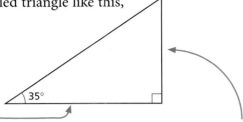

we say this side is **adjacent** to the 35° angle and this side is **opposite** to the 35° angle.

When the angle is 35° you can use the rule: adjacent side —$\boxed{\times\,0.7}$— opposite side

This is another way of stating the rule: $\dfrac{\text{opposite side}}{\text{adjacent side}} = 0.7$

A4 Write a rule in each of these ways for the angle 50°.

A5 Do this drawing experiment to find rules for other angles.

| Draw a horizontal line 10 cm long with a vertical line at the right-hand end of it. | Draw angles of 10°, 20°, 30°, ..., 60°, 70°. |

Measure the opposite side for each triangle and record your results in a table like this.

Angle	Adjacent side	× ?	Opposite side
10°	10 cm		
20°	10 cm		
30°	10 cm		
40°	10 cm		

Work out the numbers that go in here and write them in your table.

A6 Use results from your table to find the missing lengths here.

(a)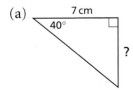

(b)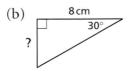

(c)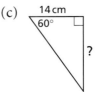

(d)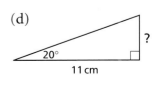

The number you multiply the adjacent side by to get the opposite side is called the **tangent** of the angle.

So the tangent of 35° is about 0.7
and the tangent of 50° is about 1.2.

'tan' is short for tangent, so we can say tan 35° = 0.7
and   tan 50° = 1.2.

A scientific calculator gives very accurate values of tangents.

Try keying in $\boxed{3}$ $\boxed{5}$ $\boxed{\text{tan}}$ or $\boxed{\text{tan}}$ $\boxed{3}$ $\boxed{5}$. Is 0.7 close to what you get?

A7 Use your calculator to find tan 10°, tan 20°, ..., tan 70°, to 3 d.p.
Compare these results with what you got in A5.

A8 Find tan 65° on your calculator and use the result to find the missing lengths here.
Give each length to two significant figures.

(a)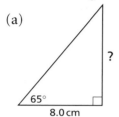

(b)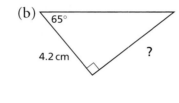

(c)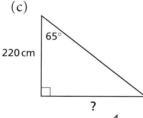

A9 (a) What is special about this right-angled triangle?

(b) How big is angle *a*?

(c) Explain how you can get the tangent of *a* without using a calculator.
Now check with a calculator.

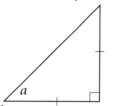

The angles in the following drawings have been drawn accurately.
For each drawing,

(a) write an estimate of the missing length

(b) then use the tangent key on your calculator to get an answer to 1 d.p.

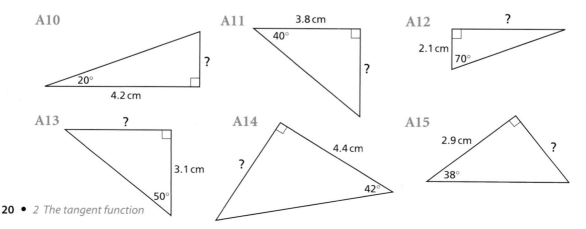

A10

A11   3.8 cm   40°

A12   ?   2.1 cm   70°

A13   ?   3.1 cm   50°

A14   4.4 cm   ?   42°

A15   2.9 cm   ?   38°

B Finding an adjacent side

Earlier you found that tan 35° is 0.7 and that
you could use this to find an opposite side.

How long is the **adjacent** side here?

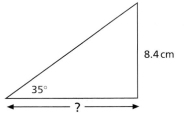

B1 Find each of these adjacent sides, using tan 35° = 0.7.

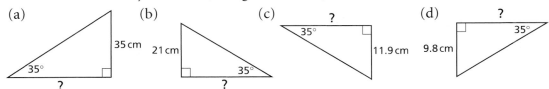

B2 Find each of these adjacent sides, using tan 50° = 1.2.

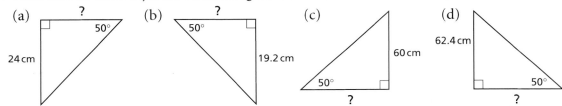

B3 The angles in the following drawings have been drawn accurately.
For each one,

(i) write an estimate of the length of the missing adjacent side

(ii) then find its length to 1 d.p. using the calculator's value of the tangent

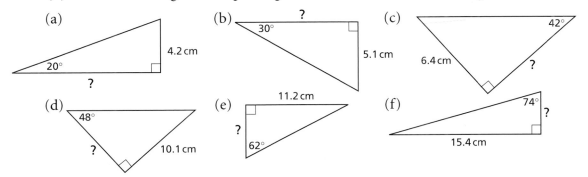

B4 Find the missing sides here to 1 d.p., using a calculator.
Be careful: some are opposite sides and some are adjacent sides.

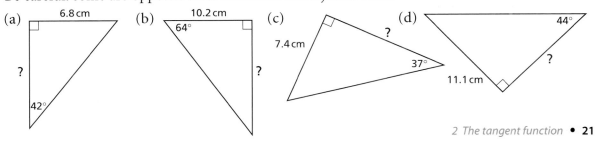

C Finding an angle

Before there were calculators, people used tables of tangents.

This simple table gives tangents to two decimal places.

a	$\tan a$	a	$\tan a$	a	$\tan a$	a	$\tan a$	a	$\tan a$	a	$\tan a$	a	$\tan a$	a	$\tan a$	a	$\tan a$
1	0.02	11	0.19	21	0.38	31	0.60	41	0.87	51	1.23	61	1.80	71	2.90	81	6.31
2	0.03	12	0.21	22	0.40	32	0.62	42	0.90	52	1.28	62	1.88	72	3.08	82	7.12
3	0.05	13	0.23	23	0.42	33	0.65	43	0.93	53	1.33	63	1.96	73	3.27	83	8.14
4	0.07	14	0.25	24	0.45	34	0.67	44	0.97	54	1.38	64	2.05	74	3.49	84	9.51
5	0.09	15	0.27	25	0.47	35	0.70	45	1.00	55	1.43	65	2.14	75	3.73	85	11.43
6	0.11	16	0.29	26	0.49	36	0.73	46	1.04	56	1.48	66	2.25	76	4.01	86	14.30
7	0.12	17	0.31	27	0.51	37	0.75	47	1.07	57	1.54	67	2.36	77	4.33	87	19.08
8	0.14	18	0.32	28	0.53	38	0.78	48	1.11	58	1.60	68	2.48	78	4.70	88	28.64
9	0.16	19	0.34	29	0.55	39	0.81	49	1.15	59	1.66	69	2.61	79	5.14	89	57.29
10	0.18	20	0.36	30	0.58	40	0.84	50	1.19	60	1.73	70	2.75	80	5.67	90	?

C1 Find these in the table.

    (a) $\tan 23°$ (b) The angle whose tangent is 0.78   (c) $\tan 52°$

    (d) The angle whose tangent is 4.01   (e) The angle whose tangent is 0.29

    (f) $\tan 6°$ (g) The angle whose tangent is 0.05   (h) $\tan 86°$

    (i) The angle whose tangent is 0.96   (j) The angle whose tangent is 28.64

C2 (a) Work out the tangent of angle a here.

    (b) Use the table to find the angle with this tangent.
        (In other words, find angle a.)

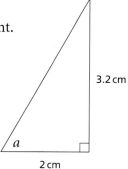

3.2 cm

a

2 cm

C3 What happens in the table of tangents as
the angle a gets closer to 90°?
Why?

C4 For each of these, work out the tangent of the angle,
then use the table to find the angle.

(a)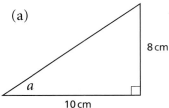

(b)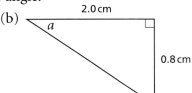

(c)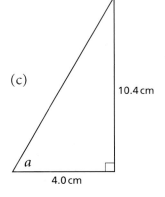

You know that the the tangent of an angle can be expressed like this.

$$\tan a = \frac{\text{opposite side}}{\text{adjacent side}}$$

C5 For each of these, work out the tangent of the angle,
then use the table to find the angle.

(a)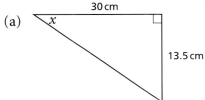

(b)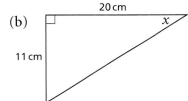

(c)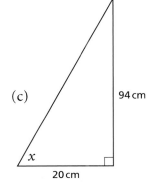

You can find 'the angle whose tangent is …' very accurately on a scientific calculator.
The key sequence you use depends on the type of calculator.

Find out how to find 'the angle whose tangent is …' on your calculator.
Experiment with different values.
Use the table opposite to see whether your method makes sense.

C6 Use your calculator to find the angles in C4 and C5 more accurately.
Give the angles to 2 d.p.

C7 For each of these, work out the tangent of the angle,
then use your calculator to find the angle to 2 d.p.

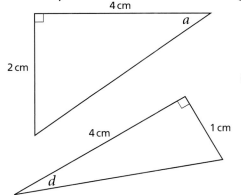

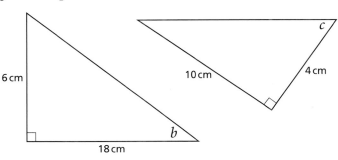

C8 For each of these, write down an estimate of the angle.
Then work out the angle to 2 d.p. using your calculator.
Write down all the stages of your working.

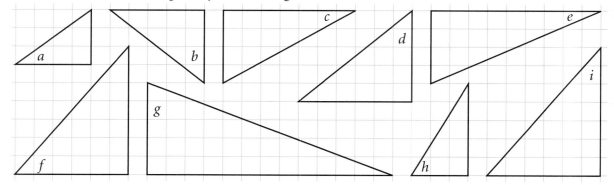

C9 Work out these angles to 2 d.p. using your calculator.
Write down all the stages of your working.

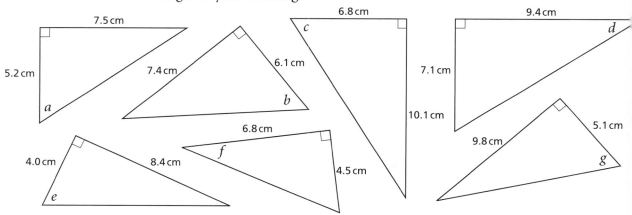

D *Mixed questions and problems*

D1 Find the missing lengths and angles here.

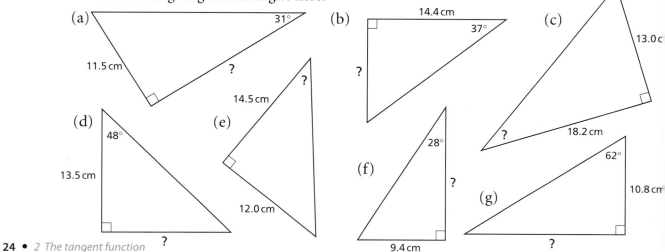

D2 This is part of the symmetrical end wall of a house. Calculate, to the nearest degree, the angle a that the roof makes with the horizontal.

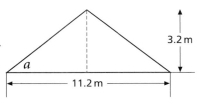

D3 This is the end wall of a shed. Calculate, to the nearest degree, the angle that the roof makes with the horizontal.

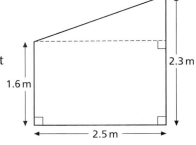

D4 This is a symmetrical trapezium. Calculate the angle marked b, to the nearest 0.1°.

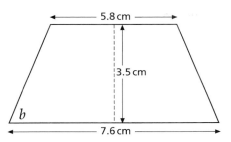

D5 It is possible to get from a farmhouse F to a windmill W by going 5.3 km north then 6.6 km east.

(a) What is the bearing of W from F?

(b) What is the bearing of F from W?

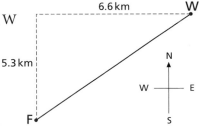

D6 The diagonals of rectangle ABCD cross at O. Find the angles x and y.

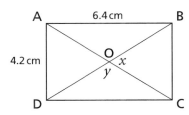

D7 The lengths of the diagonals of this rhombus are shown. Calculate the angles at the vertices of the rhombus, to the nearest 0.1°.

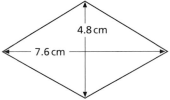

D8 This is an equilateral triangle.

    (a) How long is the dotted line (to 2 d.p.)?

    (b) What is the area of the equilateral triangle (to 2 d.p.)?

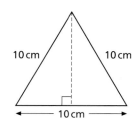

*D9 This is a regular octagon. O is the centre.

    Calculate

    (a) angle AOC

    (b) length OB (to 2 d.p.)

    (c) the area of the whole octagon (to 1 d.p.)

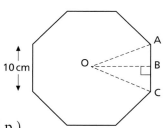

*D10 This is a regular hexagon with sides 2 cm long.

    Explain why the area of the hexagon is $(6 \tan 60°) \, \text{cm}^2$.

*D11 Find the missing length here, to 1 d.p.

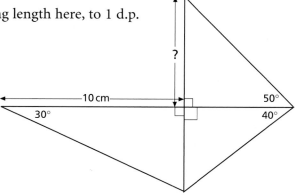

Summary

The tangent of an angle can be thought of as a multiplier ...

... or as a ratio: $\tan a = \dfrac{\text{opposite}}{\text{adjacent}}$

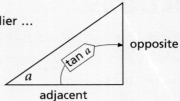

Test yourself

T1 Which is the opposite side to angle a in each of these?

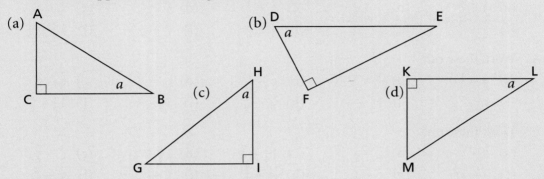

T2 Which is the adjacent side in each of the triangles in T1?

T3 Find the missing quantity in each of these. Give each answer to 1 d.p.

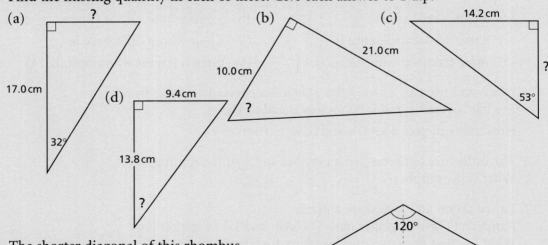

T4 The shorter diagonal of this rhombus
is 10 cm long.
How long is the longer diagonal?

3 Fractions

You will revise calculations with fractions, including

◆ addition, subtraction and multiplication

◆ reciprocals and division

A Addition and subtraction

A1 Work these out.

(a) $\frac{1}{2} + \frac{1}{3}$ (b) $\frac{1}{4} + \frac{1}{6}$ (c) $\frac{2}{3} + \frac{1}{5}$ (d) $\frac{3}{4} + \frac{1}{5}$ (e) $\frac{1}{3} + \frac{3}{10}$

(f) $\frac{2}{3} + \frac{3}{4}$ (g) $\frac{3}{8} + \frac{2}{3}$ (h) $\frac{3}{5} + \frac{1}{6}$ (i) $\frac{4}{5} + \frac{5}{8}$ (j) $1\frac{1}{3} + \frac{4}{5}$

A2 Work these out.

(a) $\frac{3}{4} - \frac{1}{3}$ (b) $\frac{4}{5} - \frac{1}{4}$ (c) $\frac{7}{8} - \frac{2}{3}$ (d) $\frac{7}{10} - \frac{2}{3}$ (e) $\frac{5}{8} - \frac{2}{5}$

(f) $1\frac{1}{2} - \frac{5}{6}$ (g) $2\frac{1}{4} - \frac{5}{8}$ (h) $\frac{17}{20} - \frac{1}{3}$ (i) $1\frac{1}{8} - \frac{1}{3}$ (j) $1\frac{1}{3} - \frac{3}{8}$

A3 Work these out.

(a) $\frac{1}{8} + \frac{5}{6}$ (b) $\frac{5}{6} - \frac{1}{8}$ (c) $1\frac{1}{4} - \frac{2}{5}$ (d) $1\frac{1}{4} + \frac{2}{5}$ (e) $\frac{3}{8} + \frac{5}{6}$

(f) $3 - 1\frac{3}{5}$ (g) $\frac{3}{4} - \frac{1}{5}$ (h) $\frac{11}{12} - \frac{2}{3}$ (i) $\frac{7}{8} + \frac{5}{6}$ (j) $\frac{2}{7} + \frac{4}{5}$

A4 From the fractions in this list find

$$\frac{1}{6} \quad \frac{1}{5} \quad \frac{1}{4} \quad \frac{1}{3} \quad \frac{2}{5} \quad \frac{2}{3} \quad \frac{3}{4} \quad \frac{4}{5} \quad \frac{5}{6}$$

(a) a pair whose sum is $\frac{9}{20}$ (b) a pair whose sum is $\frac{11}{12}$

(c) a pair whose difference is $\frac{7}{15}$ (d) a pair whose difference is $\frac{2}{3}$

(e) three fractions whose sum is $1\frac{1}{4}$ (f) three fractions whose sum is $1\frac{11}{30}$

A5 Dawn and Eve run a race. After Dawn has covered $\frac{3}{5}$ of the distance, Eve has covered $\frac{2}{3}$ and is 50 metres ahead of Dawn.

How many metres does Eve still have to run?

A6 The difference between $\frac{1}{6}$ of a number and $\frac{1}{7}$ of the number is 5. What is the number?

A7 Rama, Gavin and Jake share a pizza.
Rama and Gavin together have $\frac{5}{8}$. Gavin and Jake together have $\frac{2}{3}$.
What fraction do Rama and Jake have together?

B Multiplying fractions

If the area of this square is 1, … 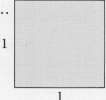 … this area is $\frac{1}{3} \times \frac{1}{4} = \frac{1}{12}$.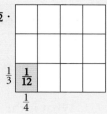

B1 Write a multiplication for each of these diagrams.
The first is done as an example.

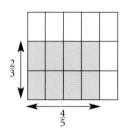

$\frac{2}{3} \times \frac{4}{5} = \frac{8}{15}$

(a)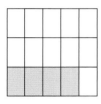

(b)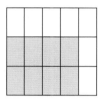

(c)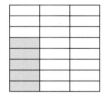

(d)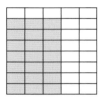

(e)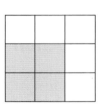

(f)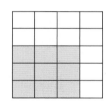

To multiply fractions, multiply the numerators and the denominators. $\frac{3}{4} \times \frac{3}{5} = \frac{3 \times 3}{4 \times 5} = \frac{9}{20}$

Often a calculation can be simplified by cancelling common factors. $\frac{1}{3\cancel{6}} \times \frac{\cancel{4}^2}{5} = \frac{1 \times 2}{3 \times 5} = \frac{2}{15}$

B2 Work these out. **Give each answer in its simplest form.**

(a) $\frac{3}{4} \times \frac{2}{3}$ (b) $\frac{5}{6} \times \frac{3}{4}$ (c) $\frac{3}{8} \times \frac{2}{5}$ (d) $\frac{3}{10} \times \frac{2}{3}$ (e) $\frac{5}{6} \times \frac{7}{10}$

(f) $\frac{1}{3} \times \frac{7}{8}$ (g) $\frac{4}{5} \times \frac{5}{6}$ (h) $\frac{5}{8} \times \frac{2}{3}$ (i) $\frac{4}{9} \times \frac{3}{8}$ (j) $\frac{3}{10} \times \frac{5}{6}$

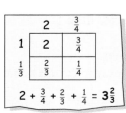

B3 Here are two ways to work out $1\frac{1}{3} \times 2\frac{3}{4}$.

Do each of these by both methods and check that they give the same result.

(a) $1\frac{1}{2} \times 2\frac{1}{2}$ (b) $2\frac{1}{3} \times 3\frac{2}{3}$ (c) $(1\frac{3}{4})^2$

C Reciprocals

The **reciprocal** of a number is $\dfrac{1}{\text{the number}}$.

A number × its reciprocal = 1.

The reciprocal of 5 is $\frac{1}{5}$, because $5 \times \frac{1}{5} = 1$.

The reciprocal of $\frac{2}{3}$ is $\frac{3}{2}$, because $\frac{2}{3} \times \frac{3}{2} = 1$.

C1 What is the reciprocal of $\frac{1}{5}$?

C2 Write down the reciprocal of (a) 2 (b) $\frac{1}{2}$ (c) $\frac{3}{8}$ (d) $\frac{9}{2}$ (e) $\frac{2}{3}$

C3 (a) Write $1\frac{1}{4}$ as an improper ('top heavy') fraction.
   (b) Write down the reciprocal of $1\frac{1}{4}$.

C4 As a decimal, $\frac{5}{8} = 0.625$.
   Without using a calculator, work out, as a decimal, the reciprocal of 0.625.

C5 (a) If A stands for a number, what is the reciprocal of the reciprocal of A?
   (b) Which number has no reciprocal?

D Dividing by a fraction

$4 \div \frac{1}{3}$ means 'how many $\frac{1}{3}$s in 4'.

| 0 | 1 | 2 | 3 | 4 |

$\frac{1}{3}$ $\frac{1}{3}$ $\frac{1}{3}$ $\frac{1}{3}$ $\frac{1}{3}$ $\frac{1}{3}$ $\frac{1}{3}$ $\frac{1}{3}$ $\frac{1}{3}$ $\frac{1}{3}$ $\frac{1}{3}$ $\frac{1}{3}$

$4 \div \frac{1}{3} = 12$

$\boxed{\div \frac{1}{3}}$ is equivalent to $\boxed{\times 3}$.

$4 \div \frac{2}{3}$ means 'how many $\frac{2}{3}$s in 4'.

| 0 | 1 | 2 | 3 | 4 |

$\frac{2}{3}$ $\frac{2}{3}$ $\frac{2}{3}$ $\frac{2}{3}$ $\frac{2}{3}$ $\frac{2}{3}$

$4 \div \frac{2}{3} = 6$

$\boxed{\div \frac{2}{3}}$ is equivalent to $\boxed{\times 3}\boxed{\div 2}$ or $\boxed{\times \frac{3}{2}}$.

Dividing by a number is equivalent to multiplying by its reciprocal.

In the following questions, give the answer in its **simplest form**.
If the answer is an improper fraction, leave it like that. (For example, don't change $\frac{3}{2}$ to $1\frac{1}{2}$.)

D1 (a) $3 \div \frac{1}{4}$ (b) $2 \div \frac{1}{5}$ (c) $10 \div \frac{2}{3}$ (d) $3 \div \frac{3}{5}$ (e) $2 \div \frac{3}{4}$

D2 (a) $\frac{3}{8} \div \frac{2}{5}$ (b) $\frac{4}{5} \div \frac{2}{3}$ (c) $\frac{3}{4} \div \frac{2}{5}$ (d) $\frac{1}{3} \div \frac{3}{8}$ (e) $\frac{3}{4} \div \frac{4}{5}$

E *Mixed questions*

Where appropriate, leave answers as improper fractions.

E1 Copy and complete this addition table.

+	$\frac{1}{4}$	$\frac{3}{10}$
$\frac{1}{2}$	$\frac{3}{4}$	
		$\frac{11}{30}$

E2 Copy and complete this multiplication table.

×	$\frac{1}{4}$	
$\frac{1}{3}$	$\frac{1}{12}$	$\frac{4}{15}$
		$\frac{2}{5}$

E3 20 pizzas are shared between a group of children so that each gets $\frac{2}{3}$ of a pizza. How many children are in the group?

E4 To solve the equation $4x = 36$, you divide both sides by 4.
This is the same as multiplying both sides by the reciprocal of 4, or $\frac{1}{4}$.
Solve the equation $\frac{3}{4}x = \frac{2}{5}$, by multiplying both sides by the reciprocal of $\frac{3}{4}$.

E5 Solve (a) $\frac{2}{3}x = \frac{1}{5}$ (b) $\frac{3}{5}x = \frac{2}{3}$ (c) $\frac{3}{8}x = \frac{1}{2}$ (d) $\frac{2}{5}x = \frac{5}{6}$

E6 People went one by one into a kitchen where there was a jug of water.
The first person drank half of the water.
The second drank half of what was left.
The next drank half of what was left, and so on.

What fraction of the water was left after the fifth person had had a drink?

E7 Re-do question E6, but this time the first person drinks $\frac{1}{3}$ of the water and the others each drink $\frac{1}{3}$ of what is left.

Test yourself

T1 Work these out.
   (a) $\frac{7}{12} - \frac{1}{5}$ (b) $\frac{1}{3} + \frac{7}{8}$ (c) $\frac{2}{5} + \frac{5}{8}$ (d) $\frac{5}{6} - \frac{1}{8}$

T2 What is the reciprocal of   (a) 4   (b) $\frac{4}{7}$ (c) $3\frac{1}{2}$

T3 Work these out.
   (a) $\frac{5}{12} \div \frac{1}{4}$ (b) $\frac{1}{4} \div \frac{5}{12}$ (c) $\frac{7}{8} \div \frac{5}{6}$ (d) $\frac{3}{10} \div \frac{2}{9}$

T4 Solve these.
   (a) $\frac{1}{5}x = \frac{3}{4}$ (b) $\frac{3}{4}x = \frac{1}{6}$ (c) $\frac{2}{3}x = \frac{3}{8}$ (d) $\frac{5}{6}x = \frac{4}{5}$

4 Indices

You should know what is meant by common multiples and factors,
the lowest common multiple (LCM) and the highest common factor (HCF) of two numbers.

This work will help you

- ◆ work with powers and positive indices, including prime factorisation
- ◆ find lowest common multiples and highest common factors using prime factorisation
- ◆ use the rules for multiplying and dividing powers
- ◆ work with negative indices
- ◆ simplify algebraic expressions that involve indices

A *Positive indices*

In an expression like 2^5, the raised number '5' is called the **index**.

Indices (more than one index) are used as a mathematical shorthand.

Examples

$2 \times 2 \times 2 \times 2 \times 2 = 2^5$ $3 \times 3 \times 3 \times 3 = 3^4$

The value of 5^3 is $5 \times 5 \times 5 = 125$

We say 3^4 as 'three to the power four'.

Powers of three can be written as $3^1,\ \ 3^2,\ \ 3^3,\ \ 3^4, \dots$
                or $3,\ \ \ 9,\ \ \ 27,\ \ \ 81, \dots$

A1 Write these in shorthand form using indices.

    (a) $2 \times 2 \times 2 \times 2 \times 2 \times 2 \times 2 \times 2$ (b) $4 \times 4 \times 4 \times 4 \times 4 \times 4 \times 4 \times 4 \times 4 \times 4 \times 4$

A2 Calculate the value of   (a) 2^4 (b) 4^3 (c) 3^5

A3 (a) List all the powers of two that are whole numbers less than 100.

    (b) What is the value of 'three to the power two'?

A4 There is one cell in a flask in a laboratory.
The number of cells doubles every 15 minutes.

    (a) How many cells are there after 2 hours?

    (b) How long does it take the number of cells to increase to 2^{17}?

A5 Decide if the following statements are true or false.

    (a) $5^2 = 5 \times 2$ (b) $3^2 > 2^3$ (c) $2^6 < 5^2$ (d) $3^4 < 6^2$

A6 Choose the correct symbol, $<$, $>$ or $=$, for each box below.

(a) 4^3 ■ 3^4 (b) 7^2 ■ 2^7 (c) 2^5 ■ 5^2 (d) 9^1 ■ 1^9

A7 What is the smallest whole-number value of n so that $3^n > 100$?

A8 Solve the following equations.

(a) $3^x = 27$ (b) $x^5 = 1$ (c) $10^x = 1\,000\,000$ (d) $2^x = 128$

A9 Most calculators have a special key for working out powers.

Find this key on your calculator.

(It might look like one of these: $\boxed{x^y}$ or $\boxed{\wedge}$)

(a) Use this key on your calculator to work out

(i) 2^3 (ii) 10^4 (iii) 3^6

Check your answers are correct without using this key.

(b) Arrange the following numbers in order of size, smallest first.

2^{31}, 7^{10}, 3^{20}, 16^6, 100^2

A10 What is the smallest power of three that is greater than a million?

A11 Solve the following equations.

(a) $x^7 = 823\,543$ (b) $x^8 = 6561$ (c) $x^9 = 262\,144$ (d) $6^x = 10\,077\,696$

A12   'As I was going to St Ives I met a man with seven wives; each wife had seven sacks; each sack had seven cats and each cat had seven kittens.'

If each kitten caught one mouse each day and each cat caught seven mice each day, how many mice would the cats and kittens catch in seven weeks?

A13 The last digit of 3^4 is 1 because $3^4 = 81$.

(a) Copy and complete this table.

n	1	2	3	4	5	6	7	8	9	10
Last digit of 3^n	3	9		1						

(b) What are the last digits of   (i) 3^{36} (ii) 3^{101} (iii) 3^{199}

A14 (a) Investigate the last digits of powers of the form n^5, for example 1^5, 2^5, 3^5, …

(b) What are the last digits of   (i) 299^5 (ii) 305^5 (iii) 411^5

*A15 Find the last digit of each of these. Give reasons for each answer.

(a) 2^{100} (b) 9^{999} (c) 5^{432} (d) 4^{120} (e) $6^{1\,000\,000}$

B *Prime factorisation*

Prime factorisation

There are many ways to write 84 as a product of factors.

For example, $84 = 4 \times 21$

$84 = 12 \times 7$

$84 = 2 \times 6 \times 7$

$84 = 2 \times 2 \times 3 \times 7$

> Euclid proved that there is only one way to write a number as a product of its prime factors. His proof is based on an algorithm for finding the highest comm...

$2 \times 2 \times 3 \times 7$ is called the **product of prime factors** or **prime factorisation** of 84.
We can use index notation to write it as $2^2 \times 3 \times 7$.

There are different ways to work out that 96 is $2 \times 2 \times 2 \times 2 \times 2 \times 3$ (or $2^5 \times 3$).

Factor trees	Repeated division	Factor pairs

Factor trees

96
8   12
2 4   2 6
   2 2   2 3

Repeated division

```
2 | 96
2 | 48
2 | 24
2 | 12
2 |  6
3 |  3
      1
```

Factor pairs

$96 = 4 \times 24$

$= 2 \times 2 \times 4 \times 6$

$= 2 \times 2 \times 2 \times 2 \times 2 \times 3$

B1 Find the prime factorisation of each of these numbers and write it using index notation.

(a) 45 　　　　 (b) 150 　　　　 (c) 48 　　　　 (d) 126 　　　　 (e) 120

B2 The prime factorisation of 24 is $2^3 \times 3$.

(a) Without doing any calculating, decide which of these are multiples of 24.

| $2^3 \times 3 \times 5$ | $2^3 \times 3 \times 7^2$ | $2^3 \times 5$ | $2^4 \times 3$ | $2^3 \times 3^2$ | $2^2 \times 3 \times 5$ |

(b) Check your answers by calculating.

B3 The prime factorisation of 875 is $5^3 \times 7$.

(a) Without doing any calculating, decide which of these are factors of 875.

| 3 | 7 | $5^3 \times 7^2$ | 5×7 | 5×7^3 | $5^2 \times 7$ |

(b) Check your answers by calculating.

B4 The prime factorisation of 275 is $5^2 \times 11$.
The prime factorisation of 1155 is $3 \times 5 \times 7 \times 11$.

Without calculating, decide which of these are **common factors** of 275 and 1155.

| 5 | 3 | 11 | 5×3 | 5×11 | 5^2 | $5^2 \times 11$ |

B5 The prime factorisation of 36 is $2^2 \times 3^2$.
The prime factorisation of 126 is $2 \times 3^2 \times 7$.

Without calculating, decide which of these are **common multiples** of 36 and 126.

| 2×3^2 | $2^3 \times 3^2 \times 7$ | $2^2 \times 3^7 \times 7$ | $2^2 \times 3^2 \times 7 \times 13$ | 2×3 | $2 \times 3^2 \times 7^2$ |

Finding the LCM (lowest common multiple) using prime factorisation

Example   Find the LCM of 15 and 20.

First find the prime factorisation of each number.

$15 = 3 \times 5$
$20 = 2 \times 2 \times 5$

The lowest number which is a multiple of 15 **and** a multiple of 20 is $2 \times 2 \times 3 \times 5$, which is **60**. So the LCM of 15 and 20 is 60.

Finding the HCF (highest common factor) using prime factorisation

Example   Find the HCF of 84 and 120.

First find the prime factorisation of each number.

$84 = 2 \times 2 \times 3 \times 7$
$120 = 2 \times 2 \times 2 \times 3 \times 5$

The highest number which is a factor of 84 **and** a factor of 120 is $2 \times 2 \times 3$, which is **12**. So the HCF of 84 and 120 is 12.

B6 Use prime factorisation to find the LCM of
(a) 12 and 20   (b) 14 and 15   (c) 45 and 165   (d) 42 and 350

B7 Use prime factorisation to find the HCF of
(a) 72 and 180   (b) 90 and 525   (c) 165 and 154   (d) 104 and 234

B8 Ten friends swim at the local pool on 1 January.
They make a new year's resolution.

The first person is going to swim every day, the second person every second day, the third every third day and so on.

How many days later do ten people again swim on the same day?

B9 Helen wants to make a patchwork quilt from squares.
The quilt is to measure 204 cm by 374 cm and the width of each square is to be a whole number of centimetres.

What is the largest size she can use for the squares?

*B10 The HCF of two numbers is 20. The LCM of the same two numbers is 420.

Both the numbers are over 50. What are the numbers?

C *Multiplying and dividing*

C1 Find the missing numbers in these calculations.

(a) $2^3 \times 2^2 = (2 \times 2 \times 2) \times (2 \times 2) = 2^\blacksquare$

(b) $\dfrac{2^5}{2^2} = \dfrac{2 \times 2 \times 2 \times 2 \times 2}{2 \times 2} = 2^\blacksquare$

(c) $5^3 \times 5^4 = (5 \times 5 \times 5) \times (5 \times 5 \times 5 \times 5) = 5^\blacksquare$

(d) $\dfrac{4^3}{4^2} = \dfrac{4 \times 4 \times 4}{4 \times 4} = 4^\blacksquare$

(e) $3^5 \times 3 = (3 \times 3 \times 3 \times 3 \times 3) \times 3 = 3^\blacksquare$

(f) $\dfrac{7^4}{7} = \dfrac{7 \times 7 \times 7 \times 7}{7} = 7^\blacksquare$

C2 Write down the numbers missing from these calculations.

(a) $3^2 \times 3^3 = 3^\blacksquare$

(b) $4^2 \times 4^4 = 4^{\blacksquare\blacksquare}$

(c) $\dfrac{2^7}{2^3} = 2^\blacksquare$

(d) $8 \times 8^7 = 8^\blacksquare$

(e) $6^3 \times 6^9 = 6^{\blacksquare\blacksquare}$

(f) $\dfrac{5^4}{5^3} = 5^\blacksquare$

(g) $2^\blacksquare \times 2^5 = 2^{11}$

(h) $7^5 \times 7^\blacksquare = 7^6$

(i) $\dfrac{11^{10}}{11^\blacksquare} = 11^7$

C3 (a) (i) Write down a rule for multiplying powers of the same number. Explain why your rule works.

(ii) Copy and complete $2^m \times 2^n = 2^\blacksquare$.

(b) (i) Write down a rule for dividing powers of the same number. Explain why your rule works.

(ii) Copy and complete $\dfrac{5^m}{5^n} = 5^\blacksquare$.

C4 Find four pairs of equivalent expressions.

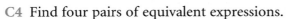

Ⓐ $2^5 \times 2^2$ **Ⓑ** 2^9 **Ⓒ** $2^9 \times 2$ **Ⓓ** 2^7 **Ⓔ** $\dfrac{2^5}{2}$ **Ⓕ** 2^{10} **Ⓖ** 2^4 **Ⓗ** $\dfrac{2^{12}}{2^3}$

C5 Write the answers to these using indices.

(a) $3^4 \times 3^3$

(b) $10^5 \times 10^6$

(c) $8^4 \times 8$

(d) $7 \times 7^9 \times 7^2$

(e) $\dfrac{4^{10}}{4^2}$

(f) $\dfrac{3^9}{3^6}$

(g) $\dfrac{5^3}{5^2}$

(h) $\dfrac{6^5}{6}$

C6 Simplify each of these.

(a) $p^4 \times p^4$

(b) $q^3 \times q^{10}$

(c) $r \times r^9$

(d) $s^3 \times s^5 \times s^2$

(e) $\dfrac{w^7}{w^3}$

(f) $\dfrac{x^5}{x^2}$

(g) $\dfrac{y^7}{y}$

(h) $\dfrac{z^9}{z^8}$

C7 Copy and complete

(a) $2^\blacksquare \times 2^4 = 2^{12}$

(b) $\dfrac{3^9}{3^\blacksquare} = 3$

(c) $a^2 \times a^\blacksquare \times a^4 = a^8$

(d) $\dfrac{b^\blacksquare}{b} = b^7$

C8 Which of these statements is false?

Ⓐ $2^5 \times 3^4 \times 2^2 = 2^7 \times 3^4$ **Ⓑ** $5^2 \times 6^3 \times 5^4 \times 6 = 5^6 \times 6^4$ **Ⓒ** $2^2 \times 3^5 = 6^7$

C9 Which of these statements are identities?

Ⓐ $x^2 \times x^3 = x^6$ **Ⓑ** $k^2 \times k^8 = k^{10}$ **Ⓒ** $m^4 \times m \times m^2 = m^7$ **Ⓓ** $\dfrac{b^{10}}{b^5} = b^2$

C10 Copy and complete each grid. Each space should contain an expression using an index or one of the signs × or ÷.

Each row (from left to right) and each column (from top to bottom) should show a true statement.

(a) 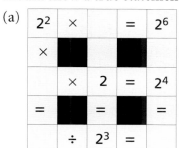 (b) 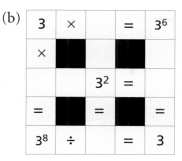 (c)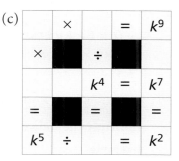

C11 Copy and complete these.

(a) $3^2 \times 5^3 \times 5^4 \times 3^6 = 3^\blacksquare \times 5^\blacksquare$ (b) $2 \times 9^2 \times 2^5 \times 9^3 = 2^\blacksquare \times 9^\blacksquare$

(c) $4^7 \times 3^\blacksquare \times 4 \times 3^2 = 3^{10} \times 4^\blacksquare$ (d) $3^4 \times 11^\blacksquare \times 3^\blacksquare \times 11^5 = 3^5 \times 11^8$

C12 Simplify these.

(a) $10^2 \times 3^4 \times 10^3 \times 3^5$ (b) $2^2 \times 5^3 \times 2^9$ (c) $5^9 \times 7 \times 7^6 \times 5$

C13 Copy and complete these.

(a) $(5^3)^2 = 5^3 \times 5^3 = 5^\blacksquare$ (b) $(2^4)^3 = 2^4 \times 2^4 \times 2^4 = 2^\blacksquare$

(c) $(3^2)^5 = 3^2 \times 3^2 \times ..... = 3^\blacksquare$ (d) $(k^2)^3 = k^2 \times k^2 \times k^2 = k^\blacksquare$

(e) $(h^9)^2 = ... = h^\blacksquare$ (f) $(m^6)^3 = ... = m^\blacksquare$

C14 Copy and complete these.

(a) $(3^4)^2 = 3^\blacksquare$ (b) $(5^3)^6 = 5^\blacksquare$ (c) $(k^3)^5 = k^\blacksquare$

(d) $(2^5)^\blacksquare = 2^{10}$ (e) $(7^\blacksquare)^4 = 7^{12}$ (f) $(n^5)^\blacksquare = n^{20}$

C15 Copy and complete these.

(a) $(2^5)^n = 2^\blacksquare$ (b) $(3^x)^\blacksquare = 3^{4x}$ (c) $(k^h)^5 = k^\blacksquare$

C16 Copy and complete the rule $(a^m)^n = a^\blacksquare$.
Explain how you know your rule is correct.

C17 Write each of these numbers as a power of three.

(a) 9   (b) 9^5 (c) 27^4 (d) 81^2 (e) 27^n

C18 Without using a calculator, arrange the following numbers in order of size, smallest first.
$2^{31},\ 4^{15},\ 8^{11},\ 16^7,\ 32^5$

C19 Which of the numbers $2^4,\ 4^2,\ 2^7,\ 2^2$ is the same as half of 4^4?

To **multiply** powers of the same number, **add** the indices ($a^m \times a^n = a^{m+n}$).

Example

$$k^4 \times k^2$$

$= (k \times k \times k \times k) \times (k \times k)$
$= k \times k \times k \times k \times k \times k$ $\qquad = k^{4+2}$
$= k^6$ $\qquad\qquad\qquad\qquad = k^6$

To **divide** powers of the same number, **subtract** the indices ($\frac{a^m}{a^n} = a^{m-n}$).

Example

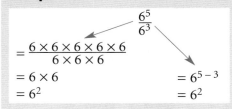

$$\frac{6^5}{6^3}$$

$= \dfrac{6 \times 6 \times 6 \times 6 \times 6}{6 \times 6 \times 6}$

$= 6 \times 6$ $\qquad\qquad = 6^{5-3}$
$= 6^2$ $\qquad\qquad\quad = 6^2$

To raise a power to a further power, **multiply** the indices ($(a^m)^n = a^{mn}$).

Example

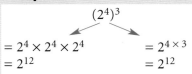

$$(2^4)^3$$

$= 2^4 \times 2^4 \times 2^4$ $\qquad = 2^{4 \times 3}$
$= 2^{12}$ $\qquad\qquad\qquad = 2^{12}$

D *More complex multiplications and divisions*

Examples

$2p^2 \times 3p^5 = 2 \times 3 \times p^2 \times p^5$
$\qquad\qquad\quad = 6p^7$

$(5q^3)^2 = 5q^3 \times 5q^3$
$\qquad\quad = 5 \times 5 \times q^3 \times q^3$
$\qquad\quad = 25q^6$

$$\dfrac{2m^9}{4m^2} = \dfrac{m^9}{2m^2} = \dfrac{m^7}{2}$$

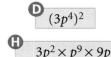

$$\dfrac{6n^3}{9n^5} = \dfrac{2n^3}{3n^5} = \dfrac{2}{3n^2}$$

D1 Find four pairs of equivalent expressions.

A $\quad 3p^6 \times 9p^2$

B $\quad 9p^8$

C $\quad 27p^{12}$

D $\quad (3p^4)^2$

E $\quad 27p^6$

F $\quad 27p^8$

G $\quad (3p^2)^3$

H $\quad 3p^2 \times p^9 \times 9p$

D2 Simplify these.

  (a) $n^2 \times 5n^9$ (b) $2n \times 3n^2$ (c) $7n^5 \times 3n^8$ (d) $5n^2 \times 2n^3 \times 3n^4$

  (e) $(4n)^2$ (f) $(2n)^3$ (g) $(5n^2)^3$ (h) $(2n^3)^5$

D3 Find four pairs of equivalent expressions.

A $\dfrac{6x^8}{3x^2}$

B $\dfrac{x^6}{2}$

C $2x^4$

D $\dfrac{2}{x^4}$

E $2x^6$

F $\dfrac{12x^4}{6x^8}$

G $\dfrac{8x^9}{4x^5}$

H $\dfrac{5x^7}{10x}$

D4 Simplify these.

  (a) $\dfrac{6m^9}{3m^2}$ (b) $\dfrac{2m^8}{10m^6}$ (c) $\dfrac{2m^3}{m^7}$ (d) $\dfrac{8m^6}{12m^3}$ (e) $\dfrac{15m^5}{10m^7}$

E Getting lower

- How do you think these patterns continue?

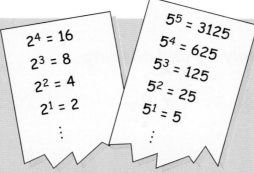

$2^4 = 16$
$2^3 = 8$
$2^2 = 4$
$2^1 = 2$
⋮

$5^5 = 3125$
$5^4 = 625$
$5^3 = 125$
$5^2 = 25$
$5^1 = 5$
⋮

- What do you think are the values of 7^0, 4^{-2} and 3^{-1}?
- What is the reciprocal of 2^5 written in the form 2^n?

E1 Find three matching pairs.

 A 2^{-12}
 B $\frac{1}{8^3}$
 C $^-24$
 D $\frac{1}{3^8}$
 E $\frac{1}{2^{12}}$
 F 8^{-3}
 G 3^{-8}

E2 Find four matching pairs.

 A 3^{-2}
 B 2^{-3}
 C 4^{-2}
 D 6^{-1}
 E $^-6$
 F $\frac{1}{6}$
 G $\frac{1}{16}$
 H $\frac{1}{8}$
I $\frac{1}{9}$

E3 What is the value of 3^0?

E4 What are the reciprocals of the following written in the form 3^n?
  (a) 3^4
  (b) 3
  (c) 3^0
  (d) 3^{-7}
  (e) 9^2

E5 2^{-4} is equivalent to the fraction $\frac{1}{16}$.
  Write these as fractions.
  (a) 3^{-3}
  (b) 7^{-1}
  (c) 9^{-2}
  (d) 4^{-3}
  (e) 11^{-1}

E6 What is the decimal value of
  (a) 5^{-2}
  (b) 2^{-6}
  (c) 10^{-2}
  (d) 4^{-3}
  (e) 16^{-1}

E7 What is the decimal value of each of these, correct to two significant figures?
  (a) 9^{-1}
  (b) 3^{-4}
  (c) 7^{-2}
  (d) 11^{-1}
  (e) 6^{-2}

E8 Solve the following equations.
  (a) $12^{-1} = \frac{1}{x}$
  (b) $2^x = \frac{1}{32}$
  (c) $4^x = 0.0625$
  (d) $x^{-3} = 0.001$

  (e) $4^x = \frac{1}{4}$
  (f) $10^x = 1$
  (g) $x^{-2} = \frac{1}{49}$
  (h) $x^{-3} = \frac{1}{8}$

*E9 Do you think there is a solution to the equation $2^x = 0$?

- $a^{-m} = \dfrac{1}{a^m}$
- $a^0 = 1$

F *More multiplying and dividing*

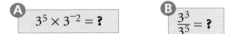

A $3^5 \times 3^{-2} = ?$

B $\dfrac{3^3}{3^5} = ?$

C $(2^3)^{-2} = ?$

D $6^{-1} \times 6^{-2} = ?$

E $\dfrac{7^2}{7^2} = ?$

F $2^{-5} \times 2^3 = ?$

G $(3^{-1})^{-2} = ?$

H $\dfrac{5^3}{5^{-2}} = ?$

F1 Write the answers to these using indices.
- (a) $3^4 \times 3^{-3}$
- (b) $10^5 \times 10^{-6}$
- (c) $8^{-4} \times 8^{-1}$
- (d) $7^{-2} \times 7^9 \times 7^{-4}$
- (e) $\dfrac{4^1}{4^3}$
- (f) $\dfrac{3^{-2}}{3^3}$
- (g) $\dfrac{5^3}{5^{-2}}$
- (h) $\dfrac{6^{-5}}{6^{-3}}$

F2 Simplify each of these.
- (a) $p^4 \times p^{-5}$
- (b) $q^3 \times q^{-3}$
- (c) $2r^{-7} \times 3r^9$
- (d) $s^4 \times s^{-10} \times 5s$
- (e) $\dfrac{w^2}{w^7}$
- (f) $\dfrac{x^5}{x^{-2}}$
- (g) $\dfrac{12y^{-7}}{3y}$
- (h) $\dfrac{z^{-1}}{z^{-8}}$

F3 Simplify each of these.
- (a) $(2^{-4})^2$
- (b) $(3^2)^{-1}$
- (c) $(5^{-2})^{-3}$
- (d) $(a^2)^{-5}$
- (e) $(x^{-3})^3$

F4 Copy and complete these.
- (a) $2^{\blacksquare} \times 2^{-2} = 2^6$
- (b) $\dfrac{3^{-1}}{3^{\blacksquare}} = 3^{-5}$
- (c) $a^2 \times a^{\blacksquare} \times a^{-4} = a^{-3}$
- (d) $\dfrac{b^{\blacksquare}}{b^{-3}} = b^5$
- (e) $(5^{\blacksquare})^3 = 5^{-12}$
- (f) $(2^{\blacksquare})^{-5} = 2^{10}$

F5 Copy and complete each grid. Each space should contain an expression using an index or one of the signs $\times$ or $\div$.

Each row (from left to right) and each column (from top to bottom) should show a true statement.

(a)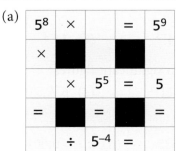

(b)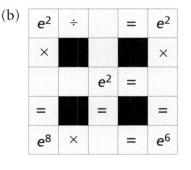

(c)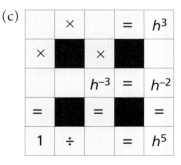

G *True, iffy, false*

G1 In each statement below, m and n can be any integer.

Decide if each statement is *always true*, *sometimes true* or *never true*.
Give reasons for each answer.

(a) $13^n \times 13^{-n} = 1$ (b) $(m + n)^2 = m^2 + n^2$ (c) n^3 is a prime number

(d) $(0.5)^n = 2^{-n}$ (e) $5^{2n} = 10^n$ (f) $2n^2$ is a square number

(g) $(2^m)^n = (2^n)^m$ (h) $2^n < n^3$ (i) $5^n \div 5^{-n} = 25^n$

(j) $2^{2n} = 4^n$ (k) $3^n = 9^m$ (l) $2^n < 0$

G2 In each statement below, a can be any integer, fraction or decimal except 0.

Decide if each statement is *always true*, *sometimes true* or *never true*.
Give reasons for each answer.

(a) $a^4 + a^4 + a^4 = 3a^4$ (b) $2a^2 + 3a^3 = 5a^5$ (c) $5a^2 = 5a \times 5a$

(d) $3a^{-2} = \dfrac{1}{3a^2}$ (e) $6a^{-3} = \dfrac{6}{a^3}$ (f) $\dfrac{6a^{-1}}{2a} = \dfrac{3}{a^2}$

(g) $(4a)^2 = 4a^2$ (h) $5a^2 = 5 \times a \times a$ (i) $a^0 \times a^0 \times a^0 \times a^0 > 2$

(j) $(2a)^{-3} = \dfrac{8}{a^3}$ (k) $(3a)^{-2} = \dfrac{1}{9a^2}$ (l) $2a^2 = a$

G3 Decide if each of the following statements is true or false.

(a) If $x^2 = x$ then the only possible value for x is 1.

(b) If $a^2 = 49$ then the only possible value for a is 7

(c) If $0 < k < 1$ then $k^4 < k^3$

(d) If $^-1 < k < 1$ then $k^4 < k^3$

Test yourself

T1 Is the statement $5^3 > 2^7$ true or false?

T2 Evaluate each of these as a whole number or as a fraction.

(a) 2^5 (b) 3^{-2} (c) 4^{-1} (d) 7^0

T3 Solve the following equations.

(a) $3^x = 81$ (b) $x^{-5} = \dfrac{1}{32}$ (c) $4^x = 0.25$ (d) $2^x = 1$

T4 (a) Find the prime factorisation of 56 and write it using index notation.

(b) Use prime factorisation to find the LCM of 56 and 44.

(c) Use prime factorisation to find the HCF of 56 and 126.

T5 Find the value of n for each of these statements.

(a) $5^6 \times 5^3 = 5^n$ (b) $2^n \times 2^3 = 2^{10}$ (c) $3^8 \times 3^n = 3^3$

(d) $6^n \times 6^2 = \dfrac{1}{6^3}$ (e) $(2^n)^{-4} = \dfrac{1}{2^8}$ (f) $(7^{-9})^n = 7^{-18}$

(g) $\dfrac{5^{12}}{5^3} = 5^n$ (h) $\dfrac{3^4}{3^n} = 3$ (i) $\dfrac{9^n}{9^{-8}} = 9^3$

T6 Simplify each of these and write your expression in the form a^n.

(a) $a^5 \times a^3$ (b) $\dfrac{a^{12}}{a}$ (c) $a^4 \times a^{-1}$ (d) $\dfrac{a^6}{a^7}$

(e) $(a^2)^4$ (f) $(a^{-3})^2$ (g) $(a^2)^{-1}$ (h) $\dfrac{a^3}{a^{-2}}$

T7 Simplify these.

(a) $3x \times 4x^3$ (b) $5x^4 \times x^{-1}$ (c) $2x^5 \times 7x^5$ (d) $2x^{-3} \times 5x$

(e) $5x^{-2} \times 3x^2$ (f) $(2x)^5$ (g) $\dfrac{9x^3}{3x}$ (h) $\dfrac{6x}{15x^2}$

Forming and solving equations

You will revise

◆ solving equations with the unknown on both sides, and with brackets

◆ forming simple equations and solving them

This work will help you

◆ solve equations involving algebraic fractions

◆ form more complex equations and solve them

A Review: unknowns on both sides

$$5 - x = x + 11$$
$$5 = 2x + 11 \quad \text{(+ x both sides)}$$
$$^-6 = 2x \quad \text{(– 11 both sides)}$$
$$^-3 = x \text{ or } \underline{x = {}^-3} \quad \text{(÷ 2 both sides)}$$

$$\tfrac{1}{2}x + 1 = \tfrac{2}{3}x - 1$$
$$x + 2 = \tfrac{4}{3}x - 2 \quad \text{(× 2 both sides)}$$
$$3x + 6 = 4x - 6 \quad \text{(× 3 both sides)}$$
$$3x + 12 = 4x \quad \text{(+ 6 both sides)}$$
$$12 = x \text{ or } \underline{x = 12} \quad \text{(– 3x both sides)}$$

A1 Solve each of these equations.
Check that your answer works in the original equation.

(a) $4a + 10 = 6a$ (b) $7b - 12 = 3b$ (c) $\tfrac{1}{2}c + 3 = c$

(d) $5d + 3 = 7d - 11$ (e) $e + 2 = 4e - 10$ (f) $5f - 12 = 6f - 15$

A2 Solve each of these equations.

(a) $10 - 2a = 2a - 14$ (b) $7b - 3 = 27 - 3b$ (c) $8 - c = 26 + 2c$

(d) $10 - 3d = 7 - 5d$ (e) $21 - e = 26 - 5e$ (f) $17 - 4f = 3f - 4$

A3 Solve each of these equations.

(a) $4 - \tfrac{1}{2}a = 6 - a$ (b) $\tfrac{3}{2}b + 2 = 3b - 4$ (c) $\tfrac{1}{4}c - 2 = 1 + \tfrac{1}{5}c$

(d) $\tfrac{1}{3}d - 10 = \tfrac{1}{7}d + 6$ (e) $\tfrac{e}{5} - 7 = 19 - \tfrac{e}{8}$ (f) $\tfrac{19f}{10} - 1 = \tfrac{7f}{13} - 1$

A4 (a) I think of a number.
I treble it and take away 10.
My answer is twice the number I started with.
What number did I start with?

(b) Abby and Becky both think of the same number.
Abby doubles her number and adds 20.
Becky multiplies her number by 5 and takes away 4.
They both get the same answer.
What number did both of them start with?

B Review: brackets

$3(x + 2) = x - 12$ (multiply out brackets)
$3x + 6 = x - 12$ (− x both sides)
$2x + 6 = {}^-12$ (− 6 both sides)
$2x = {}^-18$ (÷ 2 both sides)
$x = {}^-9$

$\dfrac{4x + 1}{3} = x - 5$ (× 3 both sides)
$4x + 1 = 3(x - 5)$ (multiply out brackets)
$4x + 1 = 3x - 15$ (− 3x both sides)
$x + 1 = {}^-15$ (− 1 both sides)
$x = {}^-16$

B1 Solve each of these equations.

(a) $3(n - 2) = n$ (b) $2(m + 3) = 24$ (c) $2(x - 3) = 1$

(d) $4(d + 2) = 2d - 7$ (e) $3f + 5 = 4(f - 3)$ (f) $2(4x - 5) = 4(x + 1)$

B2 Solve these.

(a) $2(1 - a) = 3a - 18$ (b) $3(2 - v) = 8(v - 2)$ (c) $11(2y - 5) = 0$

(d) $4(2x + 3) = 3(3x - 5)$ (e) $u - 8 = 2(1 - 2u)$ (f) $4(2r + 1) = 3(3r + 5)$

B3 Solve these.

(a) $\dfrac{3e - 2}{5} = 2$ (b) $\dfrac{f - 3}{2} = 5$ (c) $\dfrac{4f + 1}{3} = 7$

(d) $\dfrac{4j - 3}{2} = j$ (e) $\dfrac{11f + 3}{5} = 2f$ (f) $\dfrac{27 - h}{4} = 2h$

B4 Solve the following.

(a) $5(2x + 5) = 1 + 4x$ (b) $\dfrac{8 - x}{5} = 10$

B5 Solve these.

(a) $2(3x - 1) + 3(2x - 3) = 73$ (b) $3(2x + 7) - 2(2x - 1) = 17$

B6 Solve these.

(a) $\dfrac{a + 1}{2} = a - 2$ (b) $\dfrac{b + 2}{3} = b + 4$ (c) $\dfrac{2c + 3}{4} = c - 3$

(d) $\dfrac{d - 4}{2} = 2d - 5$ (e) $\dfrac{3e + 4}{2} = 2e + 1$ (f) $\dfrac{2 - 3f}{2} = 2 - 2f$

B7 (a) John thinks of a number. He takes 7 off the number.
Then he multiplies his result by 3.

His answer is twice the number he first thought of.
What number did he think of?

(b) Ria and Julia both think of the same number.
Ria takes 6 off the number and then multiplies her result by 3.

Julia adds 1 to the number and then doubles her result.

They both end up with the same answer.
What number did they both think of at the start?

C More fractions

Even with more complicated fractions, it is usually best to multiply to get rid of fractions first.

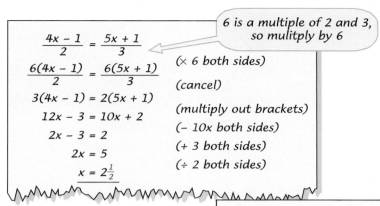

$$\frac{4x - 1}{2} = \frac{5x + 1}{3}$$

6 is a multiple of 2 and 3, so mulitply by 6

$$\frac{6(4x - 1)}{2} = \frac{6(5x + 1)}{3}$$ (× 6 both sides)

(cancel)

$$3(4x - 1) = 2(5x + 1)$$ (multiply out brackets)

$$12x - 3 = 10x + 2$$ (− 10x both sides)

$$2x - 3 = 2$$ (+ 3 both sides)

$$2x = 5$$ (÷ 2 both sides)

$$x = 2\tfrac{1}{2}$$

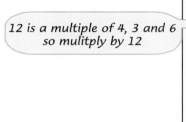

12 is a multiple of 4, 3 and 6 so mulitply by 12

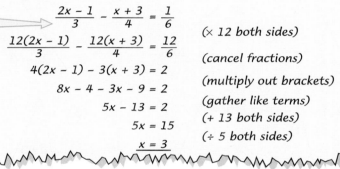

$$\frac{2x - 1}{3} - \frac{x + 3}{4} = \frac{1}{6}$$

$$\frac{12(2x - 1)}{3} - \frac{12(x + 3)}{4} = \frac{12}{6}$$ (× 12 both sides)

(cancel fractions)

$$4(2x - 1) - 3(x + 3) = 2$$ (multiply out brackets)

$$8x - 4 - 3x - 9 = 2$$ (gather like terms)

$$5x - 13 = 2$$ (+ 13 both sides)

$$5x = 15$$ (÷ 5 both sides)

$$x = 3$$

C1 Solve each of these equations.

(a) $\dfrac{x + 1}{3} = \dfrac{x - 1}{2}$ (b) $\dfrac{2x + 1}{3} = \dfrac{1}{2}$ (c) $\dfrac{3x + 5}{4} = \dfrac{x}{2}$

(d) $\dfrac{3x}{4} = \dfrac{4x - 1}{5}$ (e) $\tfrac{1}{2}(3x + 1) = \tfrac{1}{3}(5x + 3)$ (f) $\dfrac{4x + 1}{5} = \dfrac{7x - 2}{8}$

C2 Solve these.

(a) $\dfrac{5x + 2}{3} - \dfrac{3x + 1}{4} = x$ (b) $\dfrac{3x + 7}{4} - \dfrac{x}{3} = \dfrac{x}{6}$

(c) $\dfrac{1 - 2x}{5} - \dfrac{5 - 2x}{3} = 2(x + 1)$ (d) $\dfrac{3x}{4} - 2(x - 1) = \dfrac{1}{3}$

(e) $\tfrac{1}{4}(3x + 7) - 2(1 + x) = {}^{-}x$ (f) $\tfrac{1}{4}(3x - 1) + \tfrac{1}{5}(8x + 3) = \tfrac{1}{3}(7x + 1)$

C3 Solve this equation $\dfrac{3x - 5}{4} + \dfrac{12 - 11x}{6} = 4$ OCR

C4 Solve the equation $\dfrac{2x - 3}{6} + \dfrac{x + 2}{3} = \dfrac{5}{2}$ WJEC

C5 Solve the equation $4 - \dfrac{2x}{3} = 3(x + 3)$ Edexcel

D Forming equations

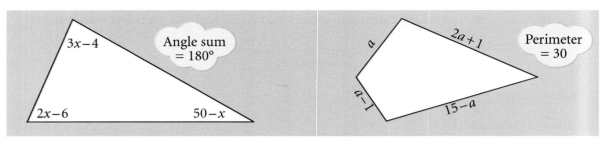

D1 Find the size of each angle in these shapes. None of the diagrams is to scale.

(a)

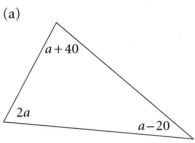

(b)

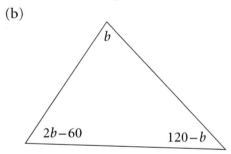

(c)

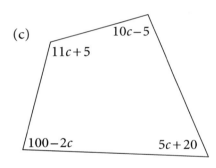

(d)

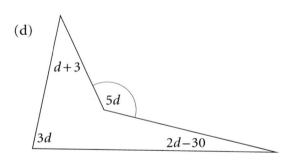

D2 The perimeter of each of these shapes is 100.
Find the length of each side.

(a)

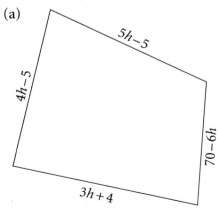

(b)

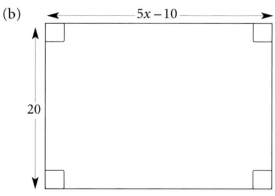

D3 The perimeter of the pentagon is 200 cm.
Work out the value of x.

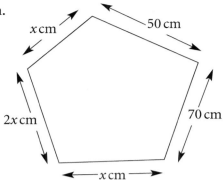

Edexcel

D4 Work out the starting number in each of these 'think of a number' puzzles.

(a)
> *I think of a number.*
> *I treble it and take off 8.*
> *My answer is double the number I first thought of.*

(b)
> *I think of a number.*
> *I double it.*
> *I take the result off 100.*
> *My answer is 1 more than the number I thought of.*

(c)
> *I think of a number.*
> *I add on half the original number.*
> *My answer is 8 more than the number I thought of.*

(d)
> *Lauren and Lucy both think of the same number.*
> *Lauren adds 4 to her number and then trebles the result.*
> *Lucy adds 99 to her number and then halves the result.*
> *They both end up with the same number.*

(e)
> *Walt and Wilf both think of the same number.*
> *Walt adds 6 to his number and then divides the result by 3.*
> *Wilf takes 10 off his number and then halves the result.*
> *They both end up with the same number.*

D5 Mary's mum is 24 years older than Mary is.
In 6 years time, Mary's mum will be exactly four times as old as Mary.

(a) Suppose that Mary's age now is m years.
Write down an expression for Mary's mum's age now.

(b) (i) Write down an expression for Mary's age in 6 years time.

(ii) Write down an expression for Mary's mum's age in 6 years.

(c) Use your answer to part (b) to write an equation in m.
Solve the equation and thus find how old Mary and her mum are now.

D6 Alex's dad is 30 years older than Alex.
In four years time, Alex's dad will be three times as old as Alex.

Work out how old Alex and his dad are now.

D7 Abbas is two years younger than Brad.
Two years ago, the sum of their ages was the same as Brad's age now.

How old are the boys now?

E Mixed questions

E1 Solve these equations.

(a) $4(x - 2) = 18$ (b) $\frac{1}{3}x + \frac{1}{2}x = 10$ (c) $7x - 4 = 2x + 11$ OCR

E2 Megan thinks of a number. She adds 15 and then doubles her result.
Olivia's starting number is 5 more than Megan's.
Olivia trebles her number and then takes off 6.

Both Megan and Olivia end up with the same number.
What numbers did each of them think of?

E3 Solve this equation $3(3x + 2) - 2(x - 3) = x + 3$ AQA(NEAB) 1997

E4 Solve these.

(a) $\frac{5}{x} = 10$ (b) $5(2x - 1) + 6x = 7 - 8x$ OCR

E5 Solve these equations.

(a) $\frac{12}{x} = 2$ (b) $\frac{12}{\sqrt{x}} = 2$ OCR

E6 Solve these.

(a) $3(x - 6) = 10 - 2x$ (b) $\frac{y + 1}{3} = \frac{1 - y}{2}$ Edexcel

E7 A rocket is fired vertically upwards with velocity u metres per second.
After t seconds the rocket's velocity, v metres per second, is given by the formula

$$v = u + gt$$

Calculate t when $u = 93.5$, $g = {}^-9.8$ and $v = 20$. AQA(NEAB) 2000

E8 Find the size of the largest
angle in the pentagon.

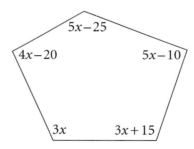

Edexcel

E9 (a) Write down the 10th term of the sequence which begins 3, 7, 11, 15, …

(b) Write down an expression for the nth term of this sequence.

(c) Show that 1997 cannot be a term in this sequence.

(d) Calculate the number of terms in the sequence 3, 7, 11, 15, …, 399.

AQA(NEAB) 1997

Test yourself

T1 Solve these and check that your answers work in the original equations.

(a) $6a + 13 = 2a + 1$ (b) $12 + 2b = 56 - 9b$ (c) $6 - 5c = 10 - 3c$

(d) $5(2d + 3) = 2(4d + 5)$ (e) $5(2e - 1) = 1$ (f) $1 + f = 1 - f$

T2 Solve each of these.

(a) $\dfrac{x + 2}{2} = \dfrac{x - 1}{3}$ (b) $\dfrac{1}{3}(1 - 2x) = \dfrac{1}{4}$ (c) $\dfrac{5x + 6}{4} = \dfrac{x}{2}$

(d) $\dfrac{x}{3} - 2(x + 1) = \dfrac{1}{4}$ (e) $\dfrac{7 - 5x}{6} = \dfrac{1 - 3x}{4}$ (f) $\dfrac{1 - 2x}{2} - \dfrac{1 - 3x}{4} = \dfrac{1 - x}{3}$

T3 The diagram shows the dimensions of a rectangle. The perimeter of the rectangle is 18 cm.

Find the value of x.

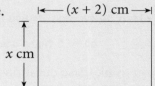

AQA(NEAB) 1998

T4 Form and solve an equation to calculate the angles of this triangle.

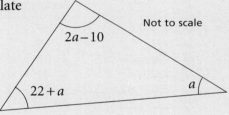

OCR

T5 (a) Write down as simply as possible an expression for the nth term of the sequence

$$100, \ 96, \ 92, \ 88, \ \ldots$$

(b) One of the terms of this sequence is $^-260$. Which term is it?

(c) Is $^-150$ a term of the sequence? Explain your answer.

T6 Bryan is twice as old as Chris.
30 years ago, Bryan was 12 times as old as Chris.
How old are they now?

T7 The sides of a regular octagon are x cm long.
Each side of a regular pentagon is 6 cm longer than each side of the octagon. The perimeter of the octagon is 3 cm longer than the perimeter of the pentagon.

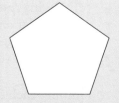

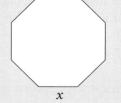

(a) Write down an equation that x satisfies.

(b) Solve the equation and hence find the length of a side of the pentagon.

WJEC

Review 1

1 Barry lives in a village. He carries out a survey of the eating habits of the people living there. He wants to find out whether meat-eating is more popular among older people than among younger people.

He gives this questionnaire to all the people in the village.

> **1 How old are you? (Please tick.)**
>
> 10 or under ____   11-20 ____   21-30 ____   31-40 ____   41-50 ____   51+ ____
>
> **2 How often, on average, do you eat meat in a week?**
>
> never ____   up to 5 times a week ____   more than 5 times a week ____

(a) Do you think these questions are satisfactory?
    If not, what improvements would you suggest?

Here is how Barry records people's responses.

Age	Tally	Frequency
10 or under	⳾⳾⳾ //	7
11-20	⳾⳾⳾ ////	9
21-30	⳾⳾⳾ ⳾⳾⳾ ⳾⳾⳾	15
31-40	⳾⳾⳾ ⳾⳾⳾ //	12
41-50	⳾⳾⳾ ⳾⳾⳾ ⳾⳾⳾ ⳾⳾⳾	20
51+	⳾⳾⳾ ⳾⳾⳾ ⳾⳾⳾ ///	18

Eating meat	Tally	Frequency
never	⳾⳾⳾ ⳾⳾⳾ ////	14
1-5	⳾⳾⳾ ⳾⳾⳾ ⳾⳾⳾ ⳾⳾⳾ ⳾⳾⳾ //	27
> 5	⳾⳾⳾ ⳾⳾⳾ ⳾⳾⳾ ⳾⳾⳾ ⳾⳾⳾ ⳾⳾⳾ ⳾⳾⳾ ⳾⳾⳾	40

(b) Explain why Barry can't tell from this record whether meat is more popular among older or younger people.

Barry decides to record people's responses in a different way, using a two-way table.

	Eating meat		
Age group	never	1-5	> 5
10 or under	1	5	1
11-20	3	6	0
21-30	7	5	3
31-40	1	0	11
41-50	2	8	10
51+	0	3	15

(c) Do his results suggest that meat-eating is less popular among people aged up to 30 than among people aged over 30?
    Support your answer using the data in this table.

2 Simplify each of these expressions as far as possible.

(a) $3(x - 5) + x(x + 2)$ (b) $n(5 - n) - 2(n + 3)$ (c) $p(p + 5) - p(p - 4)$

(d) $5(t^2 + 1) - t(7 - t)$ (e) $2a(a - 4) + a(6 - a)$ (f) $3u(4 - u) - u(3 - u)$

3 Calculate the sides and angles marked with letters.

(a)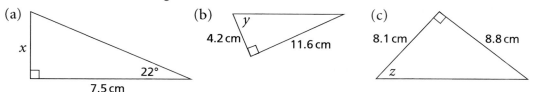
x

22°

7.5 cm

(b)
4.2 cm
y
11.6 cm

(c)
8.1 cm
8.8 cm
z

4 Marsha carried out a memory experiment with two groups of people.
Both groups were given two minutes to memorise a list of 10 words.
Then they were tested to see how many they remembered.

Here are the results for each group.

Group A										Group B									
6	5	4	7	9	3	8	9	5	7	4	9	7	5	9	10	3	6	10	9
9	10	8	10	7	9	7	6	10	8	9	8	9	5	6	7	8	9	5	7

(a) By using mean and range, show that the two groups are quite evenly matched
in their ability to remember after two minutes.

Marsha told group A that they would be tested again the next day,
but she did not tell group B.

Next day she tested both groups. Here are the results.

Group A										Group B									
3	4	7	2	6	4	3	6	4	2	4	0	2	2	1	6	2	3	3	4
5	3	5	4	3	6	8	5	5	7	0	3	2	4	2	2	1	4	5	3

(b) What do these results suggest about the effect of telling group A that
they would be tested again? Support your answer from the data.

5 (a) Write each of these numbers as a product of prime factors.

(i) 24   (ii) 40   (iii) 63   (iv) 78   (v) 225

(b) What is the highest common factor of

(i) 32 and 80   (ii) 120 and 192   (iii) 225 and 270

6 Simplify these expressions.

(a) $4x^3 \times 2x^2$ (b) $\dfrac{6x^6}{3x^2}$ (c) $5x^{-2} \times 2x^3$ (d) $6x \times 3x^{-4}$ (e) $(2x^3)^{-2}$

7 This diagram shows a sloping railway line.
Calculate the distance labelled d.

14°
28 m
d

8 (a) Write, as an improper fraction, the reciprocal of $\frac{3}{8}$.

(b) As a decimal, $\frac{3}{8} = 0.375$.
Without using a calculator, work out the reciprocal of 0.375 to three decimal places.

9 What is the lowest common multiple of
   (a) 12 and 15
   (b) 30 and 45
   (c) 80 and 120
   (d) 80 and 112

10 Solve each of these equations.
   (a) $\dfrac{x+2}{2} = \dfrac{x-4}{4}$
   (b) $\frac{1}{3}(2x + 5) = \frac{1}{2}(5 - 2x)$

11 There are 240 children in a school.
   $\frac{1}{20}$ of the children are left-handed girls.
   $\frac{1}{8}$ of the girls are left-handed.
   $\frac{3}{8}$ of the left-handed children are girls.

	Girls	Boys
Left-handed		
Right-handed		

   Copy and complete the two-way table to show the number in each section.

12 Work out the starting number in each of these 'think of a number' puzzles.
   (a)
   > I think of a number.
   >
   > I halve it and add 9.
   >
   > My answer is double the number I first thought of.

   (b)
   > I think of a number.
   >
   > I treble it.
   >
   > I take the result off 50.
   >
   > My answer is 10 less than the number I thought of.

13 In a village, the ratio of males to females is $3:5$.
   What fractions go in the gaps in the following statements?
   (a) The number of males is … of the number of females.
   (b) Females make up … of the population of the village.
   (c) Half of the males are over fifty, so males over fifty make up … of the population.

14 Form and solve an equation to calculate the angles in this quadrilateral.

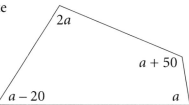

15 A landowner divides the area of her land between orchards and fields in the ratio $6:1$.
   The orchard area is divided between apples and plums in the ratio $2:1$.

   What fraction of the total area of land is
   (a) apple orchards
   (b) plum orchards
   (c) fields

16 Simplify each of these expressions as far as possible.
   (a) $p(1 - q) - q(1 - p)$
   (b) $a^2bc \times ab^3c^4$
   (c) $a^2(b + 3c) - c(a^2 - 2b)$
   (d) $(2x^2y)^3$
   (e) $x(3 - y^2) + 4xy(x^2 - y)$
   (f) $(3a^2b)^{-2}$

17 (a) Calculate the value of $x^3 - 3x$ when $x = 2.5$ and when $x = 2.6$.

   (b) By trial and improvement find, to two decimal places, the value of x for which $x^3 - 3x = 9$.

18 In this diagram, the sides of a pentagon have been extended to form a five-pointed star.

   (a) Explain why $p = a + b - 180°$.

   (b) Write down similar expressions for q, r, s and t.

   (c) Show that the sum of the angles p, q, r, s and t is

$$2S - 900°$$

where S is the sum of the interior angles of the pentagon.

   (d) A pentagon can be split into three triangles. So what is the sum of the interior angles of a pentagon?

   (e) Hence write down the sum of the angles p, q, r, s and t.

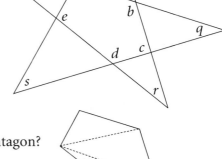

19 The points A ($^-$1, 5), B (1, 1) and C (5, 3) are three vertices of a square.

Find

   (a) the coordinates of D, the fourth vertex

   (b) the coordinates of the centre of the square

   (c) the length of each side of the square, to two decimal places

   (d) the gradient of AC

20 A museum has a collection of 34 old cars. Some have three wheels and some four.
The number of three-wheeled cars is n.
The total number of wheels on all the cars is 125.

   (a) Write an expression in terms of n for the number of four-wheeled cars.

   (b) Use the information about the total number of wheels to form an equation and solve it to find n.

Challenge

The number 32 050 000 000 has seven terminal zeros: 32 050 **000 000**.

How many terminal zeros are there in the result of the following calculation?

$$1 \times 2 \times 3 \times 4 \times ... \times 98 \times 99 \times 100$$

(The answer is not 11.)

6 Rates

You will revise

◆ the relationships between speed, time and distance

◆ distance–time graphs

This work will help you solve problems involving various rates.

A Speed: review

A train takes 5 hours to cover the distance of 410 miles from London to Edinburgh.

The **average speed** of the train is found by dividing the distance by the time.

Average speed $= \frac{410}{5} = 82$ m.p.h.

Speed calculations can also be done by informal methods, for example:

40 miles in $2\frac{1}{2}$ hours $\to$ 80 miles in 5 hours $\to$ 16 m.p.h.

80 miles in 1 hour 40 minutes $\to$ 80 miles in 100 minutes $\to$ 8 miles in 10 minutes
$\to$ 48 miles in 60 minutes $\to$ 48 m.p.h.

A1 Calculate the average speed, in m.p.h., of a train that travels

(a) 78 miles in 2 hours   (b) 144 miles in 3 hours   (c) 35 miles in $\frac{1}{2}$ hour

(d) 27 miles in $1\frac{1}{2}$ hours   (e) 14 miles in 15 minutes   (f) 49 miles in $3\frac{1}{2}$ hours

(g) 18 miles in 40 minutes   (h) 42 miles in 1 hour 10 minutes

A2 Calculate the average speed, in m.p.h., of a car that travels

(a) 6 miles in 5 minutes   (b) 35 miles in 25 minutes

(c) 39 miles in 1 hour 5 minutes

A3 A coach leaves Bristol at 10:45 and arrives in Birmingham at 13:00.
The distance from Bristol to Birmingham is 90 miles.

Calculate the average speed of the coach.

A4 This diagram shows a coach journey.

London	89 miles	Peterborough	51 miles	Lincoln
O		●		O
09:20		11:20		12:50

(a) Calculate the average speed of the coach between London and Peterborough.

(b) Calculate the average speed between Peterborough and Lincoln.

(c) Calculate the average speed between London and Lincoln.

B *Distance–time graphs*

A coach leaves a motorway service area and travels at a constant speed of 50 m.p.h.
After 1 hour it has travelled 50 miles, after 2 hours 100 miles, and so on.

This **distance–time graph** shows its journey.

The **gradient** of the graph shows the speed.

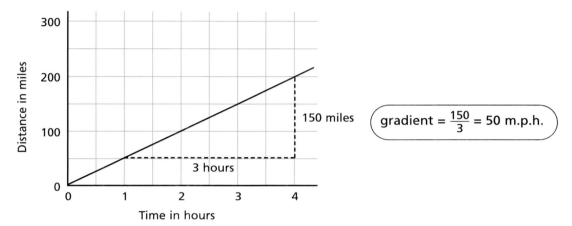

150 miles

$$\text{gradient} = \frac{150}{3} = 50 \text{ m.p.h.}$$

B1 The solid line is the distance–time graph of a model car.

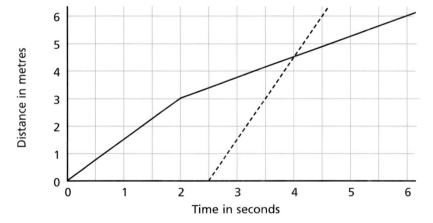

(a) What is the speed of the car during the first 2 seconds?

(b) What happens to the speed of the car 2 seconds from the start?

(c) What is the speed after that?

The dotted line is the distance–time graph of another model car.

(d) What is the speed of the second car?

(e) Where and when does the second car overtake the first?

A third model car starts 4.5 seconds after the first and overtakes it 1.5 seconds later.

(f) What is the speed of the third car?

B2 This graph shows the journey of a tourist train running between two stations.

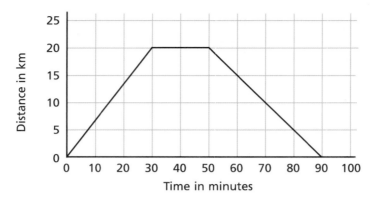

(a) How far apart are the two stations?

(b) What is the speed of the train, in km per minute, on the outward journey?

(c) What is this speed in km per hour (km/h)?

(d) What is the speed of the train, in km/h, on the return journey?

B3 This diagram shows the journeys of four ferries crossing the Channel.

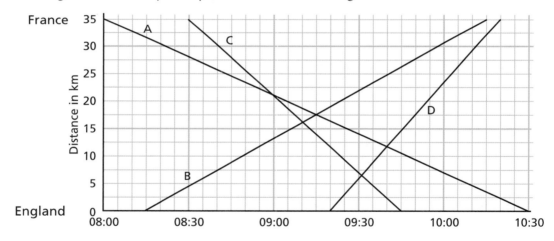

(a) Find the average speed of each ferry.

(b) Which ferry is overtaken by another going in the same direction?

(c) Peter is fishing from a stationary boat in the Channel, 10 km from England. In which order does he see the ferries?

(d) Alphonse is also fishing in the Channel.
He sees the ferries in this order: B, C, A, D.
What are his minimum and maximum distances from England?

(e) Patrice crosses the Channel in a speedboat. She leaves France at 09:10 and arrives in England at 09:50. Which ferries does she pass and in which order?

C *Calculating distances and times*

The formula **speed** = $\dfrac{\text{distance}}{\text{time}}$ can be rewritten as **distance = speed × time.**

For example, if a car travels at 40 m.p.h. for 3 hours, the distance travelled is 120 miles.

From the formula **distance = speed × time** we get **time** = $\dfrac{\text{distance}}{\text{speed}}$.

For example, if a boat travels 24 miles at 6 m.p.h. it takes 4 hours for the journey.

When you use these formulas, the units have to be consistent.
For example, if the distance is in km and the time in minutes, the speed will be in km/min.

Example

Calculate the distance travelled in 25 minutes by a car travelling at 24 km/h.

> Either convert the speed to km per minute or the time to hours.
>
> 24 km/h = $\dfrac{24}{60}$ km/min = $\dfrac{2}{5}$ km/min $\qquad$ 25 minutes = $\dfrac{25}{60}$ hour = $\dfrac{5}{12}$ hour
>
> distance = $\dfrac{2}{5} \times 25$ = **10 km** $\qquad\qquad$ distance = $24 \times \dfrac{5}{12}$ = **10 km**

C1 Calculate the distance travelled in
   (a) 4 hours at 18 m.p.h.
   (b) 22 hours at 40 km/h
   (c) 15 seconds at 6 m/s
   (d) 30 minutes at 44 m.p.h.
   (e) 15 minutes at 28 km/h
   (f) 4 minutes at 3 m/s
   (g) 20 minutes at 15 m.p.h.

C2 Calculate the time taken to travel
   (a) 150 miles at 25 m.p.h.
   (b) 300 m at 20 m/s
   (c) 280 km at 80 km/h
   (d) 120 miles at 24 miles per minute
   (e) 3 km at 150 m/s

C3 A hydrofoil travels a distance of 48 km at a speed of 36 km/h.
   Calculate the journey time
   (a) in hours, using fractions
   (b) in hours and minutes

C4 A boat travels a distance of 10 miles at a speed of 24 m.p.h.
   Calculate the journey time
   (a) in hours
   (b) in minutes

C5 Joanna set out from home at 10 a.m. By 11:15 a.m. she had travelled 55 miles.

   If she maintains the same average speed for another 45 minutes,
   how much further will she travel in this time?

D *Time on a calculator*

Example 1

A ferry travels a distance of 51.6 km at a speed of 24 km/h.
How long does the journey take, in hours and minutes?

Time taken = $\dfrac{distance}{speed}$ = $\dfrac{51.6}{24}$ = 2.15 hours

Convert the decimal part of an hour to minutes: 0.15 hour = 0.15 × 60 minutes = 9 minutes

So journey time = **2 hours 9 minutes**

Example 2

A train travels a distance of 44.5 km in 25 minutes.
What is its average speed in km/h?

Convert 25 minutes to a decimal of an hour: 25 minutes = $\dfrac{25}{60}$ hour = 0.4166… hour

Average speed = $\dfrac{distance}{time}$ = $\dfrac{44.5}{0.4166…}$ = 107 km/h (to the nearest 1 km/h)

Give the answers to the following questions to a reasonable degree of accuracy.

D1 Karl took 46 minutes to jog a distance of 6.4 km.
Calculate his average speed in km/h.

D2 Bharat's private aircraft flies at 256 m.p.h.
How long does it take to cover a distance of 145 miles?
Give your answer in minutes.

D3 Joanna's motor boat travels at 16 km/h.
How far does it travel in 35 minutes?

D4 Holly drove from Oxford to Bristol to deliver a parcel and then immediately
back to Oxford.

The distance from Oxford to Bristol is 74 miles.
Holly left Oxford at 0940, delivered the parcel at 1130 and arrived back in
Oxford at 1440.

What was Holly's average speed

(a) between Oxford and Bristol   (b) between Bristol and Oxford

(c) for the whole journey

E *Other rates*

Speed, measured in miles per hour or metres per second and so on, is an example of a **rate**. Other examples are:

the rate at which water flows from a pipe, measured in litres per second

the rate at which a photocopier makes copies, measured in copies per minute

the rate at which a household uses water, measured in litres per day

'5 litres **per** second' means '5 litres in every second'.

$$\text{Rate in litres per second} = \frac{\text{number of litres}}{\text{number of seconds}}$$

The main source of difficulty in calculating with rates is in deciding whether to multiply or divide, or which way round to divide.
It sometimes helps to simplify the numbers.

Example

Water comes out of a tap at the rate of 0.7 litre/second.
How long will it take to fill an 8 litre bottle?

Make up a similar question with 'easy' numbers.

Suppose the rate is 2 litre/second and the bottle holds 8 litres. The time taken is 4 seconds. You get it by dividing 8 by 2.

Time = $\frac{8}{0.7}$

= 11.428... seconds

A reasonable degree of accuracy here is **11 seconds**

Give the answers to the following questions to a reasonable degree of accuracy and state the units in each case.

E1 Serena is paid by the hour. She works for $4\frac{1}{2}$ hours and earns £24.30.
What is her hourly rate of pay?

E2 A furnace burns fuel at a rate of 5.5 kg per hour.
How long will 40 kg of fuel last?

E3 How long will it take to pump 45 litres of petrol at the rate of 1.2 litres per second?

E4 It take 2 minutes 50 seconds to put 520 litres of fuel into a tank.
What is the rate of delivery of the fuel in litres per second?

E5 Alicia's average pulse rate is 70 beats per minute.
How long does it take her heart to beat 1 million times?

E6 A pump operating at 7.5 litres per second took 1 hour 25 minutes to pump out the water from a flooded basement.
How much water was in the basement before pumping?

E7 An empty water tank has a leak.
Water flows into a tank at a rate of 2.4 litres per second.
After $2\frac{1}{2}$ minutes the quantity of water in the tank is 300 litres.

At what rate does the water leak from the tank?

E8 A small pump can empty a pool in 20 minutes working at a rate of
450 litres per minute.
A large pump could empty the same pool in 12 minutes.
How long would it take to empty the pool if both pumps worked together?

*E9 The fuel consumption of cars is measured in **litres per 100 km.**

(a) On a test run of 40 km, a car used 3.8 litres of fuel.
Calculate the fuel consumption of the car in litre/100 km.

(b) The fuel consumption of a van is 13.6 litre/100 km.
Calculate

(i) the amount of fuel the van uses to travel 450 km

(ii) the distance that the van can travel on 66 litres of fuel

Test yourself

T1

A ●————————————————— ● B
$3\frac{3}{4}$

The diagram shows two junctions on a motorway.
Each junction is labelled with a letter in a circle.
The distance from junction A to junction B is $3\frac{3}{4}$ miles.
Sophie drove from junction A to junction B in 5 minutes.

Work out her average speed, in miles per hour, for this part of the journey.   Edexcel

T2 A tiger runs at a speed of 50 kilometres per hour for 9 seconds.
How many metres does the tiger run?   AQA(SEG) 2000

T3 A train travels a distance of 165 miles in 1 hour 36 minutes.
Find the average speed of the train, to an appropriate degree of accuracy,
stating clearly the units of your answer.

T4 A printer takes 35 minutes to print 80 copies of a leaflet.

(a) What is the rate of printing

(i) in copies per minute   (ii) in copies per hour

(b) How long will the printer take to print 500 copies of the leaflet?
Give your answer to a reasonable degree of accuracy.

Distributions and averages

You will revise

◆ median and range

◆ calculating or estimating the mean of a frequency distribution

◆ stem-and-leaf tables

This work will help you

◆ group continuous data and represent it graphically by a bar chart or frequency polygon

◆ calculate a moving average and use it to detect a trend

A Review

A1 A coach company kept a record of the numbers of passengers in coaches from Manchester to Sheffield one day. Here are the numbers.

   34   27   41   48   30   33   29   44   40   36   28   33   45

(a) What is the median number of passengers in a coach?

(b) What is the range of the number of passengers?

Here are the numbers for coaches from Manchester to Liverpool.

   26   29   38   42   27   29   25   37   38   41   33   32

(c) What are the median and range of this set of data?

(d) Write a couple of sentences comparing the two sets of data.

(e) Calculate the mean number of passengers in each of the two sets of coaches.

(f) If you compare using the means, do you come to the same conclusion as with the medians?

A2 This frequency chart comes from a survey of the number of people in cars travelling past a checkpoint.

(a) How many cars were there in the survey?

(b) How many people were there altogether in the cars?

(c) Calculate the mean number of people per car.

(d) What is the median number in a car?

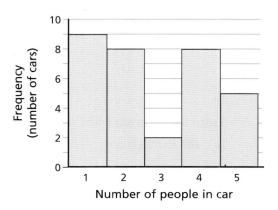

A3 A survey of the number of people in each household in a street resulted in the data shown on the right.

(a) How many households are there in the street?

(b) How many people altogether live in the street?

(c) Calculate the mean number of people per household.

Number of people	Frequency
1	5
2	8
3	12
4	18
5	9
6	4
7	2

A4

Number of children	Frequency
0	13
1	7
2	20
3	9
4	1

This data comes from a survey of the number of children (under 16) in the families in another street.

Calculate the mean number of children per family.

Stem-and-leaf tables

Here are the ages of the members of a club.

45	22	19	26	18	30	48	55	27	38
33	61	47	33	28	40	41	37	56	20
47	53	22	54	27	31	25	24	58	39

To make a stem-and-leaf table of the ages, use the tens digit as the 'stem' and the units digit as the 'leaf'.

Here the first two numbers, 45 and 22, have been entered.

```
1 |
2 | 2
3 |
4 | 5
5 |
6 |
```

Now all the numbers have been entered.

```
1 | 9 8
2 | 2 6 7 8 0 2 7 5 4
3 | 0 8 3 3 7 1 9
4 | 5 8 7 0 1 7
5 | 5 6 3 4 8
6 | 1
```

Finally the 'leaves' are put in order of size.

```
1 | 8 9
2 | 0 2 2 4 5 6 7 7 8
3 | 0 1 3 3 7 8 9
4 | 0 1 5 7 7 8
5 | 3 4 5 6 8
6 | 1
```

A5 From the stem-and-leaf table above, find

(a) the median age   (b) the range of the ages   (c) the modal age group

A6 (a) Make a stem-and-leaf table for this set of examination marks.

(b) Find the median mark and the range of the marks.

Marks out of 80

56	39	47	28	66	72	24	35	47	58
63	70	30	49	56	44	68	41	55	47
69	31	71	63	55	39	28	44	51	70

A7 (a) Make a stem-and-leaf table for this set of data. Use the units digit as the stem and the tenths digit as the leaf.

(b) Find the median weight and the range of the weights.

Weights, in kg, of 30 newborn babies

2.5	3.2	1.7	2.6	3.8	4.2	3.6	2.8	1.9	0.8
4.0	2.7	3.1	3.0	3.3	2.8	1.8	2.6	1.7	2.7
3.5	3.5	2.9	3.3	2.2	3.6	1.5	1.7	2.8	3.7

A8 Write a couple of sentences comparing the boys' and girls' test marks shown in this double stem-and-leaf table.

Boys		Girls
5 5 4 3 3 2	4	1 3 6
8 6 6 4 2 2 1 0	5	2 5 5 7 8
7 7 4 2 2 1	6	1 3 3 5 7 8 9 9
5 4 2	7	0 2 4 4 6 7 8

Mid-interval values

This table gives information about the numbers of passengers in buses leaving a bus station.

The numbers of passengers have been grouped 0–9, 10–19, 20–29, ... so we do not know exactly how many there were in each bus.

So we cannot calculate the mean number accurately. But we can **estimate** it using **mid-interval values**.

The mid-interval value for the interval 0–9 is 4.5.

We count all the buses in this interval as if they had 4.5 passengers each, and similarly for the other intervals.

Number of passengers	Mid-interval value	Frequency
0–9	4.5	5
10–19	14.5	8
20–29		11
30–39		9
40–49		5
50–59		3

A9 Complete the table above and calculate an estimate of the mean number of passengers per bus.

A10 This table comes from a survey of the numbers of children in the primary school classes in a town.

Calculate an estimate of the mean number of children in a class.

Number of children	Frequency
1–10	4
11–20	18
21–30	26
31–40	12

A11 This frequency table shows the distribution of students' marks in a test. The test was marked out of 60.

Marks	1–10	11–20	21–30	31–40	41–50	51–60
Frequency	0	3	12	37	29	44

(a) Which class interval contains the median mark?

(b) Calculate an estimate of the mean mark.

B *Continuous data*

Discrete and continuous quantities

The number of people in a car is an example of a **discrete** quantity.
It can be 0, 1, 2, 3, … but not numbers in between, such as 2.3.

Weight is an example of a **continuous** quantity. A person's weight could be,
for example, 46.783… kg (if the weighing machine were accurate enough).

Describing class intervals

This dot plot shows the weights of a group of newborn babies.

Weight in kg

To group the data we could use class intervals such as 0–1, 1–2, 2–3, and so on.
But there is a problem: what do we do about a weight of, say, exactly 3 kg?

Suppose we decide to put 3 kg in the interval 2–3.
Then we can describe the interval like this.

$2 < \text{weight} \leq 3$ or $2 < w \leq 3$ (where w stands for weight in kg)

The frequency table looks like this.

Weight (w kg)	$0 < w \leq 1$	$1 < w \leq 2$	$2 < w \leq 3$	$3 < w \leq 4$	$4 < w \leq 5$
Frequency	2	4	7	6	3

Frequency polygon

In a frequency polygon, the frequency is plotted at the mid-interval value.

This is the frequency polygon for the babies' weights.

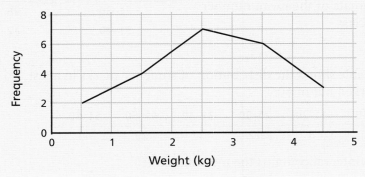

B1 This dot plot shows the times, in seconds, for which a group of people could hold their breath.

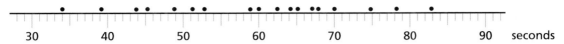

(a) Copy and complete this frequency table for the data.

(b) Which is the modal interval?

Time (t seconds)	Frequency
30 < t ≤ 40	2
40 < t ≤ 50	
50 < t ≤ 60	
60 < t ≤ 70	
70 < t ≤ 80	
80 < t ≤ 90	

B2 Here are the heights in metres of two groups of students.
Each set of data is given in height order.

Girls	1.51	1.53	1.54	1.55	1.56	1.58	1.58	1.60	1.62	1.63
	1.63	1.66	1.67	1.67	1.69	1.70	1.70	1.71	1.73	1.75
Boys	1.57	1.57	1.60	1.62	1.66	1.68	1.70	1.70	1.70	1.70
	1.72	1.74	1.75	1.77	1.79	1.82	1.82	1.83	1.83	1.86

(a) Find the median height of the girls and the median height of the boys.

(b) Find the range of the girls' heights and the range of the boys' heights.

(c) Copy and complete this table.

(d) Draw, on the same axes, a frequency polygon for the girls and for the boys.

(e) Write a couple of sentences comparing the girls' and boys' heights.

Height (h m)	Frequency (girls)	Frequency (boys)
1.5 < h ≤ 1.6	8	
1.6 < h ≤ 1.7		
1.7 < h ≤ 1.8		
1.8 < h ≤ 1.9		

Height (h cm)	Frequency
120 < h ≤ 130	3
130 < h ≤ 140	5
140 < h ≤ 150	8
150 < h ≤ 160	7
160 < h ≤ 170	4

B3 This table shows the distribution of the heights of a group of women.

(a) Draw a frequency polygon for this data.

(b) Calculate an estimate of the mean height of the women.

B4 This table shows the distribution of the lengths of some snakes.

Calculate an estimate of the mean length of the snakes.

Length (L cm)	Frequency
0 < L ≤ 50	4
50 < L ≤ 100	10
100 < L ≤ 150	17
150 < L ≤ 200	9

C Different groupings

The same raw data can be grouped into class intervals in different ways, resulting in different frequency tables and charts.

Here for example are different ways of grouping a set of heights.

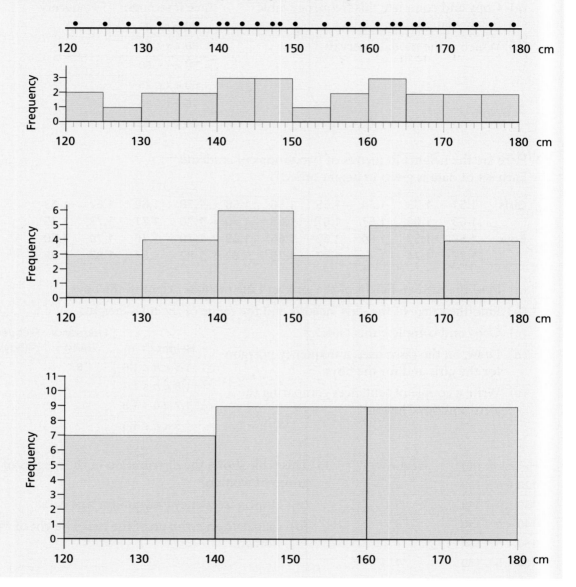

C1 (a) Which of the three frequency charts above best shows the overall 'shape' of the distribution?

(b) Draw a rough sketch of the frequency chart you would get by using only two class intervals, 120–150 and 150–180.

C2 (a) Make a frequency table for the data below. Choose your own class intervals.

Weights, in kg, of fish caught in a competition

| 2.3 | 3.5 | 0.8 | 8.4 | 10.6 | 5.5 | 1.7 | 22.4 | 3.7 | 12.0 | 7.5 | 2.2 | 24.8 | 9.4 | 17.2 | 4.1 |
| 11.6 | 4.1 | 10.7 | 20.9 | 3.1 | 14.4 | 6.2 | 15.8 | 8.4 | 23.2 | 4.0 | 3.5 | 2.1 | 0.6 | 1.9 | 4.9 |

(b) Draw a frequency chart for the data.

(c) From your frequency table, calculate an estimate of the mean weight of the fish.

(d) Now calculate the mean of the actual weights, and compare your estimate with it.

D Trends

Saveway profits rise steadily

The profits of supermarket giant Saveway show a steady upward trend over a seven-year period, the company reported yesterday.

'One must be careful about predicting the future from the past,' said a company spokesman, 'but we believe that the trend will

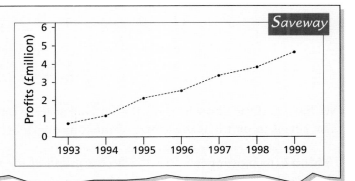

What's happening to unemployment in Marby-on-Sea?

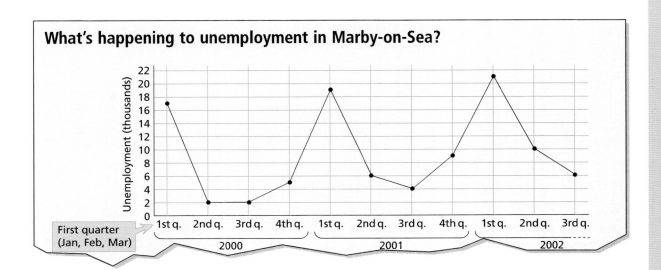

One way to see a trend in the unemployment figures would be to take the mean for each year and see how it changes over time. But we would have to wait for a whole year's figures before we could compare with previous years.

To get a more frequent update we can use a **moving average**. We use overlapping periods of four quarters, like this, and find the mean for each period.

1st	2nd	3rd	4th	
	2nd	3rd	4th	1st
		3rd	4th	1st	2nd

This is called a **4-point moving average**.

The following example uses the data for the Marby-on-Sea graph.

Year	2000				2001				2002		
Quarter	1	2	3	4	1	2	3	4	1	2	3
Unemployment (000)	17	2	2	5	19	6	4	9	21	10	6

$$\frac{17 + 2 + 2 + 5}{4} = 6.5$$

$$\frac{2 + 2 + 5 + 19}{4} = 7 \quad \text{... and so on.}$$

We have to decide how to show the moving average on the graph.
The rule is to plot it at the centre of the four quarters it covers.

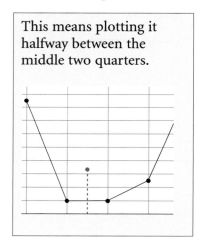

This means plotting it halfway between the middle two quarters.

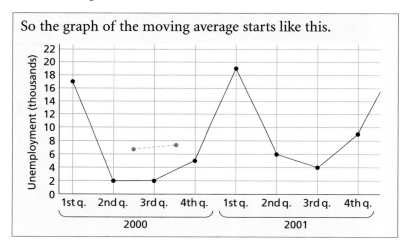

So the graph of the moving average starts like this.

In the table we can show the moving average like this:

Year	2000				2001				2002		
Quarter	1	2	3	4	1	2	3	4	1	2	3
Unemployment (000)	17	2	2	5	19	6	4	9	21	10	6
4-point moving average			6.5	7							

D1 (a) Complete the calculation above of the moving average.

(b) Describe the trend in unemployment.

A short cut

You can calculate the next value of the moving average by adjusting the value before it.

The first moving average calculation is $\dfrac{17 + 2 + 2 + 5}{4}$

The next is $\dfrac{2 + 2 + 5 + 19}{4}$

To get from the total $17 + 2 + 2 + 5$ to the total $2 + 2 + 5 + 19$, we have taken away the 17 and replaced it by the 19.

The total changes by $19 - 17$, or 2.
So the moving average changes by $\frac{2}{4}$ or **0.5**.

The first value is 6.5, so the next value will be 7.

D2 This table shows the numbers of visitors to a seaside town.

Year	1998			1999				2000	
Quarter	2	3	4	1	2	3	4	1	2
Visitors (000)	16	23	10	7	15	21	11	3	13

(a) Calculate a 4-point moving average and show it, with the original data, on a graph.

(b) Describe the trend.

D3 This table shows the numbers of visitors to a museum.

Year	1998			1999				2000	
Quarter	2	3	4	1	2	3	4	1	2
Visitors (000)	7	9	14	22	10	10	17	25	11

(a) Calculate a 4-point moving average and show it, with the original data, on a graph.

(b) Describe the trend.

D4 This table shows the number of visitors (in thousands) at a tourist attraction each month.

Month	04/99	05/99	06/99	07/99	08/99	09/99	10/99	11/99	12/99	01/00
Number	18	19	29	33	31	23	10	6	9	5

Month	02/00	03/00	04/00	05/00	06/00	07/00	08/00	09/00	10/00	11/00
Number	7	14	20	22	25	28	27	20	12	7

(a) Choose an appropriate moving average and calculate it.

(b) Describe the trend in the data.

Test yourself

T1 Here are the heights in centimetres of a group of students.

155	147	162	174	150	166	142	139	148	152
175	170	154	182	180	148	158	166	166	177
160	163	172	158	155	170	170	181	159	161

Copy and complete this frequency table.

Height h in cm	Frequency
$130 < h \leq 140$	
$140 < h \leq 150$	

T2 This table shows the distribution of the weights of the apples picked from a tree.

Weight w in grams	Frequency
$0 < w \leq 20$	7
$20 < w \leq 40$	10
$40 < w \leq 60$	14
$60 < w \leq 80$	20
$80 < w \leq 100$	18
$100 < w \leq 120$	4

(a) How many apples have weights in the interval $40 < w \leq 100$?

(b) Draw a frequency polygon to show the distribution.

(c) Calculate an estimate of the mean weight of the apples.

T3 The total rainfall figures, in millimetres, for the past 7 years in Egypt are shown below.

      27   24   31   30   28   15   29

Find the five-yearly moving averages.   OCR

T4 This table shows the quarterly profits of a garden centre.

Year	1998				1999				2000	
Quarter	1	2	3	4	1	2	3	4	1	2
Profits (000)	32	58	75	41	30	63	77	40	36	65

(a) Calculate a 4-point moving average.

(b) Draw a graph showing the original data and the moving average.

(c) Comment on the trend.

8 Changing the subject 1

You should know how to solve equations like $\frac{6t}{5} = 12$, $\frac{a}{5} - 2 = 10$, $2k^2 = 18$ or $28 - 3a = 7$.

This work will help you rearrange a formula where the new subject appears only once.

A Review

A1 The formula connecting the number of black beads (b) and the number of white beads (w) in these patterns is $b = 4w + 2$.

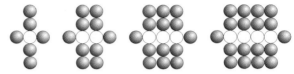

(a) Find b when $w = 20$.
(b) Find b when $w = 62$.
(c) Find w when $b = 50$.
(d) Find w when $b = 550$.

A2 The quantities v, u, a and t are connected by the formula $v = u + at$.

(a) You are told that $t = 3$, $u = 12$ and $v = 30$.

    (i) Substitute these values into the formula to get an equation in which a is unknown.

    (ii) Solve the equation to find the value of a.

(b) Find the value of u when $v = 17$, $a = 1\frac{1}{2}$ and $t = 3$.

(c) Find the value of t when $v = 20$, $u = 14$ and $a = 5$.

(d) Find the value of a when $v = 21$, $u = 25$ and $t = 8$.

A3 The formula connecting C with p and q is $C = \frac{p}{5} - 3q$.

(a) Find C if $p = 200$ and $q = 12$.
(b) Find p if $C = 9$ and $q = 2$.
(c) Find q if $C = 2$ and $p = 55$.
(d) Find q if $C = 3$ and $p = 180$.

A4 The formula connecting f, x, y and z is $f = \frac{xy}{z}$.

(a) Find f when $x = 10$, $y = 20$ and $z = 8$.
(b) Find x when $f = 1$, $y = 25$ and $z = 8$.
(c) Find y when $f = 20$, $x = 5$ and $z = 6$.
(d) Find z when $x = 3$, $y = 8$ and $f = 6$.

A5 The formula for the volume V of a pyramid with height h and a square base of side length b is $V = \frac{1}{3}b^2h$.

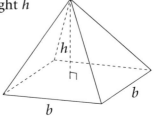

(a) Find V when $b = 3$ and $h = 5$.
(b) Find h when $b = 2$ and $V = 24$.
(c) Find b when $h = 6$ and $V = 50$.
(d) Find b when $h = \frac{1}{2}$ and $V = 24$.

B *Moving letters*

In the bead pattern in question A1, the formula $b = 4w + 2$ connects the variables b and w.

We can find w when $b = 50$
like this.

$$50 = 4w + 2$$
$$48 = 4w$$
$$12 = w$$

Take 2 from both sides

Divide both sides by 4

We can work with the letters themselves
instead of numbers.

$$b = 4w + 2$$
$$b - 2 = 4w$$
$$\frac{b - 2}{4} = w, \text{ or } w = \frac{b - 2}{4}$$

Each time, we do the same thing to both sides of the formula.
We call this 'rearranging the formula to make w the subject'.

Now we can find the value of w easily for different values of b.

B1 This bead pattern has formula $b = 3w + 4$.

(a) Rearrange the formula to make w the subject.

(b) What is the value of w when $b = 70$?

(c) What is the value of w when $b = 271$?

(d) Check that the values of b and w
in parts (b) and (c) fit the original formula.

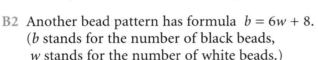

B2 Another bead pattern has formula $b = 6w + 8$.
(b stands for the number of black beads,
w stands for the number of white beads.)

(a) Rearrange the formula to make w the subject.

(b) How many white beads are there when there are 350 black beads?

(c) How many white beads are there when there are 440 black beads?

(d) Check that your answer to part (c) fits the original formula.

B3 (a) Copy and complete this working to make x
the subject of the formula $y = 5x - 6$.

(b) Use your new formula to find x when $y = 129$.

(c) Use the value of x you found in part (b)
to check that your rearrangement is correct.

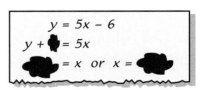

$$y = 5x - 6$$
$$y + \blacklozenge = 5x$$
$$\blacksquare = x \text{ or } x = \blacksquare$$

B4 Rearrange each of these formulas to make the bold letter the subject.

(a) $b = 8\boldsymbol{w} + 7$ (b) $u = 5\boldsymbol{v} - 2$ (c) $g = 6\boldsymbol{d} - 12$ (d) $y = 12 + 3\boldsymbol{x}$

(e) $t = 3\boldsymbol{b} - 5$ (f) $f = 3\boldsymbol{d} + 8$ (g) $h = \boldsymbol{k} - 5$ (h) $w = 7\boldsymbol{d} + 1$

B5 (a) Copy and complete this working to make x the subject of the formula $y = 16 - 3x$.

(b) Use suitable values of x and y to check that your rearrangement is correct.

(c) Find x when $y = 5.5$.

(d) Find x when $y = {}^-36.5$.

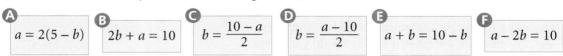

$$y = 16 - 3x$$
$$y + 3x = 16$$
$$3x = 16 - $$
$$x = \underline{16 - }$$

Add 3x to both sides

Take y from both sides

B6 (a) Make r the subject of the formula $s = 10 - 5r$.

(b) (i) Find r when $s = {}^-65$.

(ii) Use this value of r to check that your rearrangement is correct.

B7 Which four of the following are correct rearrangements of $a = 10 - 2b$?
(b need not be the subject of the rearrangement.)

A $a = 2(5 - b)$ **B** $2b + a = 10$ **C** $b = \dfrac{10 - a}{2}$ **D** $b = \dfrac{a - 10}{2}$ **E** $a + b = 10 - b$ **F** $a - 2b = 10$

B8 Rearrange each of these formulas to make the bold letter the subject.

(a) $s = 8 - 7\boldsymbol{t}$ (b) $t = 8\boldsymbol{u} - 7$ (c) $g = 66 - 5\boldsymbol{h}$ (d) $y = 12 - 3\boldsymbol{x}$

(e) $h = 4\boldsymbol{b} - 5$ (f) $m = 3 - 8\boldsymbol{n}$ (g) $y = 15 - 5\boldsymbol{x}$ (h) $w = 7\boldsymbol{b} - 10$

B9 Here are eight formulas.
Find four matching pairs of equivalent formulas.

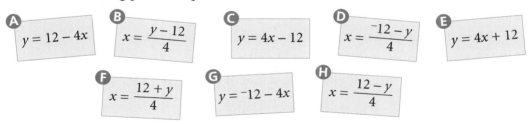

A $y = 12 - 4x$ **B** $x = \dfrac{y - 12}{4}$ **C** $y = 4x - 12$ **D** $x = \dfrac{{}^-12 - y}{4}$ **E** $y = 4x + 12$

F $x = \dfrac{12 + y}{4}$ **G** $y = {}^-12 - 4x$ **H** $x = \dfrac{12 - y}{4}$

B10 (a) Copy and complete this working to make a the subject of the formula $3b - 2a = 30$.

(b) Use suitable values of a and b to check that your rearrangement is correct.

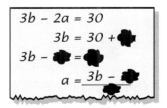

$$3b - 2a = 30$$
$$3b = 30 + $$
$$3b - \ \ = $$
$$a = \underline{3b - }$$

B11 Rearrange each of these formulas to make the bold letter the subject.

(a) $2a + 3\boldsymbol{b} = 30$ (b) $5s - \boldsymbol{t} = 40$ (c) $t = 60 - 12\boldsymbol{g}$ (d) $f = 12 + 3\boldsymbol{b}$

(e) $y = 8\boldsymbol{x} + 12$ (f) $r = 5\boldsymbol{s} - 20$ (g) $2a = 3\boldsymbol{b} - 5$ (h) $2v = 7\boldsymbol{u} - 10$

(i) $x + \boldsymbol{y} = 35$ (j) $d - 4\boldsymbol{j} = 8$ (k) $6k - 8\boldsymbol{j} = 45$ (l) $4\boldsymbol{w} = 7z - 1$

C *More letters*

Some formulas only have letters in them, for example $v = u + at$.
(u stands for the initial speed in $m\,s^{-1}$, v for the final speed in $m\,s^{-1}$,
a for the acceleration in $m\,s^{-2}$, t for the time in seconds.)

We can rearrange formulas containing only letters just as before.

Suppose we wish to make t the subject of $v = u + at$.

We would rearrange $v = 8 + 5t$ like this.

$v = 8 + 5t$ **Take 8 from both sides**

$v - 8 = 5t$

$t = \dfrac{v - 8}{5}$ **Divide both sides by 5**

We rearrange $v = u + at$ like this.

$v = u + at$ **Take u from both sides**

$v - u = at$

$t = \dfrac{v - u}{a}$ **Divide both sides by a**

Since the letters in a formula stand for numbers,
we can treat the letters just as we would numbers.

C1 (a) Rearrange the formula $v = u + at$ to give a in terms of v, u and t.
      (u, v, a and t have the same meaning as above.)

     (b) A dragster can accelerate from $20\,m\,s^{-1}$ to $90\,m\,s^{-1}$
      in 5 seconds. Use your rearrangement
      to work out its acceleration.

C2 The formula $p = m + q$ connects p, m and q.

     (a) Rearrange the formula to make m the subject.

     (b) What is the value of m when $p = 12$ and $q = 15$?

     (c) Check that this value of m and the values of p and q fit in the original formula.

C3 Copy and complete each of these to make m the subject.

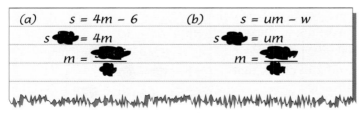

(a) $s = 4m - 6$ *(b)* $s = um - w$

$s \; \blacksquare = 4m$ $s \; \blacksquare = um$

$m = \dfrac{\blacksquare}{\blacksquare}$ $m = \dfrac{\blacksquare}{\blacksquare}$

C4 Rearrange each of these formulas to make the bold letter the subject.

     (a) $b = k\boldsymbol{w} + l$ (b) $l = b\boldsymbol{v} - h$ (c) $g = s\boldsymbol{d} - q$ (d) $y = m\boldsymbol{x} + c$

     (e) $t = 3\boldsymbol{b} - f$ (f) $f = n\boldsymbol{d} + 8$ (g) $h = \boldsymbol{k} - l$ (h) $w = a\boldsymbol{d} + 1$

     (i) $g = 6\boldsymbol{j} + k$ (j) $a = c\boldsymbol{d} + ef$ (k) $2j = 5\boldsymbol{k} + l$ (l) $ak = b\boldsymbol{j} - m$

C5 (a) Copy and complete this working to make h
the subject of the formula $s = f - kh$.

(b) Use your new formula to find h
when $f = 30$, $s = 2$ and $k = 7$.

(c) Use the value of h you found in part (b)
to check that your rearrangement is correct.

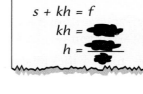

C6 (a) Copy and complete this working to give p
in terms of a, b, k and q.

(b) Find some values that fit the original formula.
Check that they fit your rearrangement.

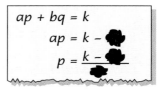

C7 The temperature in a cold store is given by the formula

$$T = T_0 - \alpha h$$

T_0 stands for the starting temperature in °C;
T stands for the temperature in °C
in the store after h hours;
α is a constant and is the rate of cooling in degrees per hour.

T_0 is a single symbol - a useful way of writing 'T at the start'.

(a) Rearrange the formula to make α the subject.

(b) What rate of cooling is needed to cool a store
from 22°C to ⁻18°C in 5 hours?

C8 The mass of a brass bar in kilograms is given by

$$W = aV_1 + bV_2$$

W is the mass of the brass bar in kg;
a is the density of zinc in kilograms per cubic metre ($kg\,m^{-3}$);
V_1 is the volume of zinc in m^3;
b is the density of copper in $kg\,m^{-3}$; V_2 is the volume of copper in m^3.

(a) Rearrange the formula to give V_2 in terms of the other variables.

(b) What volume of copper needs to be added to $0.05\,m^3$ of zinc
to make a brass bar with mass 1000 kg? Give your answer to 2 s.f.
(Density of zinc $= 7100\,kg\,m^{-3}$; density of copper $= 8900\,kg\,m^{-3}$)

C9 Which of these are correct rearrangements of $ax - by = c$?

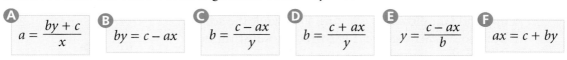

A $a = \dfrac{by + c}{x}$ B $by = c - ax$ C $b = \dfrac{c - ax}{y}$ D $b = \dfrac{c + ax}{y}$ E $y = \dfrac{c - ax}{b}$ F $ax = c + by$

C10 Rearrange each of these formulas to make the bold letter the subject.

(a) $na + m\mathbf{b} = k$ (b) $fs - \mathbf{t} = j$ (c) $t = k - f\mathbf{g}$ (d) $f = 2a - m\mathbf{b}$

(e) $H = H_0 - 8\mathbf{x}$ (f) $r = 5\mathbf{s} - 20$ (g) $a_1 = a_2\mathbf{b} - c$ (h) $v = k_1\mathbf{u} - k_2$

(i) $ax + b\mathbf{y} = r^2$ (j) $cd - 4\mathbf{j} = e$ (k) $ak - b\mathbf{j} = 45$ (l) $ef = 2z\mathbf{w} - c$

D Fractions and squares

You can rearrange formulas with fractions in them.
Just think of what you would do if the letters were numbers.

For example, make k the subject of the formula $a = \dfrac{k}{c} - j$.

Think of the other letters as numbers.

$10 = \dfrac{k}{4} - 5$ **Add 5 to both sides**

$10 + 5 = \dfrac{k}{4}$ **Multiply both sides by 4**

$k = 4 \times (10 + 5)$

Now do the same with the letters.

$a = \dfrac{k}{c} - j$ **Add j to both sides**

$a + j = \dfrac{k}{c}$ **Multiply both sides by c**

$k = c(a + j)$

D1 (a) Copy and complete this working to give g in terms of p and r.

(b) Find some values of g, p and r that fit the original formula.
Check that they fit your rearrangement.

$p = \dfrac{g}{5} + 3r$

$p - \blacklozenge = \dfrac{g}{5}$

$\blacklozenge(p - \blacklozenge) = g$

D2 Rearrange each of these formulas to make the bold letter the subject.

(a) $f = 3 + \dfrac{\boldsymbol{g}}{2}$ (b) $h = \tfrac{1}{2}\boldsymbol{k} - 5j$ (c) $g = \dfrac{\boldsymbol{s}}{2a} - t$ (d) $l = h - \dfrac{\boldsymbol{e}}{d}$

D3 Both of these are wrong.
Explain what each mistake is and write out correct rearrangements.

(a) Make k the subject.	(b) Make h the subject.
$h = j + \tfrac{1}{2}k$	$s = 5 - \dfrac{h}{2}$
$h - j = \tfrac{1}{2}k$	$s + \dfrac{h}{2} = 5$
$h - j + \tfrac{1}{2}k = \tfrac{1}{2}k + \tfrac{1}{2}k = k$	$s + h = 10$
$k = h - j + \tfrac{1}{2}k$ ✗	$h = 10 - s$ ✗

D4 Copy and complete this working to make R the subject.
Check your rearrangement is correct by substituting values in the original formula and in your own.

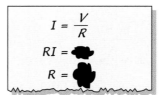

$I = \dfrac{V}{R}$

$RI = \blacklozenge$

$R = \blacklozenge$

D5 Rearrange each of these formulas to make the bold letter the subject.

(a) $Y = \dfrac{\boldsymbol{F}s}{A}$ (b) $V = \dfrac{L\boldsymbol{H}W}{\pi}$ (c) $V = \dfrac{LHW}{\boldsymbol{\pi}}$ (d) $x_1 y_1 = \dfrac{\boldsymbol{x_2} y_2}{6d}$

D6 A cube has edges of length b.
Its total surface area, A, is $6b^2$.

(a) Copy and complete this working
to make b the subject of the formula.

(b) What is the length of the edge of a cube
with total surface area $10\,\text{cm}^2$?
(Give your answer to 3 s.f.)

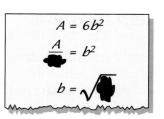

$A = 6b^2$

$\dfrac{A}{\text{⬤}} = b^2$

$b = \sqrt{\text{⬤}}$

D7 Here are some formulas connecting the lengths a and b.
Find four pairs of equivalent formulas.

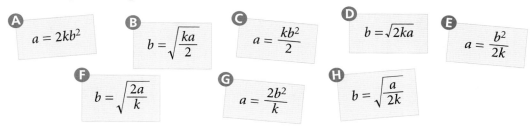

A $a = 2kb^2$

B $b = \sqrt{\dfrac{ka}{2}}$

C $a = \dfrac{kb^2}{2}$

D $b = \sqrt{2ka}$

E $a = \dfrac{b^2}{2k}$

F $b = \sqrt{\dfrac{2a}{k}}$

G $a = \dfrac{2b^2}{k}$

H $b = \sqrt{\dfrac{a}{2k}}$

D8 The area of a circle is given by the formula $A = \pi r^2$.

(a) Rearrange this formula to give r in terms of A and π.

(b) Work out the radius of a circle with area
(i) $100\,\text{cm}^2$ (ii) $5.00\,\text{m}^2$ (iii) $1.00\,\text{mm}^2$ (iv) $600\,\text{km}^2$

Give each answer to an appropriate degree of accuracy.

D9 A stone is dropped from the top of a cliff. Its distance (D metres)
below the top of the cliff after t seconds is given by $D = 4.9t^2$.

(a) Rearrange this formula to give t in terms of D.

(b) How long will it take a stone to hit the ground from the top of a cliff $50\,\text{m}$ high?

D10 Make the bold letter the subject of each of these.

(a) $V = \pi r^2 h$ (b) $V = \pi r^2 \mathbf{h}$ (c) $A = 4\pi r^2$ (d) $V = \tfrac{1}{3}\pi r^2 h$

D11 Write each formula in terms of the bold letter.

(a) $A = \dfrac{\pi \mathbf{d}^2}{4}$ (b) $g = \sqrt{\dfrac{a\mathbf{b}}{6}}$ (c) $a^2\mathbf{b}^2 = 1$ (d) $a^2 + \mathbf{b}^2 = 1$

D12 Which of these are correct rearrangements of the formula $e = s(a - b)$?

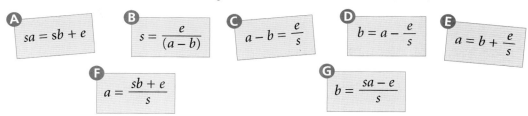

A $sa = sb + e$

B $s = \dfrac{e}{(a - b)}$

C $a - b = \dfrac{e}{s}$

D $b = a - \dfrac{e}{s}$

E $a = b + \dfrac{e}{s}$

F $a = \dfrac{sb + e}{s}$

G $b = \dfrac{sa - e}{s}$

D13 Copy and complete this working to make h the subject of the formula.

Check your rearrangement is correct by substituting suitable values in the original and your rearrangement.

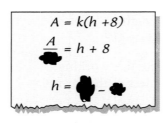

D14 (a) Make d the subject of $f = 25(d - l)$.

(b) Make l the subject of $f = 25(d - l)$.

D15 One of these formulas does not have an equivalent here. Which is it?

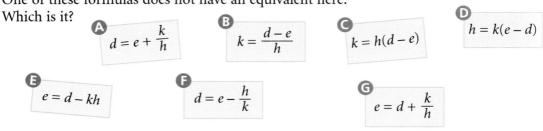

Ⓐ $d = e + \dfrac{k}{h}$

Ⓑ $k = \dfrac{d - e}{h}$

Ⓒ $k = h(d - e)$

Ⓓ $h = k(e - d)$

Ⓔ $e = d - kh$

Ⓕ $d = e - \dfrac{h}{k}$

Ⓖ $e = d + \dfrac{k}{h}$

D16 The area A of a trapezium can be written $A = \dfrac{(a + b)h}{2}$.

(a) Rearrange this formula to make h the subject.

(b) Work out the height of a trapezium whose area is $100 \, \text{cm}^2$ and whose parallel sides are $6.5 \, \text{cm}$ and $7.8 \, \text{cm}$. Give your answer to a sensible degree of accuracy.

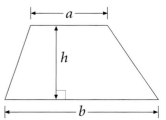

D17 A rope of length L m is formed into a circle.
The area $A \, \text{m}^2$ enclosed by the rope is given by the formula

$$A = \dfrac{L^2}{4\pi}$$

(a) Rearrange this formula to make L the subject.

(b) How long must a rope be for it to enclose an area of $50 \, \text{m}^2$ when formed into a circle? Give your answer in metres, to one decimal place.

(c) What is the total length of rope needed to make two circles each of area $25 \, \text{m}^2$? Is this more or less than what is needed to make one circle of area $50 \, \text{m}^2$?

Test yourself

T1 Rearrange each of these formulas to make the bold letter the subject.
    (a) $a = 6\boldsymbol{r} + 8$ (b) $b = 4\boldsymbol{s} - 6$ (c) $c = 12 + 5\boldsymbol{t}$ (d) $d = 8 - \boldsymbol{u}$
    (e) $e = 12 - 4\boldsymbol{v}$ (f) $2f = 15 + 2\boldsymbol{w}$ (g) $3g + 4\boldsymbol{x} = 12$ (h) $4h - 5\boldsymbol{y} = 10$

T2 Rearrange the equation $x + 2y = 6$ to make y the subject. Edexcel

T3 Rearrange each of these formulas to make the bold letter the subject.
    (a) $n = k + g\boldsymbol{t}$ (b) $s = a\boldsymbol{t} - l$ (c) $y_1 = g\boldsymbol{y_2} + h$
    (d) $m = h - \boldsymbol{u}$ (e) $j = ab - u\boldsymbol{v}$ (f) $A_1 f = A_2 g + v\boldsymbol{w}$
    (g) $kg + j\boldsymbol{x} = s^2$ (h) $jh - x\boldsymbol{y} = b$

T4 You are given the formula $a = bc + d$.
    (i) Calculate a when $b = \frac{2}{5}$, $c = 3$ and $d = {}^{-}0.8$.
    (ii) Rearrange the formula to give b in terms of a, c and d. AQA 2002

T5 Rearrange each of these formulas to make the bold letter the subject.
    (a) $y = \frac{\boldsymbol{x}}{8} + 6$ (b) $f = ab - \frac{1}{3}\boldsymbol{k}$ (c) $g = \frac{\boldsymbol{h}}{2c} - 6j$ (d) $S = \frac{A\boldsymbol{F}M}{2\pi}$

T6 You are given the formula $p = \frac{2}{3}n^2$.

Rearrange the formula to give n in terms of p. AQA 2001

T7 Make d the subject of the following formula.
$$h = \sqrt{t - d}$$
WJEC

T8 The volume of a cone, in cm^3, is given by $V = \frac{\pi r^2 h}{3}$.

r is the base radius and h is the height, both in cm.
    (a) Rearrange the formula to make r the subject.
    (b) Use your rearrangement to work out the base radius of a cone of volume 100 cm^3 and height 8 cm. Give your answer to three significant figures.

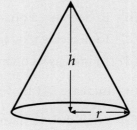

9 Increase and decrease

You should know

- ◆ that the decimal equivalent of, for example, 43% is 0.43
- ◆ how to express one quantity as a percentage of another
- ◆ how to increase a quantity by, for example, 8% by multiplying by 1.08, and how to decrease it by 8% by multiplying by 0.92

This work will help you

- ◆ calculate the overall percentage change given two successive percentage changes
- ◆ calculate the final amount when money is invested at a given rate of interest
- ◆ calculate the original value given a percentage change and the final value

A Review

A1 Write down the decimal equivalent of   (a) 72%   (b) 6%   (c) 14.5%   (d) 3.5%

A2 In a class of 31 children, 13 are absent with flu.
What percentage of the class (to the nearest 1%) are absent with flu?

A3 Express each of these as a percentage, to the nearest 0.1%.
(a) 26 out of 85   (b) 49 out of 65   (c) 64 out of 818   (d) 21.3 out of 34.7

A4 What is missing in each of these statements?
(a) To increase a quantity by 4%, multiply it by ……
(b) To decrease a quantity by 7%, multiply it by ……
(c) To decrease a quantity by ……%, multiply it by 0.88.
(d) To increase a quantity by ……%, multiply it by 1.125.

A5 (a) Increase £43 by 13%.   (b) Increase £74 by 7%.   (c) Decrease £38 by 16%.

A6 (a) What do you have to multiply £58 by to get £66.70?
(b) What is the percentage increase when a price of £58 is increased to £66.70?

A7 What is the percentage increase
(a) from £26 to £33.28   (b) from £48 to £50.40   (c) from £140 to £178.50

A8 (a) What do you have to multiply £64 by to get £60.16?
(b) What is the percentage decrease when a price of £64 is decreased to £60.16?

A9 What is the percentage decrease
(a) from £36 to £30.60   (b) from £120 to £86.40   (c) from £320 to £264

A10 A garage increases its prices by 4% in September.
In the following January, prices go up by 6%.

A car cost £7645 before the first price increase.
What does it cost after the second increase, to the nearest pound?

A11 The population of a town is expected to rise by 7% during the next twelve months.
During the following twelve months a fall of 7% is expected.
If the present population is 28 470, what is it expected to be after 24 months?

A12 In June, 2640 people paid £1.20 each to ride on a miniature railway.
In July the fare went up to £1.50 and 2250 people rode on the railway.
Calculate, to the nearest 0.1%,

(a) the percentage increase in the fare

(b) the percentage decrease in the number of people riding on the railway

(c) the percentage change in the total amount of money paid, stating whether
it was an increase or a decrease

A13 Instone's Olde Fashioned Ginger Beer used to be sold in 1.5 litre bottles
costing £1.85. It is now sold in 2 litre bottles costing £2.65.
Calculate, to the nearest 0.1%,

(a) the percentage increase in the quantity in a bottle

(b) the percentage increase in the price of a bottle

(c) the percentage change in the price per litre

A14 A bedspread is a square 2.5 m by 2.5 m.
After washing, its length and breadth have both shrunk by 15%.
Calculate the percentage reduction in the area of the bedspread.

B *Successive percentage changes*

A hardware shop increases all its prices
by 15% in April, …

… and increases them again
by 20% in September.

Buy now!

Prices up 15% from
1st April!

Buy now!

Prices up 20% from
1st September!

What is the overall percentage change?

Example

The population of a town grows by 12% in one year and by 15% in the next year.
What is the overall percentage increase?

This diagram shows the two increases.

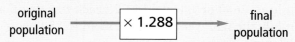

Multiplying by 1.12 and then by 1.15 is equivalent to multiplying by 1.12 × 1.15, or **1.288**.

original population —— × 1.288 ——→ final population

This is equivalent to a **28.8% increase**.

B1 Calculate the overall percentage change for each of these.
  (a) A 25% increase then a 20% increase   (b) A 4% increase then a 15% increase
  (c) A 5% increase then a 30% increase   (d) A 16.5% increase then a 13.5% increase

B2 Calculate the overall percentage change when a 15% decrease is followed
  by a 20% decrease.

B3 Calculate the overall percentage change for each of these.
  (a) A 25% decrease then a 20% decrease   (b) A 3% decrease then a 10% decrease
  (c) A 6% decrease then a 40% decrease   (d) A 9.5% decrease then a 12.5% decrease

B4 Calculate the overall percentage change for each of these, stating whether
  it is an increase or a decrease.
  (a) A 30% increase then a 20% decrease   (b) A 20% decrease then a 30% increase
  (c) A 12% decrease then a 12% increase   (d) A 14.5% decrease then a 20.5% increase

B5 Over the last three years, the wren population in a wood increased by 20% in the first
  year, increased by 15% in the second year and decreased by 25% in the third year.
  Calculate the overall percentage change.

B6 The edges of a cube are each increased in length by 10%.
  Calculate the percentage increase in
  (a) the volume of the cube   (b) the surface area of the cube

*B7 A club wants its membership to increase by 25% over a period of two years.
  After one year it has gone up by 10%. By what percentage must it go up
  over the second year to reach the target?

*B8 Between 1998 and 2000, the lapwing population on an island went up by 51.2%.
  It increased by 35% between 1999 and 2000.
  What was the percentage increase between 1998 and 1999?

C Compound interest

Aunt Ethel wanted to save some money to give to her favourite niece in 10 years time. She considered three ways of saving the money.

A Put £1000 in a jar on the mantelpiece and at the end of each year put in an extra £100.

B Put £1000 into a bank account earning 8% interest every year.

C Put £1000 in a safe, and add £1 after one year, £2 after two years, £4 after three years, £8 after four years, … doubling the amount each year.

What would Aunt Ethel's niece like her to do?

C1 Rajesh puts £500 into a building society account which pays 5% interest p.a. (per annum, i.e. per year)

Copy and complete this table, which shows the amount in the account at the end of each year for 4 years. (Round up to the nearest penny.)

Number of years	Amount
0	£ 500.00
1	£ 525.00
2	
3	
4	

C2 Calculate the final amount when

(a) £800 is invested at 7% p.a. for 3 years   (b) £650 is invested at 4% p.a. for 5 years

(c) £1200 is invested at 2.5% p.a. for 4 years (d) £800 is invested at 3.75% p.a. for 3 years

---

Using the power key on a calculator

If £400 is invested at 6% p.a. for 5 years, the final amount can be calculated like this.

$$400 \times 1.06 \times 1.06 \times 1.06 \times 1.06 \times 1.06 = 400 \times 1.06^5$$

Do 1.06 x^y 5 or 1.06 ^ 5

C3 Calculate the final amount when

(a) £400 is invested at 3% p.a. for 8 years   (b) £750 is invested at 8% p.a. for 9 years

(c) £1500 is invested at 4.5% p.a. for 10 years (d) £300 is invested at 5.5% p.a. for 7 years

C4 Jacqui invests £2000 in an account which pays interest at 4% p.a. Find how many complete years she will have to leave the money in the account for it to become at least £2500.

C5 £5000 is invested in an account paying 6% p.a. interest. For how many complete years will it have to be left to become at least

(a) £5500   (b) £6000   (c) £10 000

C6 Rachael opens a building society account on 1 January 2000.
She puts £1000 into the account to start with, and adds an extra £500 at
the end of each year.

The building society pays interest at the
rate of 6% p.a.

Continue the calculation and find
the amount on 1 January 2004.

Amount on 01/01/00	£1000.00
Interest for year 2000 +	60.00
Investment, 31/12/00 +	500.00
Amount on 01/01/01	1560.00
Interest for year 2001 +	93.60
Investment, 31/12/01 +	500.00
Amount on 01/01/02	2153.60

C7 Sean borrows £2000 from a bank on 1 January.
He agrees to pay back £500 at the end of each
month.

The bank charges interest at 2% per month on
the outstanding amount of the loan.

Amount on 1 January	£2000.00
Interest, January +	40.00
Repayment, 31 Jan −	500.00
Amount on 1 February	1540.00
Interest, February +	30.80
Repayment, 28 Feb −	500.00
Amount on 1 March	1070.80

(a) Continue the calculation until the loan is
fully repaid. (The final repayment will be
less than £500.) When is it finally repaid?

(b) How much is the last repayment?

C8 A hospital physiotherapy department gives ultraviolet treatment.
Every patient having the treatment receives a dose of 1 minute 9 seconds on day 1.
Each day the dose is increased by a percentage which depends on the patient's skin type,
as shown in the table below.

Skin type	Percentage increase per day
1 Always burns	10%
2 Tans with care but burns easily	15%
3 Tans easily and rarely burns	20%
4 Always tans, never burns	25%

(The dose is increased until it reaches a maximum of 46 minutes 18 seconds, when it
is kept constant from then on.)

(a) Janine has skin of type 3. Calculate her dose on day 3.

(b) Karl has skin type 4. On which day will his dose first go above 3 minutes?

(c) Rita has skin type 2. On day 14 her dose is 4 minutes 0 seconds.
What is her dose on day 16?

C9 The population of newts in a pond is decreasing by 7% a year.
There are 428 newts in the pond now. How many will there be in 5 years time?

C10 The value of a secondhand Ford Gerbil decreases by 18% every year.
What is the percentage decrease in its value over a period of 3 years?

C11 A credit card company charges interest at a rate of 2% per month.
Calculate the overall percentage rate of interest for 12 months, to the nearest 0.1%.

C12 Another credit card company's monthly interest rate is 1.5%.
Calculate the annual interest rate, to the nearest 0.1%.

*C13 A loan company charges interest at the rate of 60% per half year.
What is the annual interest rate?

*C14 The Sharks Loans Company is considering different ways of charging interest.

Option A	Charge 78% per year
Option B	78% ÷ 2 = 39%, so charge 39% per six months
Option C	78% ÷ 4 = 19.5%, so charge 19.5% per three months
Option D	78% ÷ 12 is 6.5%, so charge 6.5% per month
Option E	78% ÷ 52 is 1.5%, so charge 1.5% per week

(a) Calculate the equivalent yearly interest rate for each option.

(b) Sharks also considers the option of a daily rate of 78% ÷ 365.
Calculate the equivalent yearly rate for this option.
You could also investigate the option of an hourly rate of 78% ÷ (365 × 24).

D *Percentage in reverse*

Problem 1

Jackie bought a printer for £149.95 from a computer shop.
This price included VAT at 17.5%.

What was the cost of the printer excluding VAT?

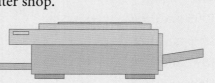

Problem 2

A shop reduces the price of a jacket by 15% in a sale.
The sale price is £39.95.

How much did the jacket cost before the sale?

Worked example

A rail fare goes up by 8%. The new fare is £37.80.
What was the old fare?

The old fare is multiplied by 1.08 to get the new fare.

So the new fare has to be divided by 1.08 to get the old fare.

Old fare = £37.80 ÷ 1.08 = **£35**

$$? \xrightarrow{\times 1.08} 37.80$$

$$35 \xleftarrow{\div 1.08} 37.80$$

D1 The price of a theatre ticket goes up by 6% to £13.25. What was it before the increase?

D2 The cost of hiring a digger goes up by 15% to £96.60. What was it before?

D3 A restaurant adds a 12.5% service charge to customers' bills.
The total cost of a meal, including the service charge, is £18.90.
What did the meal cost before the service charge was added?

D4 The price of a coat goes down by 14% to £27.95. What was it before?

D5 Calculate the original price of an article which costs
    (a) £29.50 after an 18% increase
    (b) £39.60 after a 12% reduction
    (c) £88.80 after a 7.5% reduction
    (d) £50.50 after a 37.5% reduction

E *Mixed questions*

E1 The graph shows the profits
    of a small business.
    (10k = 10 000)

    Calculate the percentage
    increase in the profits
    between

    (a) 1995 and 1996
    (b) 1996 and 1997
    (c) 1997 and 1998
    (d) 1998 and 1999
    (e) 1996 and 1998
    (f) 1996 and 1999
    (g) 1995 and 1998

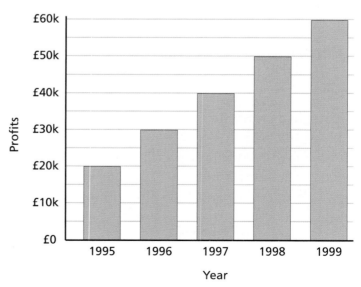

E2 Each year that Jane leaves her money in a bank account the amount grows by 4%.
    By what percentage does it grow in 3 years? Give your answer to the nearest 0.1%.

E3 Calculate the missing entries in this table.

Old price	£26.50	£30.50	(c)	(d)	£48.40	£31.50
Percentage change	up 24%	down 18%	up 16%	down 26%	(e)	(f)
New price	(a)	(b)	£16.82	£13.69	£65.34	£28.98

E4 Kolleeg's Corn Flakes are sold in standard packs of 500 g for £1.20.
During a promotion, the quantity in a pack is increased by 20%,
but the cost stays the same.

Calculate the percentage reduction in the cost per kilogram
during the promotion.

E5 Lord Elaudi is thinking of increasing the entrance fee to his castle by 15%.
His advisers tell him that this will reduce the number of visitors by 14%.
Should he increase the fee? Explain your answer.

Test yourself

T1 Between 1960 and 1990 the population of a town increased by 34%.
Between 1990 and 2000 the population increased by 8%.

Calculate the percentage increase in the population between 1960 and 2000,
to the nearest 0.1%.

T2 Zuhair put £3000 into a bank account which paid 5.5% p.a. interest.
He left the money in the account for 4 years.
Calculate the amount at the end of this period.

T3 The number of visitors to a theme park dropped by 18% between 1998 and 1999.
Improvements were made to the park and the number of visitors rose by 15%
between 1999 and 2000.

Calculate the overall percentage change in the number of visitors between 1998
and 2000.

T4 (a) A shop reduces its prices by 12% in a sale.
What was the original price of a sweater which costs £14.30 in the sale?

(b) A garage increases the prices of cars by 4%.
What was the original price of a car which costs £5876 after the increase?

Review 2

☒ 1 Find the average speed for these journeys.

(a) 115 miles in 5 hours   (b) 245 miles in $3\frac{1}{2}$ hours

2 Borchester Film Club sends out leaflets designed to attract new members, especially young members. Before the leaflets were sent out, the age distribution of the members was as shown in this table.

Age group	15–24	25–34	35–44	45–54	55–64	65–74
Number of members	18	28	25	14	11	7

After the leaflets were sent, the age distribution was as follows.

Age group	15–24	25–34	35–44	45–54	55–64	65–74
Number of members	28	40	29	18	16	8

Calculate an estimate of the mean age before and after the leaflets were sent out.

3 The bullet train in Japan covers the 120 miles between Hiroshima and Kokura at an average speed of 164 mph. How long does it take, to the nearest minute?

4 Rearrange each of these formulas to make the bold letter the subject.

(a) $s = 3(\mathbf{r} - 4)$ (b) $d = \frac{2\mathbf{c} + 5}{3}$ (c) $b = 7 - 4\mathbf{a}$ (d) $y = 10 + \frac{\mathbf{x}}{4}$

5 The population of a town is planned to increase by 15% over the next three years and by 10% over the three years after that.

The present population is 24 600. Calculate, to the nearest hundred, the planned population in six years time.

6 The table below shows the takings of a garden centre for three-monthly periods.

Year	2002				2003			
Months	Jan–Mar	Apr–Jun	Jul–Sep	Oct–Dec	Jan–Mar	Apr–Jun	Jul–Sep	Oct–Dec
Takings (£000)	160	250	420	210	140	240	380	220

(a) Calculate a four-point moving average for this data.

(b) Comment on the trend in the takings.

7 A pump operates at a rate of 1.8 litres per second. Calculate, to the nearest minute, the time it will take to empty a tank containing 2540 litres of water.

8 Rearrange each of these formulas to make the bold letter the subject.

(a) $g = a\mathbf{f} - b$ (b) $g = af - \mathbf{b}$ (c) $t = \frac{\mathbf{p} + qs}{r}$ (d) $t = \frac{p + qs}{\mathbf{r}}$

9 Calculate the length AB in this diagram.

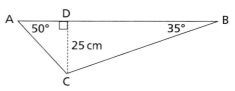

10 A company is doing badly. In March, the employees agree to a 7% reduction in their wages. In September, business picks up and they are given an 11.5% rise.

Calculate the overall percentage change between March and September, to the nearest 0.1%.

11 Which of the following scatter diagrams shows

(a) positive correlation (b) negative correlation

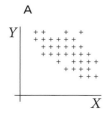
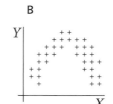
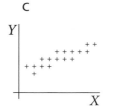
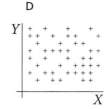

12 Rearrange the formula $q = ap^2 - 2b$ so that

(a) a is the subject (b) b is the subject (c) p is the subject

13 A company has an agreement with its employees that every year their salaries will be increased by a percentage equal to the annual rate of inflation.

When the next increase is due, the annual rate of inflation is 2.8%. Calculate, to the nearest pound, the new salary of an employee whose current salary is £26 405 a year.

14 This diagram shows a rectangle ABCD 70 cm by 40 cm. Calculate to a reasonable degree of accuracy

(a) the area of triangle ABD

(b) the length BD

(c) the length AH

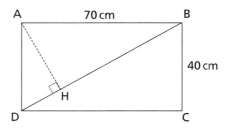

15 Jawal has two beakers, A and B. Beaker A contains 40 ml water and beaker B contains 30 ml water. Jawal pours some water from B into A. Afterwards, beaker A contains three times as much water as beaker B.

Calculate how much water was poured from B into A, showing your working.

16 This diagram shows two villages P and Q.
    Q is 31 km east and 53 km north of P.

    Calculate

    (a) the bearing of Q from P

    (b) the bearing of P from Q

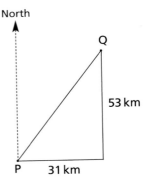

17 Brian found the nests of a colony of birds and counted the eggs in each nest.
    His results are shown in this table.

Number of eggs in nest	0	1	2	3	4	5
Number of nests	15	3	5	3	12	7

    Find

    (a) the modal number of eggs in a nest

    (b) the median number of eggs

    (c) the mean number of eggs

18 Solve each of the following equations.

    (a) $\frac{x-5}{3} + \frac{x+1}{5} = 6$

    (b) $3(x-7) - 2(4-x) = 26$

    (c) $\frac{3(x+1)}{2} - \frac{2(x-1)}{3} = 8$

    (d) $\frac{1}{2}(8+x) = 7 - x - \frac{1}{4}(18-2x)$

Challenge

Two cities, A and B, are linked by a motorway.
A coach service runs between them.
The journey in either direction takes $2\frac{1}{2}$ hours.

Coaches leave A for B every hour from 0800 until 1900.
The same is true of the service from B to A.

- If I travel on the 1100 coach from A to B, how many coaches will
  I meet that are coming in the opposite direction?

When a coach reaches either A or B it waits 30 minutes for cleaning and refuelling
and then becomes the next coach to leave that city.

Coaches that arrive after 1900 wait until the next morning and become the first coaches
to leave.

- How many coaches are needed to run the service
  (assuming that there are no breakdowns)?

10 Using area and volume

You should know how to

- ◆ find the area of any rectangle, triangle or parallelogram
- ◆ find the volume of a simple object
- ◆ use Pythagoras's theorem

This work will help you

- ◆ find and use the area of a trapezium
- ◆ find and use density
- ◆ convert between units of area and volume

A Review: parallelogram and triangle

The area of a parallelogram is **base × height**.

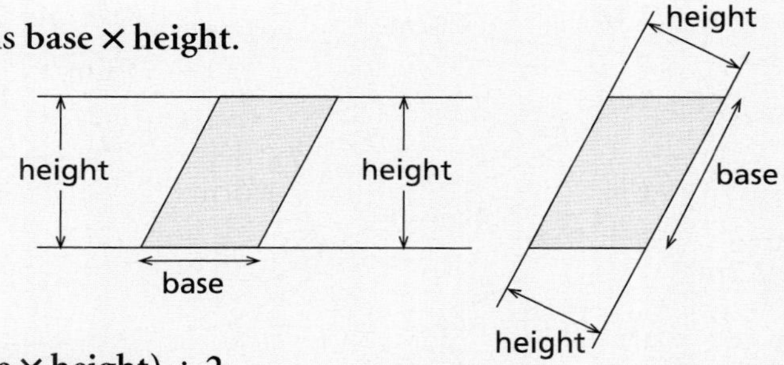

The area of a triangle is **(base × height) ÷ 2**.
You see this written as $\frac{1}{2}bh$ or $\frac{bh}{2}$.

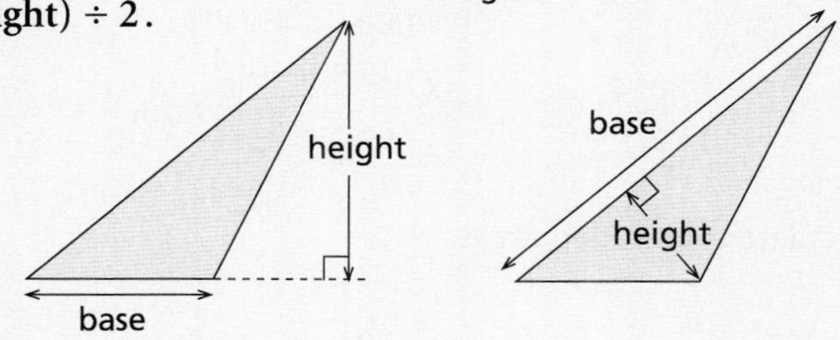

Any side of these shapes can be chosen as the 'base'.

But the height must be measured perpendicular (at right angles) to the base.

A1 Work out the areas of these shapes.

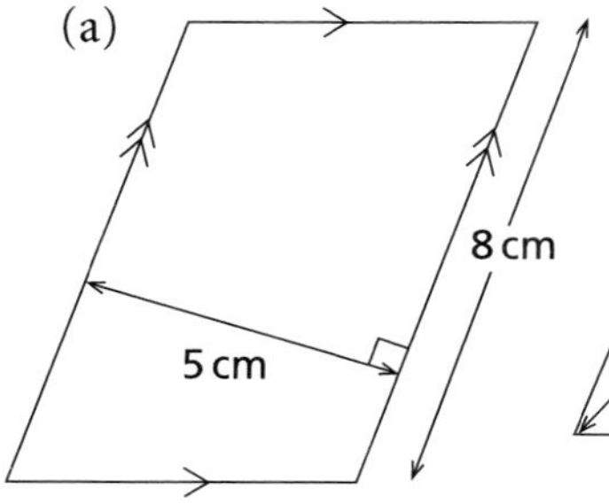

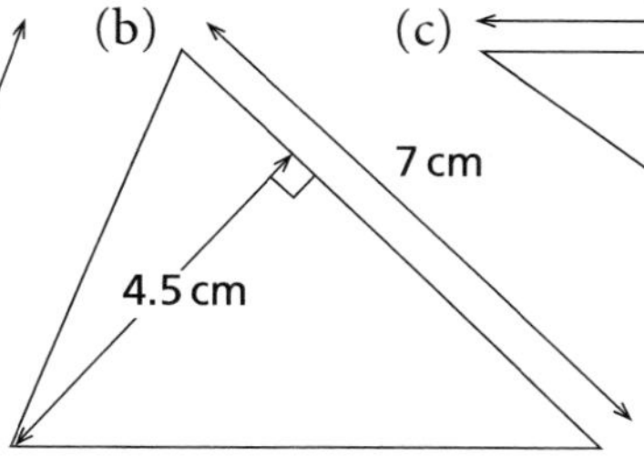

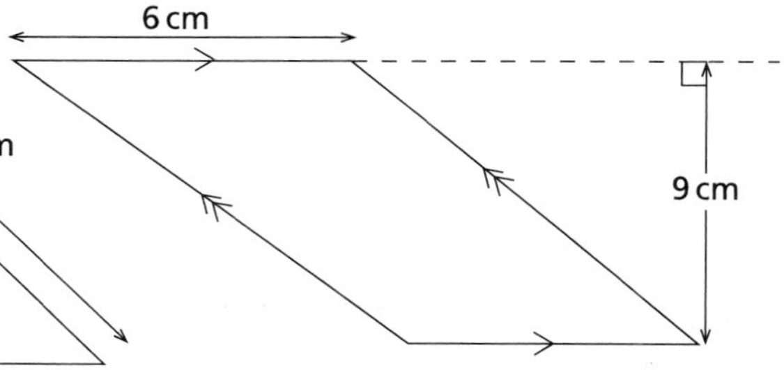

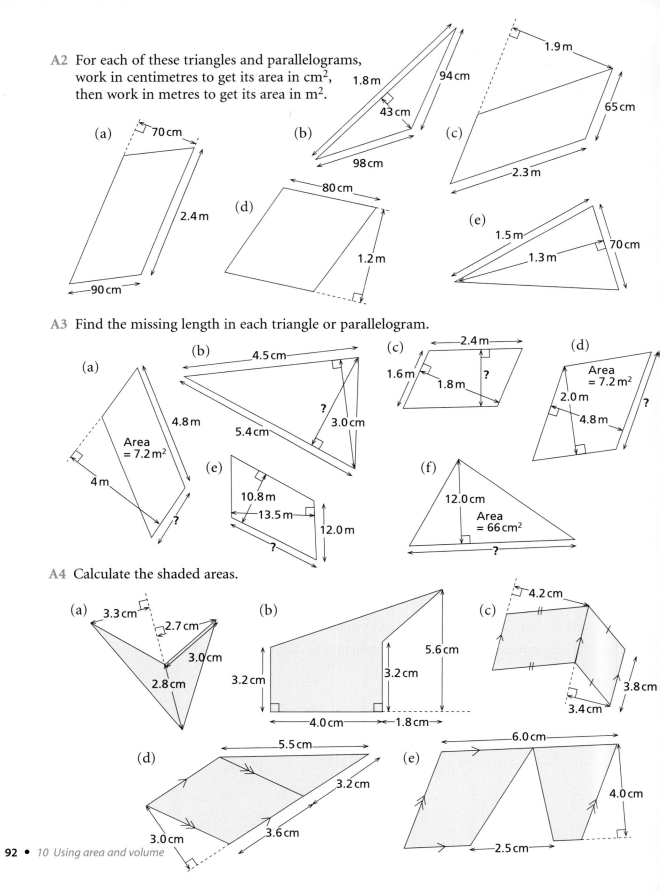

A2 For each of these triangles and parallelograms, work in centimetres to get its area in cm², then work in metres to get its area in m².

(a) 70 cm, 2.4 m, 90 cm

(b) 1.8 m, 94 cm, 43 cm, 98 cm

(c) 1.9 m, 65 cm, 2.3 m

(d) 80 cm, 1.2 m

(e) 1.5 m, 70 cm, 1.3 m

A3 Find the missing length in each triangle or parallelogram.

(a) 4.8 m, Area = 7.2 m², 4 m, ?

(b) 4.5 cm, ?, 3.0 cm, 5.4 cm

(c) 2.4 m, 1.6 m, 1.8 m, ?

(d) Area = 7.2 m², 2.0 m, 4.8 m, ?

(e) 10.8 m, 13.5 m, 12.0 m, ?

(f) 12.0 cm, Area = 66 cm², ?

A4 Calculate the shaded areas.

(a) 3.3 cm, 2.7 cm, 3.0 cm, 2.8 cm

(b) 5.6 cm, 3.2 cm, 3.2 cm, 4.0 cm, 1.8 cm

(c) 4.2 cm, 3.8 cm, 3.4 cm

(d) 5.5 cm, 3.2 cm, 3.0 cm, 3.6 cm

(e) 6.0 cm, 4.0 cm, 2.5 cm

B Using algebra

B1 Write, as simply as possible, an expression for the area of each shape.

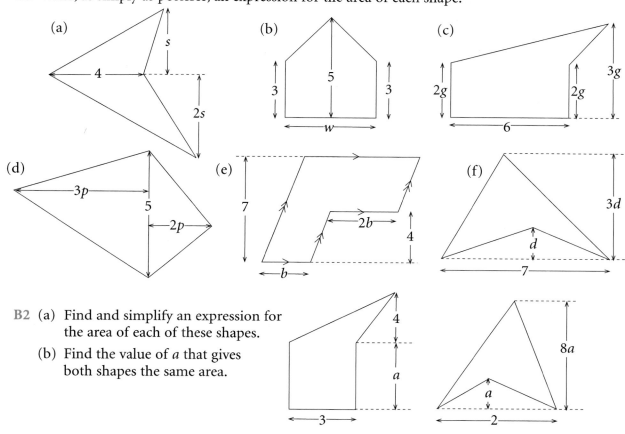

(a)

(b)

(c)

(d)

(e)

(f)

B2 (a) Find and simplify an expression for the area of each of these shapes.

(b) Find the value of a that gives both shapes the same area.

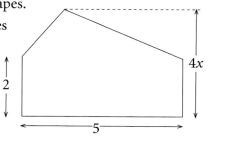

*B3** (a) Find and simplify an expression for the area of each of these shapes.

(b) Find the value of x that gives both shapes the same area.

B4 Write a formula for the area of this shape, in terms of a, b and h.

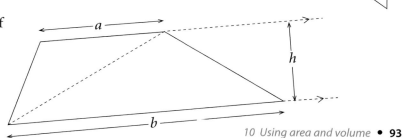

C *Trapezium*

The shape in B4, a quadrilateral with a pair of parallel sides, is called a **trapezium**.

You perhaps wrote $\frac{1}{2}ah + \frac{1}{2}bh$ as your formula.

This is correct, but the more usual ways of writing it are

$$\frac{(a+b)h}{2} \quad \text{and} \quad \frac{1}{2}(a+b)h$$

You can think of them as saying 'Add the lengths of the parallel sides, multiply by the perpendicular distance between them, then halve.'

C1 For each of these trapeziums,
work in millimetres to get its area in mm², (b)
then work in centimetres to get its area in cm².

(a)

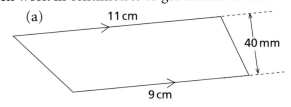

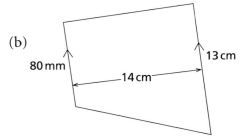

C2 Calculate the area in square units of
trapezium A + trapezium B + ... + trapezium E.

The total area of the trapeziums is approximately
equal to the area shaded green.

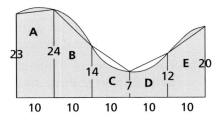

C3 This is a quadrant (quarter circle), with radius 5 cm.

(a) Find the approximate area of each strip
by measuring and using the trapezium formula.
(Treat the strip on the far right as a triangle.)

(b) Add the areas of the strips to get the
approximate area of the quadrant.

(c) Find the area of the quadrant, using the
formula for the area of a circle.
Was your approximation in part (b) too large
or too small? Can you give a reason for this?

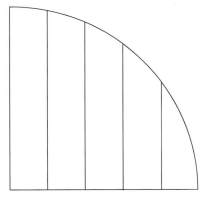

C4 (a) Find and simplify an expression for the area of each of these trapeziums.

(b) Find the value of a that gives both trapeziums the same area.

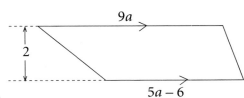

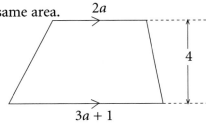

D Population density

- In which countries are people crowded together?

- In which countries are people most spread out?

- Why must you be careful when using area to make comparisons of this kind?

Working out the number of people per square kilometre is called calculating **population density**.

$$\text{Population density} = \frac{\text{population}}{\text{area}}$$

By making an estimate of area from a map and finding out about the local population, you could calculate the population density for your town or village, county or region.

Country	Population	Area (km²)
Australia	16 530 000	7 682 000
Bangladesh	104 530 000	144 000
Canada	25 950 000	9 221 000
Holland	14 760 000	37 000
India	796 600 000	3 288 000
Japan	122 610 000	378 000
Kenya	23 880 000	583 000
Libya	4 230 000	1 760 000
Russian Republic	283 680 000	22 403 000
Saudi Arabia	14 020 000	2 150 000
South Africa	29 600 000	1 223 000
United Kingdom	57 080 000	244 000
United States	246 330 000	9 373 000

D1 (a) Calculate the population density for the United Kingdom.

   (b) Which country in the table has the closest population density to that of the United Kingdom?

D2 Rwanda has an area of 26 000 km² and is estimated to have a population density of 260 people per square kilometre. Estimate the population of Rwanda.

D3 Mr Jones keeps a herd of 82 beef cattle on 160 000 m² of pasture.
His neighbour keeps 48 beef cattle on 70 000 m² of pasture.
How many more cattle could Mr Jones keep on his pasture if he used the same stocking density as his neighbour?

D4 Farm animals can only be described as 'free-range' if they are given enough space. The table shows the limits.

Bird	Maximum live weight per m²
Chicken	27.5 kg
Duck, guinea fowl or turkey	25.0 kg
Goose	15.0 kg

   (a) Assuming that the live weight of an adult chicken is about 4 kg, how many chickens could you keep in a yard with area 7.8 m²?

   (b) A farmer has 18 m² available for free-range birds. These are the birds she wants to keep, with their typical live weights.

   Duck 3 kg   Turkey 12 kg   Goose 9 kg

   She wants to keep an equal number of each type of bird. What is the most she can keep?

You might like to calculate the area of your back garden or a suitable place in the school grounds, and work out numbers of birds for your own free-range farm!

E *Solids*

Prism

The volume of a prism is **area of cross-section × length**.

The simplest prism is a cuboid.

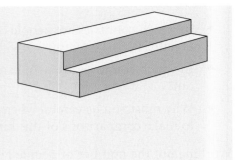

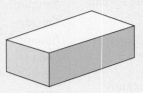

E1 For each of these prisms,
work in centimetres to get its volume in cm³,
then work in metres to get its volume in m³.

(a)

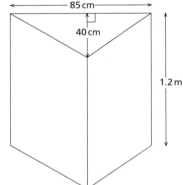

(b)

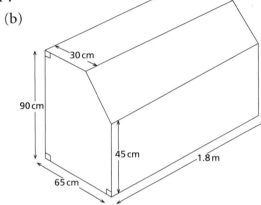

E2 These shapes are made up from cuboids. Find the volume of each one.

(a)

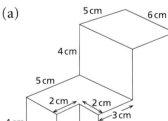

(b)

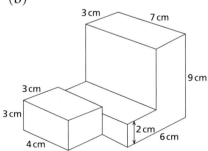

E3 The cross-section of this prism is an isosceles triangle.

(a) Use Pythagoras to work out the height h
of the triangular cross-section.

(b) Calculate the volume of the prism.

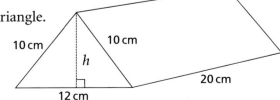

E4 The diagram shows a sketch of a garage.

(a) Calculate the height marked h.

(b) Find the volume of the garage. Give your answer to a reasonable degree of accuracy.

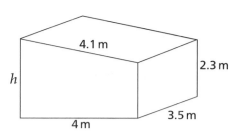

Volumes can also be measured in litres or millilitres.

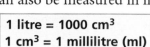

1 litre = 1000 cm³
1 cm³ = 1 millilitre (ml)

E5 (a) How many litres of water does this fish tank hold when it is completely full?

(b) How deep will the water be (to the nearest cm) when the tank contains 50 litres?

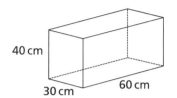

E6 Half a litre of water is poured into a tin.
The depth of the water is 8 cm.
What is the area of the base of the tin?

Surface area

To calculate the surface area of a solid object, work out the area of each face.

It may help you organise the calculation if you draw a net of the object, or draw separate sketches of the faces, and write measurements on the diagram.

E7 Find the surface area of these shapes.

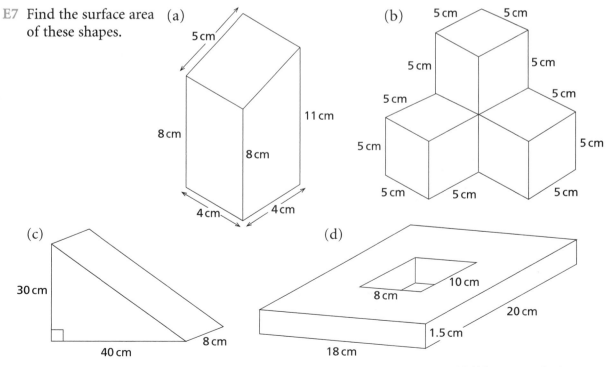

E8 For each of these cuboids, calculate
(i) the length marked x cm
(ii) the total surface area

(a)
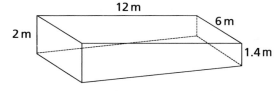
Volume 69.3 cm³
2.8 cm
4.5 cm
x cm

(b)
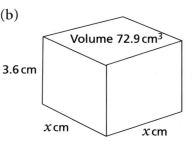
Volume 72.9 cm³
3.6 cm
x cm x cm

E9 A load of 15 m³ of liquid concrete is poured over an area of 75 m², covering it to a constant depth everywhere.
Calculate the depth of the concrete (a) in metres (b) in centimetres

E10 The diagram shows a swimming pool which slopes from a maximum depth of 2 m to a minimum depth of 1.4 m.

(a) Calculate the volume of the water in the pool when it is full.

(b) By how much does the depth of water go down if 50 m³ of water is pumped out of the pool?

12 m
6 m
2 m
1.4 m

E11 A piece of copper of volume 4.5 cm³ is made into wire with cross-sectional area 0.03 cm².
Calculate the length of the wire.

E12 The doors and the walls of this shed have all to be painted on the outside.
This tray is full of paint. Each litre of the paint should cover 5 square metres.

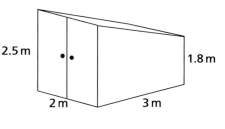
2.5 m
1.8 m
2 m 3 m

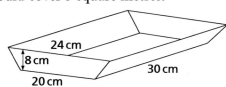
24 cm
8 cm
20 cm
30 cm

Is there enough paint to cover the shed as required? Explain your answer.

*E13 A cuboid has width 7 cm, height 8 cm and surface area 547 cm².
Calculate the length and the volume of the cuboid.

*E14
10 cm
6 cm
8 cm

The surface area of this prism is 438 cm².
Calculate the volume of the prism.

F Density

The **density** of a substance tells you how heavy a standard volume of it is.

For example, the density of insulating foam is 0.1 g/cm³ (grams per cubic centimetre), while the density of lead is 11.4 g/cm³.

$$\text{Density} = \frac{\text{mass}}{\text{volume}}$$

F1 (a) A glass ornament has mass 690 g and volume 240 cm³. Find its density in g/cm³.

   (b) A statue has volume 1.5 m³ and mass 2.4 tonnes (1 tonne = 1000 kg). Find its density in kg/m³.

F2 Use the table on the right to work out the mass of these objects.

   (a) A stone with volume 92 cm³

   (b) A wooden cuboid 22 cm by 18 cm by 11 cm

   (c) The water in a 25 m by 15 m swimming pool, which is 1.5 m deep everywhere

Material	Density
Wood	0.7 g/cm³
Stone	3.0 g/cm³
Water	1.0 g/cm³

F3 Density can help identify metals (or at least rule out fake metals).

One of these ingots is platinum, one is gold, one is silver and one is fake gold.

Work out the density of each ingot and use the table to say what each ingot is made of.

Material	Density
Platinum	21.5 g/cm³
Gold	19.3 g/cm³
Silver	10.5 g/cm³
Fake gold	9.8 g/cm³

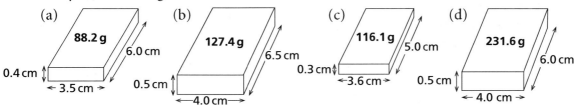

(a) 88.2 g 6.0 cm 0.4 cm 3.5 cm

(b) 127.4 g 6.5 cm 0.5 cm 4.0 cm

(c) 116.1 g 5.0 cm 0.3 cm 3.6 cm

(d) 231.6 g 6.0 cm 0.5 cm 4.0 cm

F4 An aluminium vase weighs 413 g. The density of aluminium is 2.6 g/cm3. What is the volume of the aluminium used to make the vase?

F5 A measuring cylinder has an internal base area of 6.5 cm². It is partly filled with water. When a pebble is put into it the water rises by 18 mm. The pebble has mass 32.8 g. What is its density?

For the next two questions use the densities given in question F3.

F6 An ornament consists of a gold part and a silver part joined together. The gold and silver parts weigh the same. The volume of the gold part is 8.4 cm³. What is the volume of the whole ornament?

*F7 Another ornament has a gold part and a silver part. The whole ornament weighs 363 g and has a volume of 22 cm³. By first finding the volumes, find the mass of each metal in the ornament.

G Converting units

G1 (a) Calculate the area of this rectangle in cm².

(b) Change the length and width into metres and calculate the area in m².

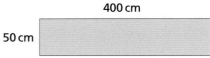

(c) How many cm² are there in 1 m²?

(To check your answer makes sense, imagine filling a metre square with centimetre squares.)

G2 How many mm² are there in 1 cm²?

1 cm

1 cm

G3 How many m² are there in 1 km²? Explain how you worked it out.

G4 Convert

(a) 3 m² to cm² (b) 7.2 m² to cm² (c) 50 cm² to m² (d) 1040 cm² to m²

(e) 0.7 m² to cm² (f) 302 cm² to m² (g) 11 km² to m² (h) 8000 m² to km²

(i) 0.09 km² to m² (j) 720 m² to km² (k) 6 000 000 m² to km²

Area can also be measured in **hectares**.

> **1 hectare (ha) = 10 000 m²**

G5 (a) A city centre square is an exact square, and has area 1 hectare. How long is each side of the square?

(b) How many hectares are there in 1 km²?

G6 A ranch has 4.8 km² of grazing land which is stocked at 8 cattle per hectare. How many cattle are there?

G7 (a) Calculate the volume of this cuboid in m³ and in cm³.

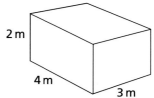

(b) How many cm³ are there in 1 m³?

(To check your answer makes sense, imagine filling a metre cube with centimetre cubes.)

G8 Convert

(a) 2.4 m³ to cm³ (b) 0.06 m³ to cm³ (c) 8000 cm³ to m³ (d) 342 000 cm³ to m³

G9 A litre of orange juice is spilt and spreads out to a depth of 1 mm. What area of floor, in m², is covered in orange juice?

G10 A 5 litre tin of paint is enough to cover an area of 24 m². How thick, in mm, should the layer of paint be when it is applied?

G11 The area of an ornamental pond is 4500 m². How many litres of water need to be added to the pond to increase the depth by 5 cm?

Test yourself

T1 The three parallelograms sketched here all have the same area.

Find the missing lengths.

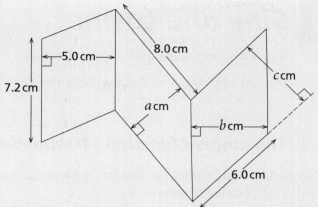

T2 The diagram shows a trapezium ABCD.

AB is parallel to DC.

AB = 4.8 m, DC = 5.2 m, AD = 1.6 m,

angle BAD = 90°, angle ADC = 90°

Calculate the area of trapezium ABCD.

State the units of your answer.

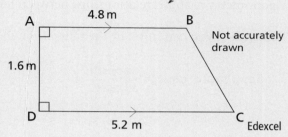

Not accurately drawn

Edexcel

T3 The diagram shows a cuboid of length 53.1 cm. The cross-section, PQRS, is such that PR = 24.7 cm and QR = 16.3 cm.

(a) Calculate the length of PQ.

(b) The density of the material from which the cuboid is made is 4.3 g/cm³. Calculate the mass of the cuboid in kilograms.

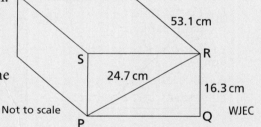

Not to scale

WJEC

T4 (a) Find an expression for the volume of each cuboid.

(b) What value of *a* gives them the same volume?

(c) What value of *a* gives them the same surface area?

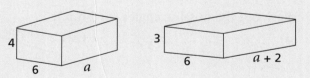

T5 The diagram represents a solid metal bar with an uniform cross-section in the form of the trapezium ABCD, in which AB = 9.3 cm and DC = 5.8 cm.

The height of the bar is 3.5 cm and the length of the bar, BE, is 14.7 cm.

The density of the metal is 5.6 g/cm³.

Calculate the weight, in kilograms, of the bar.

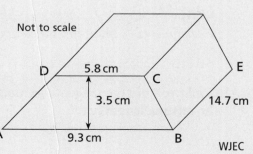

Not to scale

WJEC

11 Sine and cosine

You will revise the use of the tangent function.

This work will help you use sine and cosine to find sides and angles in right-angled triangles.

A The tangent function – a reminder

The work you did earlier on tangents is part of a branch of mathematics called trigonometry.
Trigonometry is about relationships between lengths and angles.

Remember, you can think of the tangent of an angle as a multiplier ...

... or as a ratio: $\tan a = \dfrac{\text{opposite}}{\text{adjacent}}$

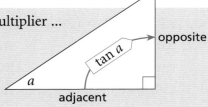

A1 Work out the missing lengths (to 1 d.p.).

   (a)   (b)   (c)

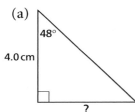
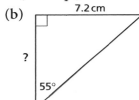
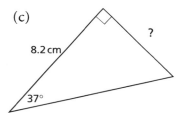

A2 Use your calculator to find (to 1 d.p.) the angle whose tangent is 1.666.

A3 Work out the missing angles (to 1 d.p.).

   (a)   (b)   (c)

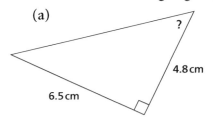
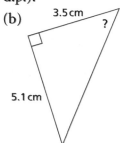
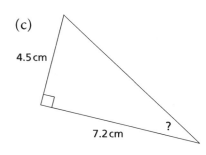

- Could you use the tangent function to work out the missing length here?

B *The sine function*

Draw a horizontal line.
Draw a quarter circle of radius 10 cm
with its centre on the line.

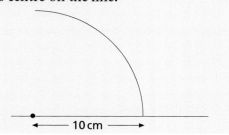

10 cm

Draw a line 40° to the horizontal line.
Label this point P.

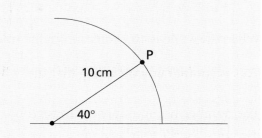

P

10 cm

40°

Use a set-square to draw a perpendicular
from P to the horizontal line.

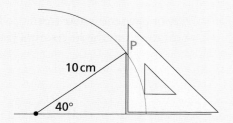

P

10 cm

40°

Measure this length.

10 cm

P

40°

Keep your drawing.

B1 Use your result to find the missing lengths here.

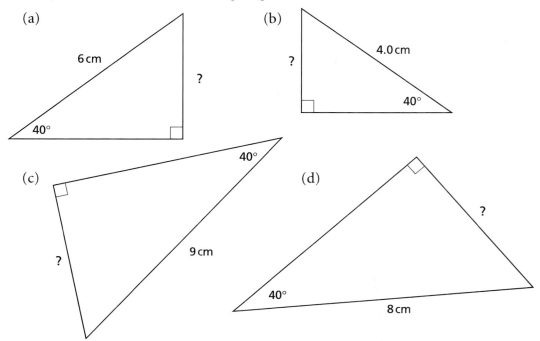

(a)

6 cm

?

40°

(b)

4.0 cm

?

40°

(c)

?

40°

9 cm

(d)

?

40°

8 cm

In a right-angled triangle like this, this side is the **hypotenuse** 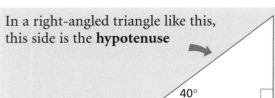 and this is the side **opposite** to the 40° angle.

When the angle is 40°, you can use the rule   hypotenuse ─[× 0.64]→ opposite side

This is another way of stating the rule: $\dfrac{\text{opposite side}}{\text{hypotenuse}} = 0.64$

The number you multiply the hypotenuse by to get the opposite side is called the **sine** of the angle.

So the sine of 40° is about 0.64.

B2 Continue your drawing to find the sines of other angles.

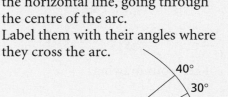 Draw lines at 10°, 20°, 30°, ..., 80° to the horizontal line, going through the centre of the arc.
Label them with their angles where they cross the arc.

Draw a perpendicular from each crossing point down to the horizontal line.

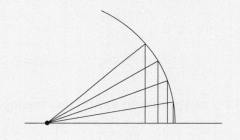

Measure the opposite side for each triangle and record your results in a table like this.

Angle	Hypotenuse ─[×?]→	Opposite side
10°	10 cm	
20°	10 cm	
30°	10 cm	
40°	10 cm	

Work out the numbers that go in here and write them in your table.

'sin' is a short way to write sine (though we still pronounce it 'sine' when written this way).

So we can write sin 40° = 0.64.

A scientific calculator gives very accurate values of sines.

Try keying in [4] [0] [sin] or [sin] [4] [0] . Is 0.64 close to what you get?

B3 Use your calculator to find sin 10°, sin 20°, sin 30°, ..., sin 80°.
Compare these results with what you got in B2.

B4 Look back at your drawing for B2.
What do you think sin 90° will be?
Does your calculator agree?

*B5 What do you think sin 100° will be?
Does your calculator agree?

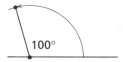

B6 (a) Find sin 75° on your calculator.

Use this result to find the missing lengths here (to 2 d.p.).

(b)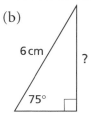

(c)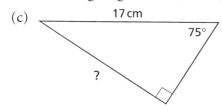

(d)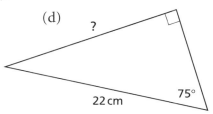

B7 The angles in the following drawings have been drawn accurately.
For each drawing,

    (i) write an estimate of the missing length

    (ii) then use the sine key on your calculator
        to get an answer to 2 d.p.

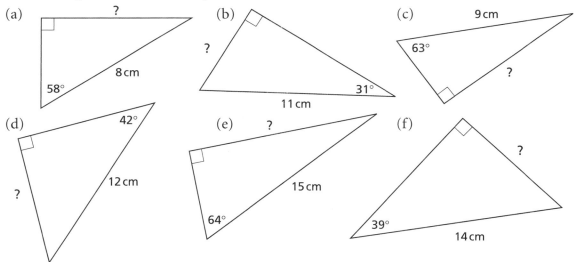

B8 The sloping section of this escalator is 12 metres long.
It slopes at an angle of 35° to the horizontal.

How high is the top of the escalator above the bottom
(to the nearest 0.1 m)?

B9 A ladder 6.4 metres long leans against a vertical wall and makes an angle of 67° with the ground.

How far is the top of the ladder from the bottom of the wall, to the nearest 0.1 m?

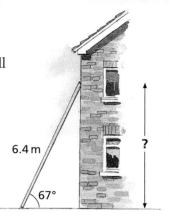

B10 This is a parallelogram.

(a) Work out the length of the perpendicular AQ, showing how you did the calculation.

(b) What is the area of the parallelogram (to the nearest 0.1 cm²)?

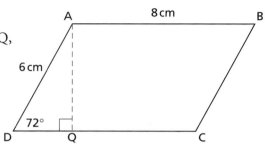

B11 Work out the area of each of these parallelograms to the nearest 0.1 cm².

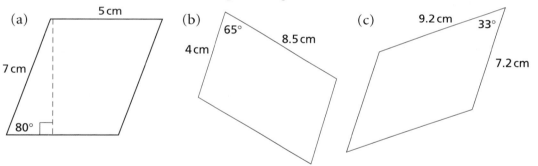

(a) 5 cm

7 cm

80°

(b) 65°

4 cm

8.5 cm

(c) 9.2 cm 33°

7.2 cm

B12 In your own words, explain how to find the area of a parallelogram if you know the acute angle between two sides and the lengths of those sides.

Now try to write your method as a formula.

B13 Work out the area of each of these triangles to the nearest 0.1 cm².

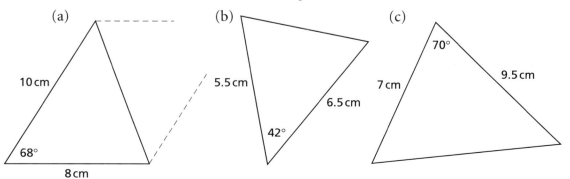

(a) 10 cm

68°

8 cm

(b) 5.5 cm

6.5 cm

42°

(c) 70°

9.5 cm

7 cm

C Using sine to find the hypotenuse

sin 53° is about 0.8.

How long is the hypotenuse here?

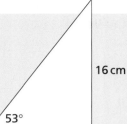

16 cm

53°

C1 Find each hypotenuse using sin 53° = 0.8.

(a)

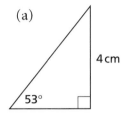

4 cm

53°

(b)

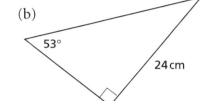

53°

24 cm

(c)

28 cm

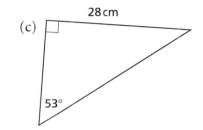

53°

C2 The angles in the following drawings have been drawn accurately.
For each drawing,

    (i)   write an estimate of the length of the missing hypotenuse

    (ii) then find its length to 1 d.p. using the calculator's value of the sine.

(a)

7.1 cm

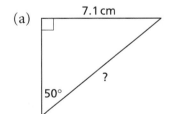

?

50°

(b)

18°

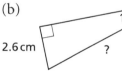

2.6 cm

?

(c)

14.6 cm

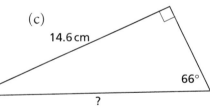

66°

?

(d)

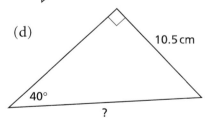

10.5 cm

40°

?

(e)

?

12.4 cm

47°

(f)

?

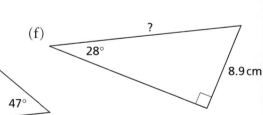

28°

8.9 cm

C3 Write a formula for finding h
if you know x and p.

$h = \ldots\ldots$

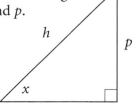

h

p

x

D *Finding an angle*

D1 This table shows approximate values of the sines of angles every 10°.

Use the table to **estimate**

(a) the angle whose sine is 0.7

(b) the angle whose sine is 0.9

(c) the angle whose sine is 0.2

Now use 'the angle whose sine is …' function on your calculator to find these angles. Your calculator will work the way it did for 'the angle whose tangent is …'

a	sin a
10	0.17
20	0.34
30	0.50
40	0.64
50	0.77
60	0.87
70	0.94
80	0.98

D2 (a) Work out the sine of angle a here.

(b) Now use your calculator to find the angle whose sine is this value (in other words find a).

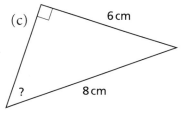

D3 For each of these, work out the sine of the angle. Then use your calculator to find the angle to 1 d.p.

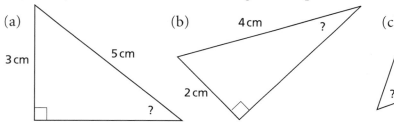

D4 Work out these angles to 1 d.p.

(a)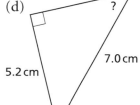

(b)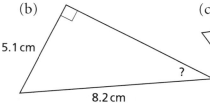

(c)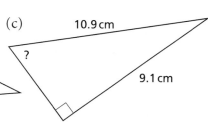

(d)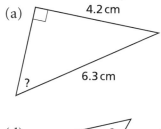

(e)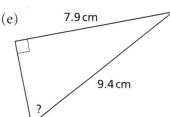

(f)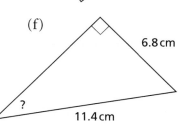

D5 What angle, to the nearest degree, does this bridge make with the horizontal when point P is 3.8 metres above road level?

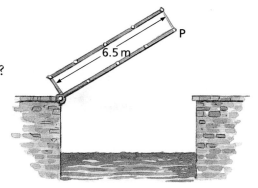

D6 A ladder 4.2 m long is placed against a vertical wall. The top of the ladder is 4.0 m above the ground. The ground is horizontal.

What angle does the ladder make with the ground?

*D7 The area of this parallelogram is 26 cm². Find angle x.

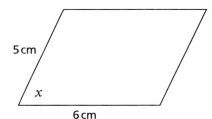

*D8 The area of this triangle is 19 cm². Find angle y.

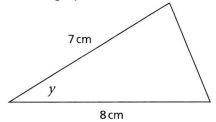

E *The cosine function*

The cosine function ('cos' for short) involves the hypotenuse again but relates it to the side **adjacent** to the angle.

The cosine of an angle can be thought of as a multiplier ...

... or as a ratio: $\cos a = \dfrac{\text{adjacent}}{\text{hypotenuse}}$

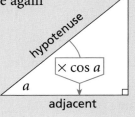

E1 Find these on your calculator and write them down to 3 d.p.

cos 10°, cos 20°, cos 30°, cos 40°, cos 50°, cos 60°, cos 70°, cos 80°

E2 (a) What do you think the cosine of 90° is? Does your calculator agree?

(b) What do you think the cosine of 0° is?

*E3 What do you think the cosine of 100° is? Does your calculator agree?

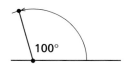

E4 Work out the length of the adjacent side to 1 d.p. in each of these triangles.

(a)

5.0 cm

43°

(b)

6.0 cm

29°

(c)

8.2 cm

73°

E5 Work out the length of the hypotenuse in each of these.

(a)

35°

7.0 cm

(b)

4.0 cm

64°

(c)

7.5 cm

32°

E6 Work out the missing lengths.

(a)

?

48°   4.0 cm

(b)

?

5.2 cm   36°

(c)

?

61°

3.8 cm

(d)

5.9 cm   29°

?

(e)

2.8 cm   70°

?

(f)

?

7.6 cm   30°

E7 Work out the missing angles.

(a)

6.2 cm

?

4.3 cm

(b)

3.9 cm

?

7.1 cm

(c)

9.5 cm

?

6.5 cm

E8 Calculate the width (marked w)
of this building.

3.8 m   3.8 m

27°   27°

w

F Mixed trigonometry questions and problems

This method may help you decide whether to use sine, cosine or tangent.

1 Make a sketch and label the sides:

hyp (hypotenuse)

opp (opposite the angle that is given or required)

adj (adjacent to the angle that is given or required)

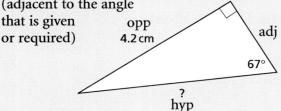

2 These are the three basic formulas.

$$\text{adj} \times \tan a = \text{opp}$$
$$\text{hyp} \times \sin a = \text{opp}$$
$$\text{hyp} \times \cos a = \text{adj}$$

Choose the formula that contains the values you know and the value you want to calculate.

F1 Find the missing lengths and angles here.

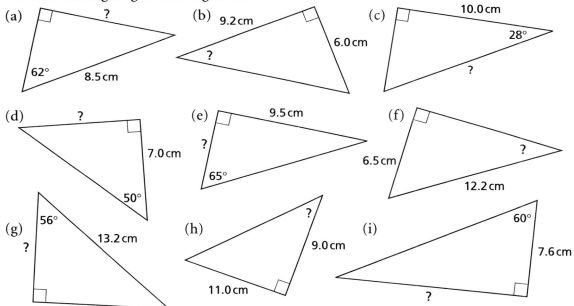

F2 (a) Work out the lengths of sides AC and BC in this triangle. Give your results to 2 d.p.

(b) Use Pythagoras to find AB from the lengths AC and BC. Show your working.
Do you get 7 cm, as shown in the diagram?

F3 Repeat question F2 choosing your own angle at vertex A and your own length AB. Does the same thing happen as before?

F4 (a) Calculate PR and QR to 2 d.p.

(b) Work out (to 2 d.p.) the tangent of 59° from the lengths PR and QR.
Do you get the same by simply keying tan 59° on your calculator and writing the result to 2 d.p.?

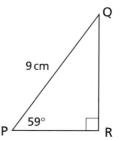

F5 Repeat question F4 choosing your own angle at vertex P and your own length PQ. Does the same thing happen as before?

F6 (a) Sketch this right-angled triangle (including its labels). What is special about it?

(b) How big is angle x? Mark the value on your sketch.

(c) Use Pythagoras to work out length DE and mark it on your sketch.

(d) Work out sin x from two of the lengths given on your sketch. Do you get the same by keying the sine of this angle on your calculator?

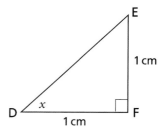

F7 (a) What special kind of triangle is triangle ABC?

(b) What are the values of angles w, x, y and z?

(c) Use Pythagoras to work out the length AN.

(d) Use your answer to (c) and the lengths in the diagram to work out these.

(i) sin 30°   (ii) cos 30°   (iii) tan 30°

(iv) sin 60°   (v) cos 60°   (vi) tan 60°

(e) Check your answers using the sin, cos and tan functions on your calculator.

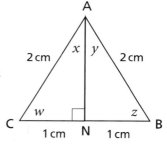

F8 (a) What special kind of triangle is this?

(b) Work out the length of the side AC. (Make a sketch and add information to it if you need to.)

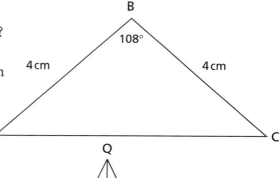

F9 Calculate these.

(a) The length QK to 2 d.p.

(b) The length PK to 2 d.p.

(c) The length KR to 2 d.p.

(d) The area of triangle PQR to 1 d.p.

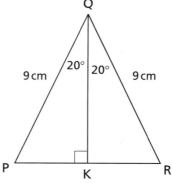

F10 (a) Calculate the area of an equilateral triangle with sides 10 cm long.

(b) What is the area of a regular hexagon with sides 10 cm long?

F11 Calculate angle x.

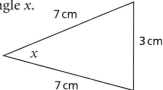

F12 Calculate the area of this triangle.

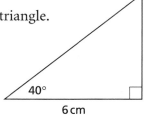

F13 A glider travels for 8 km on a bearing of 065°.

(a) How much further north is it from when it started?

(b) How much further east?

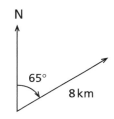

F14 A helicopter travels for 78 km on a bearing of 206°.

(a) How much further south is it from when it started?

(b) How much further west?

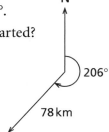

F15 A boat travels for 24 km on a bearing of 080°.
It then travels for 18 km on a bearing of 043°.
(Make a sketch.)

(a) How much further north is it from when it started?

(b) How much further east?

F16 An orienteer runs for 1.2 km on a bearing of 220°.
She then runs for 2.1 km on a bearing of 285°.

(a) How much further north or south is she from when she started?

(b) How much further west?

F17 How long are the diagonals of this rhombus?

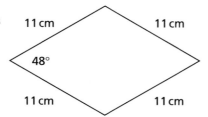

F18 The diagonals of a rhombus are 14 cm and 8 cm long. Calculate the angles at the vertices of the rhombus.

F19 This is a regular octagon. Make a large labelled sketch showing the way it has been divided up.

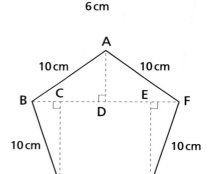

  (a) Label the dimensions of the rectangles and triangles it has been divided up into.
  (Use sine or cosine and give answers to 2 d.p.)

  (b) Find the areas of these rectangles and triangles.

  (c) Calculate the area of the whole octagon (to 1 d.p.).

*F20 This is a regular pentagon. Make a large labelled sketch showing the way it has been divided up.

  (a) Calculate the acute angles in the triangles formed by dotted lines. Mark them on your sketch.

  (b) Use sine and cosine to calculate the lengths AD, BD, DF, BC, CG, EF, EH to 2 d.p. Mark them on your sketch.

  (c) Calculate the area of the whole pentagon (to 1 d.p.).

*F21 This is the same regular pentagon as in question F20.

  (a) Use tangent to calculate the length h to 2 d.p.

  (b) Calculate the area of one of the isosceles triangles to 1 d.p.

  (c) Calculate the area of the whole pentagon to 1 d.p. Compare this answer with the one for question F20.

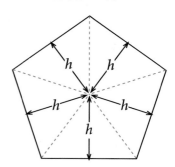

F22 A regular nonagon is drawn by spacing nine points equally around a circle of radius 10 cm.
Calculate the length of one side of the nonagon.

*F23 Nick wants to draw a regular decagon by spacing ten points equally around a circle. He wants each side to be 6 cm long.
What radius circle should he use?

Test yourself

T1 Use your calculator to find the value of each of these to 3 d.p.

   (a) sin 16°   (b) tan 8°   (c) cos 57°   (d) tan 29°   (e) sin 1°

T2 Find the values of these to 1 d.p.

   (a) The angle whose tangent is 0.04   (b) The angle whose sine is 0.56

   (c) The angle whose cosine is 0.82   (d) The angle whose tangent is 156

T3 Find the missing angles.

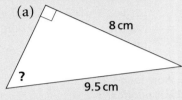
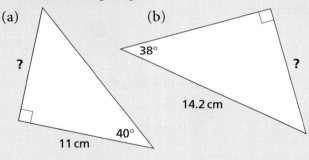
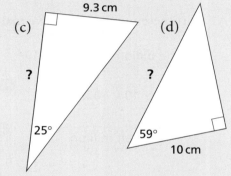

T4 Find the missing lengths.

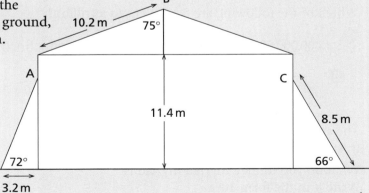

T5 Calculate the heights of the points A, B, C above the ground, each to the nearest 0.1 m.

T6 A tree 2.50 metres tall casts a shadow 4.36 metres long. Calculate the angle of elevation of the sun to the nearest degree. (The angle is marked *a* in the diagram.)

12 Large and little

This work will help you

◆ write large and small numbers in standard form

◆ calculate with numbers in standard form

A Millions and billions

How old will you be one billion seconds from now?

What were you doing a million hours ago?

If you travelled a million metres north of London, where would you be?

How high is a pile of one million 1p pieces?

If you spend £1000 each day, how long will it take you to spend a billion pounds?

A1 Sort these numbers into five matching pairs.

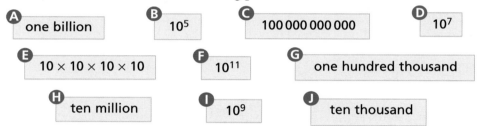

A one billion

B 10^5

C 100 000 000 000

D 10^7

E $10 \times 10 \times 10 \times 10$

F 10^{11}

G one hundred thousand

H ten million

I 10^9

J ten thousand

A2 A trillion is a thousand billion.

Which of the following are equivalent to a trillion?

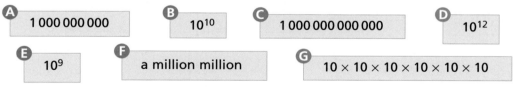

A 1 000 000 000

B 10^{10}

C 1 000 000 000 000

D 10^{12}

E 10^9

F a million million

G $10 \times 10 \times 10 \times 10 \times 10 \times 10$

A3 The ancient Greeks called ten thousand a myriad. It was the largest number they had a name for.

What do you think a myriad myriad is?
Write it in the form 10^n.

A4 The total fish population of the world's oceans has been estimated at about 10^{14} individual fish.

How many billion fish is this?

B *Down to Earth*

Large numbers can be written in different ways.

2.3 thousand	20 million	3.45×10^8
$= 2.3 \times 1000$	$= 20 \times 1\,000\,000$	$= 3.45 \times 100\,000\,000$
$= 2300$	$= 20\,000\,000$	$= 345\,000\,000$

B1 The St Lawrence river in Canada is 3130 km long.
Which of these are ways to write its length?

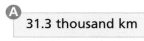

A 31.3 thousand km

B 31.3×100 km

C 3.13 thousand km

D 3.13×10^3 km

E 3.13×10 km

B2 This table shows the lengths of five well-known rivers.

In kilometres, the length of the Nile is 6670 km.
Write the length of each river in kilometres.

River	km (thousands)
Nile	6.67
Yangtze-Kiang	6.3
Zaire	4.7
Danube	2.84
Seine	0.78

B3 The total surface area of land on Earth is 149 200 000 square kilometres.

Forest covers the largest part.

Forest	39 000 000 sq km
Desert	35 300 000 sq km
Pasture	34 200 000 sq km
Icecap	15 000 000 sq km
Cultivated	14 400 000 sq km
Other	11 300 000 sq km

(a) Copy and complete this table to show the areas in millions of square kilometres.

Forest	39 million sq km
Desert	
Pasture	

(b) What is the total surface area of land in millions of square kilometres?

(c) What percentage of land on Earth is
  (i) forest   (ii) cultivated

(d) The total surface area of the Earth is 5.1×10^8 square kilometres.
What percentage of this is covered by sea?

B4 Here are some estimates of the world population for different years.

| 1900 | 1630 million | 1950 | 2520 million | 1960 | 3 billion | 1970 | 3.7×10^9 |

| 1980 | 4.5 billion | 1990 | $53 \times 100\,000\,000$ | 1995 | $5\,690\,000\,000$ |

(a) Copy and complete this table to show these figures as ordinary numbers and in billions.

(b) Draw a graph to show how the population has changed between 1900 and 1995. (Use axes that go up to 8 billion and along to 2020.)

(c) Use your graph to estimate the population for the year 2020 in billions.

Year	Population	Billions
1900	1 630 000 000	1.63
1950		
1960		
1970		
1980		
1990		

C Standard form

In 1997 the assets of Bill Gates, the world's richest man, were about $14\,800\,000\,000$ dollars. We say $14\,800\,000\,000$ is written in **ordinary form**.

All of these are ways to write this number …

- 14 800 million
- 14.8 billion
- $1\,480\,000 \times 10^4$
- 14.8×10^9
- 1.48×10^{10}
- $1.48 \times 10\,000\,000\,000$

… but only 1.48×10^{10} is in **standard form**.

In standard form, a number is written $a \times 10^n$

where • a is greater than or equal to 1 but less than 10 ($1 \leq a < 10$)
and • n is an integer

C1 Which of these numbers are written in standard form?

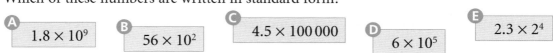

A 1.8×10^9 B 56×10^2 C $4.5 \times 100\,000$ D 6×10^5 E 2.3×2^4

C2 Write these numbers in ordinary form.

(a) 4.2×10^5 (b) 8.1×10^8 (c) 5.89×10^9

(d) 3×10^{10} (e) 7.04×10^6 (f) 8×10^{15}

C3 Write these numbers in standard form.

(a) 90 000

(b) 91 000

(c) 9 100 000 000

(d) 76 000 000 000 000

(e) 500 000 000 000 000 000

(f) 329 000 000

(g) 107 800 000 000 000

(h) 20 000 000 000 000 000 000 000

C4 This table shows the most common languages spoken in the world in 1996.

(a) Write each number of speakers in ordinary form.

(b) Copy and complete the following statement.

	Language	Number of speakers (approx.)
1	Chinese (Mandarin)	1.093×10^9
2	English	4.5×10^8
3	Hindi	3.67×10^8
4	Spanish	3.52×10^8
5	Russian	2.04×10^8

In 1996, the number of Mandarin speakers was about ? times the number of Russian speakers.

C5 This table shows the countries with the most English language speakers in 1997.

Write each number of speakers in standard form.

Country	Number of English speakers (approx)
Australia	15 200 000
Canada	17 700 000
Guyana	900 000
Ireland	3 300 000
Jamaica	2 400 000
New Zealand	3 200 000
South Africa	3 600 000
Trinidad and Tobago	1 200 000
UK	56 800 000
USA	224 900 000

C6 This table shows the organised religions with most members in 1997.

(a) Write the number of Buddhists in ordinary form.

(b) Which religion had most members?

(c) Which religion had fewest members?

Religion	Members
Baha'ism	5.84×10^6
Buddhism	3.39×10^8
Christianity	1.9×10^9
Confucianism	6.33×10^6
Hinduism	7.64×10^8
Islam	1.03×10^9
Judaism	1.35×10^7
Sikhism	2×10^7

C7 In 1996, 0.4 million tonnes of fish were caught in the Antarctic. Write this figure in standard form.

D Lost in space

A light year is about 9.46×10^{12} km.

- The Andromeda Spiral Galaxy is about 2 200 000 light years away from Earth. What is this distance in kilometres?

- The star Arcturus is about 321 700 000 000 000 km away from Earth. About how far away is Arcturus in light years?

D1 This table shows some of the brightest stars in our galaxy.

Star	Constellation	Distance (light years)
Sirius	Canis Major	8.64
Canopus	Carina	1200
Vega	Lyra	26
Rigel	Orion	900
Procyon	Canis Minor	11.4

Write the distance from Earth to each star in kilometres to three significant figures

(a) in standard form   (b) in ordinary form

D2 This table shows the stars that are closest to Earth.

Star	Distance (km)
Proxima Centauri	39 920 000 000 000
Alpha Centauri	41 150 000 000 000
Barnard's Star	56 570 000 000 000
Wolf 359	73 320 000 000 000
Lalande 21185	77 770 000 000 000

Alpha Centauri

(a) Write each distance in standard form.

(b) You are in a spaceship travelling at 40 000 kilometres per hour.

   About how many years would it take you to reach Proxima Centauri?

(c) Calculate how far away each star is in light years, correct to three significant figures.

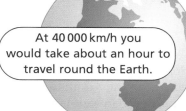

At 40 000 km/h you would take about an hour to travel round the Earth.

Here is some information about the planets in the Solar System.

Planet	Equatorial radius (km)	Mass (kg)	Mean distance from Sun (km)
Earth	6378	5.98×10^{24}	1.496×10^8
Jupiter	71942	1.90×10^{27}	7.783×10^8
Mars	3397	6.39×10^{23}	2.279×10^8
Mercury	2439	3.29×10^{23}	5.791×10^7
Neptune	25269	1.03×10^{26}	4.497×10^9
Pluto	1162	1.31×10^{22}	5.914×10^9
Saturn	60268	5.69×10^{26}	1.427×10^9
Uranus	25559	8.72×10^{25}	2.871×10^9
Venus	6052	4.87×10^{24}	1.082×10^8

D3 Copy and complete the following statements.

(a) The mass of ...... is about 18 times the mass of Mercury.

(b) The mass of Venus is about ...... times the mass of Pluto.

(c) The mass of Jupiter is about ...... times the mass of Earth.

(d) The mass of Neptune is about 160 times the mass of ......

D4 Ken wants to put models in a park to show how far away from the Sun the planets are.

Mercury is to be 1 metre from the Sun.

This is Ken's calculation for the Earth.

$(1.496 \times 10^8) \div (5.791 \times 10^7) = 2.58$ (to 2 d.p.)

Earth: 2.58 metres from Sun

(a) Explain why Ken divides by 5.791×10^7.

(b) How far will each model be from the Sun?

D5 A formula for the approximate mean density (D) of each planet in kg/m³ is

$$D = \frac{3m}{4\pi(1000r)^3}$$

where m is the mass in kg and r is the radius in km.

(a) Work out the mean density of each planet to two significant figures.

(b) An object floats in water when its density in kg/m³ is less than 1000.
Which planets would float in water (if you could find a bowl big enough!)?

E *A small world*

Standard form can be used for very small numbers.

$\vdots$

$3 \times 10^2 = 3 \times 100$	$= 300$	
$3 \times 10^1 = 3 \times 10$	$= 30$	
$3 \times 10^0 = 3 \times 1$	$= 3$	
$3 \times 10^{-1} = 3 \times \frac{1}{10} = \frac{3}{10}$	$= 0.3$	
$3 \times 10^{-2} = 3 \times \frac{1}{10^2} = \frac{3}{100}$	$= 0.03$	
$3 \times 10^{-3} = 3 \times \frac{1}{10^3} = \frac{3}{1000}$	$= 0.003$	

$\vdots$

$\vdots$

$426 = 4.26 \times 100$	$= 4.26 \times 10^2$
$42.6 = 4.26 \times 10$	$= 4.26 \times 10^1$
$4.26 = 4.26 \times 1$	$= 4.26 \times 10^0$
$0.426 = \frac{4.26}{10} = 4.26 \times \frac{1}{10}$	$= 4.26 \times 10^{-1}$
$0.0426 = \frac{4.26}{100} = 4.26 \times \frac{1}{10^2}$	$= 4.26 \times 10^{-2}$
$0.00426 = \frac{4.26}{1000} = 4.26 \times \frac{1}{10^3}$	$= 4.26 \times 10^{-3}$

$\vdots$

E1 (a) Which of these are ways to write the number 0.000 045 3?

A 0.453×10^{-5} B 4.53×10^{-5} C 4.53×10^{-4}

D 453×10^{-7} E $4.53 \div 10^5$ F $45.3 \div 1\,000\,000$

(b) What is 0.000 045 3 written in standard form?

E2 Write these numbers in ordinary form.

(a) 9.1×10^{-3} (b) 6.21×10^{-5} (c) 3.4×10^{-10}

(d) 5×10^{-7} (e) 3.01×10^{-8} (f) 1×10^{-21}

E3 Write these numbers in standard form.

(a) 0.000 000 002 9   (b) 0.000 000 428   (c) 0.092 34

(d) 0.000 000 006   (e) 0.000 000 000 001   (f) 0.000 000 000 000 000 000 08

E4 A jawa is a coin that was used in Nepal.
It is the smallest coin ever issued and weighed 0.014 grams.

Write the weight of this coin in **kilograms**

(a) in ordinary form   (b) in standard form

E5 Write 6.7×10^{-9} kg in **grams** in ordinary form.

E6 The patu marplesi spider is the smallest in the world.
It is found in Samoa and is 0.43 mm long.

Write the length of this spider in **metres**

(a) in ordinary form   (b) in standard form

E7 Write 6.03×10^{-4} metres in centimetres in ordinary form.

E8 Write 5×10^{-2} millimetres in metres in ordinary form.

E9 The weights of some living creatures are given below.

| Blue whale: 130 000 kg | Pygmy shrew: 1.5×10^{-3} kg | Giraffe: 1200 kg |

| Kitti's hog-nosed bat: 1.7 grams | House mouse: 1.2×10^{-2} kg | House spider: 10^{-4} kg |

| Bee hummingbird: 0.0016 kg | Helena's hummingbird: 2×10^{-3} kg |

(a) List the creatures in order of weight, starting with the heaviest.

(b) Copy and complete the following statements
   (i) The weight of a giraffe is ...... times the weight of a house mouse.
   (ii) ...... house spiders would weigh the same as a bee hummingbird.
   (iii) A ...... weighs the same as 800 thousand pygmy shrews.

E10 A transmission electron microscope can examine features of width 2×10^{-9} metres or more. Which of these could be detected using a transmission electron microscope?

   A: A grain of pollen that is 0.0039 cm in diameter

   B: A particle of tobacco smoke that is 5×10^{-7} metres wide

   C: The nucleus of a carbon atom about 5×10^{-15} metres in diameter

   D: A virus that is 0.000 024 cm wide

   E: A distance of 0.000 000 019 7 centimetres between two calcium atoms

F *Small numbers on your calculator*

Example

A molecule of water has a mass of about 3×10^{-23} grams.

About how many water molecules are in a drop of water with a mass of 0.25 grams?
(Give your answer in ordinary form correct to three significant figures.)

$0.25 \div (3 \times 10^{-23}) \approx 8.33 \times 10^{21}$

$\qquad\qquad\qquad = 8\,330\,000\,000\,000\,000\,000\,000\,000$

Make sure you can work this out on your calculator.

F1 Write each answer to three significant figures
   (i) in standard form   (ii) in ordinary form

   (a) $(2.4 \times 10^{-5}) \times (6.9 \times 10^{-4})$ (b) $\dfrac{2.3 \times 10^{-10}}{6}$ (c) $\dfrac{4.2 \times 10^{-5}}{2.01 \times 10^{-20}}$

F2 A molecule of water has a mass of about 3×10^{-23} grams.
   About how many water molecules are in the 5 grams of water in a teaspoon?

   Give your answer to three significant figures
   (a) in standard form   (b) in ordinary form

F3 The width of a thread of DNA is about 1×10^{-8} metres.
How many times wider is a human hair with width 0.008 cm?

These pictures show enlargements of very small things (not all to the same scale).

These tiny organisms live in water.

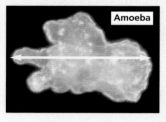
Amoeba
Length 3×10^{-4} m

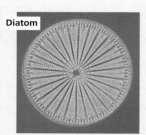
Diatom
Diameter 2.9×10^{-2} cm

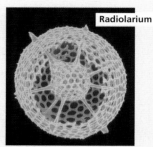
Radiolarium
Diameter 2×10^{-4} m

These are found in the human body.

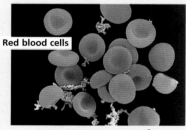
Red blood cells
Diameter 1.7×10^{-6} m

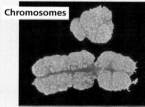
Chromosomes
Length 5×10^{-6} m

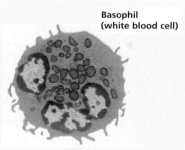
Basophil
(white blood cell)
Diameter 7.1×10^{-4} cm

These are viruses.

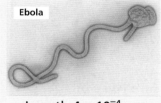
Ebola
Length 4×10^{-4} cm

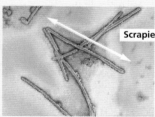
Scrapie
Length 5.5×10^{-7} m

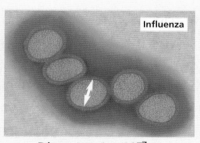
Influenza
Diameter 1×10^{-7} m

F4 How many chromosomes, placed end to end, would fit along 1 cm?

F5 This line — is 0.5 cm long. To two significant figures …
   (a) How many diatoms would fit along this line?
   (b) How many red blood cells would fit along the line?
   (c) How many scrapie viruses, placed end to end,
       would fit along the line?

F6 Measure the length of your thumbnail, to the nearest mm.
How many influenza viruses would make a line as long as your thumbnail?

F7 The diameter of a 5p coin is about 18 mm.
Copy and complete the following statements.

(a) The diameter of a 5p coin is …… times the diameter of the radiolaria.

(b) The diameter of a 5p coin is …… times the length of an ebola virus.

F8 Sue is 1.68 metres tall. How many basophils (to the nearest thousand) would make a line this long?

F9 Which has been magnified more, the red blood cell or the scrapie virus?

F10 If the influenza virus is drawn with a diameter of 2 cm, what would be the width of your thumb drawn to the same scale?

F11 Atoms consist of tiny particles.
The table shows the masses of some of these particles (to six significant figures).

Electron	$9.109\,39 \times 10^{-28}$ grams
Proton	$1.672\,62 \times 10^{-24}$ grams
Neutron	$1.674\,93 \times 10^{-24}$ grams

(a) How many electrons are equivalent in mass to one proton?
Give your answer in ordinary form correct to two significant figures.

(b) What is the difference in mass between a proton and a neutron?
Give your answer in standard form correct to one significant figure.

(c) How many protons are equivalent in mass to a grain of rice with mass 0.01 grams?
Give your answer in ordinary form correct to one significant figure.

F12 The mass of $6.022\,136 \times 10^{23}$ atoms of gold is about 197 grams.

Find the mean mass in grams of one atom of gold.
Give your answer correct to three significant figures

(a) in standard form   (b) in ordinary form

F13 The mass of about $3.011\,068 \times 10^{24}$ atoms of copper is about 0.318 kg.

Find the mean mass in grams of one atom of copper.
Give your answer in standard form correct to two significant figures.

F14 The force of attraction (in Newtons) between the Moon and the Earth,
is given by the formula $\dfrac{GmM}{r^2}$, where

$G \approx 6.7 \times 10^{-11}$ (a gravitational constant),
$M \approx 6.0 \times 10^{24}$ (the mass of the Earth in kg),
$m \approx 7.3 \times 10^{22}$ (the mass of the Moon in kg),
$r \approx 3.8 \times 10^{10}$ (the mean distance between the Earth and the Moon in metres).

Calculate this force of attraction correct to one significant figure.

G Standard form without a calculator

Examples

Calculate $(6.4 \times 10^5) \times (2 \times 10^3)$.
Give your answer in standard form.

$$(6.4 \times 10^5) \times (2 \times 10^3)$$
$$= 6.4 \times 2 \times 10^5 \times 10^3$$
$$= 12.8 \times 10^8$$
$$= 1.28 \times 10 \times 10^8$$
$$= 1.28 \times 10^9$$

Calculate $\dfrac{3 \times 10^3}{6 \times 10^5}$.
Give your answer in standard form.

$$\frac{3 \times 10^3}{6 \times 10^5}$$
$$= \frac{3}{6} \times \frac{10^3}{10^5}$$
$$= 0.5 \times 10^{-2}$$
$$= 5 \times 10^{-1} \times 10^{-2}$$
$$= 5 \times 10^{-3}$$

G1 Write these numbers in **standard form**.

(a) 26×10^6 (b) 0.05×10^9 (c) 340×10^{-5} (d) 0.2×10^{-12}

G2 Calculate the following, giving your answers in standard form.

(a) $(2 \times 10^6) \times (4 \times 10^4)$ (b) $(4.5 \times 10^9) \times 5$

(c) $(6 \times 10^{12}) \times (8 \times 10^{-5})$ (d) $(7 \times 10^3)^2$

(e) $(2.3 \times 10^{-3}) \times (4 \times 10^{-2})$ (f) $(9 \times 10^{-8}) \times (3 \times 10^4)$

G3 Calculate the following, giving your answers in standard form.

(a) $\dfrac{8.4 \times 10^8}{4}$ (b) $\dfrac{6.1 \times 10^9}{10^4}$ (c) $\dfrac{8 \times 10^8}{2 \times 10^6}$

(d) $\dfrac{2 \times 10^{13}}{8 \times 10^7}$ (e) $\dfrac{4 \times 10^5}{5 \times 10^9}$ (f) $\dfrac{6 \times 10^2}{4 \times 10^{-3}}$

G4 Calculate the following, giving your answers as **ordinary numbers**.

(a) $(4 \times 10^{-6}) \times (2 \times 10^7)$ (b) $(2.4 \times 10^5) \div 10^3$

(c) $(3.2 \times 10^8) \times (5 \times 10^{-6})$ (d) $(1.4 \times 10^5) \times (9 \times 10^{-6})$

(e) $\dfrac{1.6 \times 10^9}{8 \times 10^7}$ (f) $\dfrac{8 \times 10^5}{1.6 \times 10^6}$ (g) $\dfrac{3.9 \times 10^{-2}}{3 \times 10^3}$

G5 (a) Write $384\,216$ and $0.005\,29$ in standard form, correct to one significant figure.

(b) Use your answers to (a) to estimate

(i) $384\,216 \times 0.005\,29$ (ii) $384\,216 \div 0.005\,29$ (iii) $0.005\,29 \div 384\,216$

G6 These are additions and subtractions.
Write the answers both in ordinary form and standard form.

(a) $(5 \times 10^3) + (3 \times 10^4)$

(b) $(8 \times 10^5) + (4 \times 10^4)$

(c) $(5 \times 10^5) - (7 \times 10^3)$

(d) $(8 \times 10^{-2}) + (2 \times 10^{-3})$

(e) $(8 \times 10^{-3}) - (6 \times 10^{-4})$

(f) $(3 \times 10^{-1}) - (6 \times 10^{-2})$

Test yourself

T1 The Earth's volume is about $1\,080\,000\,000\,000$ km^3.
Write this number in standard form.

T2 The approximate area of the Australian desert is 1.47×10^6 square miles.
Write this number in ordinary form.

T3 The smallest known species of reptile is believed to be a
tiny gecko found on the island of Virgin Gorda.

The largest females found were 0.018 metres long. Write this number in standard form.

T4 The eggs of a certain parasitic wasp each weigh 2×10^{-7} grams.
Write this number in ordinary form.

T5 A light year is about 9.46×10^{12} km.

(a) The centre of the Milky Way is about 26 100 light years away.
In km, about how far away is the centre of the Milky Way

(i) in standard form

(ii) in ordinary form

(b) The spiral galaxy Andromeda is about 2.0812×10^{19} km away.
About how far away is this in light years?

T6 Calculate $(3.4 \times 10^5) \div (1.92 \times 10^{-9})$ and write the answer as
an ordinary number correct to three significant figures.

T7 Calculate the following, giving your answers as ordinary numbers.

(a) $(4.5 \times 10^3) \times (3 \times 10^4)$

(b) $\dfrac{6 \times 10^{-3}}{3 \times 10^4}$

(c) $\dfrac{2 \times 10^3}{4 \times 10^{-2}}$

T8 Calculate the following, giving your answers in standard form.

(a) $(3.2 \times 10^{-6}) \times (5 \times 10^3)$

(b) $\dfrac{4.2 \times 10^{-4}}{3}$

(c) $\dfrac{1.2 \times 10^{-2}}{4 \times 10^{-6}}$

T9 (a) (i) Write the number 5.01×10^4 as an ordinary number.

(ii) Write the number 0.0009 in standard form.

(b) Multiply 4×10^3 by 6×10^5.
Give your answer in standard form

Edexcel

13 Gradients and equations

You should know how to
◆ plot points and draw the graph of a straight line, given its equation
◆ rearrange simple formulas

This work will help you
◆ calculate and use gradients
◆ understand and use the gradient–intercept form of the equation of a line

A Gradients

> How could you measure the steepness of each line?

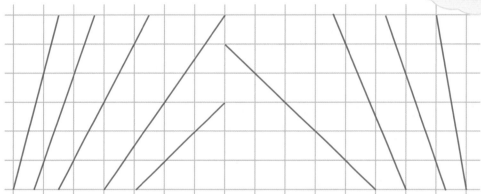

A1 Find the gradients of the lines joining
(a) (0, 1) and (4, 5) (b) (⁻1, 0) and (3, 8) (c) (0, 8) and (8, 0)
(d) (2, 1) and (6, ⁻1) (e) (⁻2, 5) and (2, ⁻1) (f) (4, 5) and (⁻2, 5)

A2 This is a sketch of a child's slide.
(a) Find the gradient of the sloping part of the slide.
(b) For safety reasons the gradient of the slide should be no greater than 0.7.
Is this a safe slide?

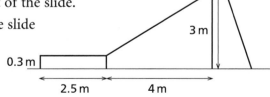

0.3 m

3 m

2.5 m 4 m

A3 (a) Find the gradient of this ramp (to 2 d.p.).
(b) A fork-lift truck can safely climb a ramp with a gradient of less than 0.17.
Can the fork-lift truck safely use this ramp?

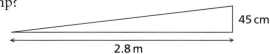

45 cm

2.8 m

B *Gradients and rates*

The gradient of a straight line is $\dfrac{\text{vertical change}}{\text{horizontal change}}$.

Each sketch shows the volume of water in a bath.

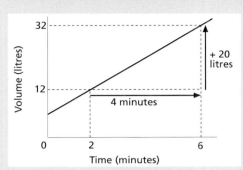

For a horizontal increase of 4 minutes there is a vertical **increase** of 20 litres.

So the **gradient** is $\dfrac{20 \text{ litres}}{4 \text{ minutes}} = 5$ litres per minute.

This is the rate of flow of the water.

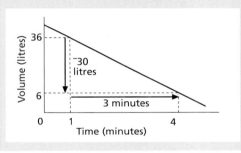

For a horizontal increase of 3 minutes there is a vertical **decrease** of 30 litres.

So the **gradient** is $\dfrac{^-30 \text{ litres}}{3 \text{ minutes}} = {}^-10$ litres per minute.

The negative gradient shows the volume of water is decreasing.

B1 What rates of flow are shown by the following graphs?

(a)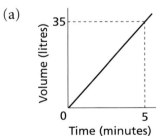

(b)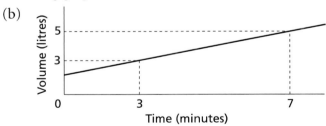

B2 This graph shows the volume of oil in a tank.

(a) Calculate the gradient of the line.

(b) What happens to the oil in the tank during these 5 minutes?

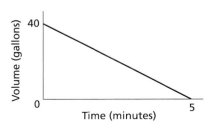

B3 This graph shows the volume of water in a tank.

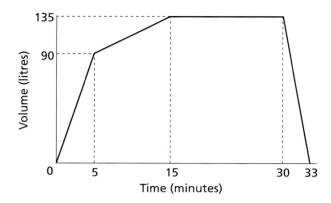

(a) Find the gradient of each straight-line segment.

(b) What do you think happened after

    (i) 5 minutes

    (ii) 15 minutes

    (iii) 30 minutes

B4 For this travel graph, find the gradient of each straight-line segment.
Describe, as fully as you can, what is happening at each stage of the journey.

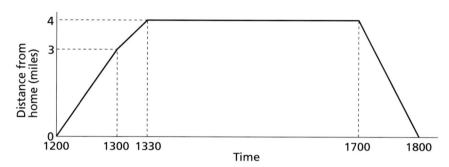

C Graphs and equations

C1 (a) Copy and complete these tables for the functions $y = 2x$, $y = 3x$ and $y = {}^-4x$.

x	$^-1$	0	1	2	3
$y = 2x$					

x	$^-1$	0	1	2	3
$y = 3x$					

x	$^-1$	0	1	2	3
$y = {}^-4x$					

    (b) On the same set of axes, draw the graphs of $y = 2x$, $y = 3x$ and $y = {}^-4x$ for values of x from $^-1$ to 3.

    (c) What is the gradient of each line?

    (d) Without drawing the graph, write down the gradient of

        (i) $y = 5x$ (ii) $y = {}^-3x$ (iii) $y = \frac{1}{2}x$ (iv) $y = x$

C2 (a) Copy and complete these tables for the functions $y = 3x$ and $y = 3x - 2$.

x	$^-1$	0	1	2	3
$y = 3x$					

x	$^-1$	0	1	2	3
$y = 3x - 2$					

    (b) On the same set of axes, draw the graphs of $y = 3x$ and $y = 3x - 2$.

    (c) What is the gradient of each line?

    (d) Where does the graph of $y = 3x - 2$ cross the y-axis?

    (e) Where would the graph of $y = 3x + 5$ cross the y-axis?

C3 (a) Where do you think the line with equation $y = 2x + 1$ crosses the y-axis?

(b) What do you think is the gradient of the line with equation $y = 2x + 1$?

C4 (a) What is the gradient of each line in the diagram on the right?

(b) Write down the equation of the red line.

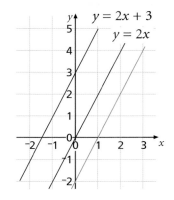

C5 What is the link between the gradient of a straight line, where it crosses the y-axis and its equation?

C6 A line has equation $y = 4x - 5$.
From the equation, write down the gradient and where the line crosses the y-axis.

C7 A line has a gradient of 8 and crosses the y-axis at $(0, 10)$.
Write down the equation of this line.

C8 Explain how you can decide, without drawing, that the lines with equations $y = {}^-5x + 1$ and $y = {}^-5x - 5$ are parallel.

C9 This sketch shows a pair of parallel lines.
The equation of the line through the origin is $y = {}^-x$.
Write down the equation of the other line.

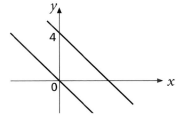

C10 Each line labelled A to E matches an equation below.

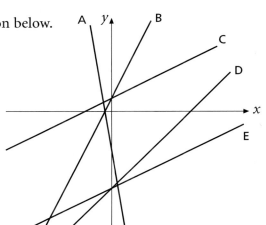

| $y = 2x + 1$ |
| $y = x - 6$ |
| $y = 0.5x - 6$ |
| $y = {}^-6x - 3$ |
| $y = 0.5x + 1$ |

Match each line to its correct equation.

A graph with equation of the form $y = mx + c$ (where m and c are constants) is a straight line.
Its gradient is m and its y-intercept is c.

Example

The y-intercept is where
the line cuts the y-axis.
It is the value of y when $x = 0$.

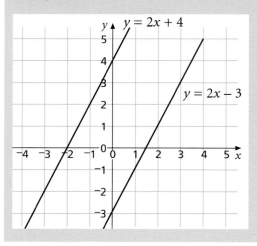

From the equations, the gradient
of each line is **2** so they are parallel.

The y-intercept of the line $y = 2x + 4$ is **4** and
the y-intercept of the line $y = 2x - 3$ is **-3**.

D *Including fractions*

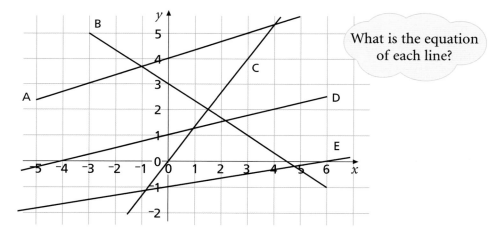

What is the equation
of each line?

D1 For each line on the right …
    (i)   Find the gradient as a fraction.
    (ii) Write down the y-intercept.
    (iii) Write down the line's equation.

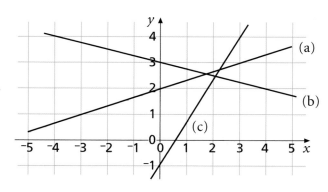

D2 A line has a gradient of $\frac{1}{3}$ and cuts the y-axis at $(0, 2)$.
Explain why it is not correct to write its equation as $y = 0.3x + 2$.

D3 Find an equation for each line sketched below.

(a)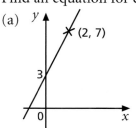

(b)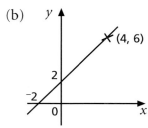

(c)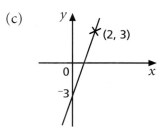

(d)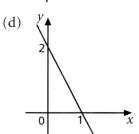

(e)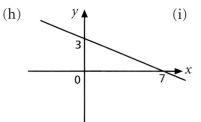

(f)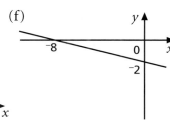

(g)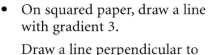

(h)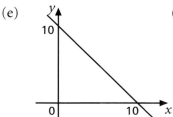

(i)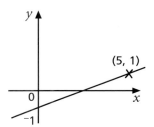

E *Perpendicular lines*

- What is the gradient of line a?

  Line b is perpendicular to line a.
  What is its gradient?

- What is the gradient of line c?

  Line d is perpendicular to line c.
  What is its gradient?

- On squared paper, draw a line
  with gradient 3.

  Draw a line perpendicular to
  this line and find its gradient.

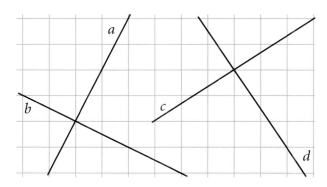

- Investigate the relationship between the gradients of perpendicular lines.

Perpendicular lines and gradients

The gradient of the line is $\frac{a}{b}$.   So the gradient of the perpendicular is $\frac{^-b}{a}$.

So, if the gradient of a line is g, then the gradient of a perpendicular is $\frac{^-1}{g}$.

E1 (a) Write down the gradient of the line whose equation is $y = 4x - 3$.

(b) What is the gradient of any line perpendicular to $y = 4x - 3$?

E2 The equation of line a is $y = \frac{3}{4}x + 3$.
What is the equation of line b, which is perpendicular to a?

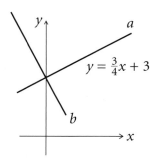

F Rearranging

• Do any of these equations give parallel lines?

A $y + 3x = 6$ **B** $x = \frac{y}{3} + \frac{2}{3}$ **C** $6x = 8 - 2y$ **D** $3y = x - 3$

E $y - 3x = 4$ **F** $2(3y - x) = 12$ **G** $2(y + 3x) = 7$ **H** $x = 3y + 5$

F1 Find the gradient and y-intercept of lines with the following equations.

(a) $y - 2x = 3$ (b) $x = y + 4$ (c) $y + 3x = 2$

(d) $2y = 4x - 3$ (e) $3y + 6x = 12$ (f) $x = 2(y - 1)$

(g) $3y + x + 2 = 0$ (h) $5y - 2x = 10$ (i) $3x = 4(1 - y)$

F2 Which of these lines is parallel to $2y - x = 3$?

A $2y + x = 5$ **B** $4y - 1 = 2x$ **C** $y = 2x + 3$ **D** $x = \frac{1}{2}y + 1$

F3 Show that $10x = 2y - 2$ and $6y - 30x = 6$ are equations for the same line.

F4 Which of these lines is perpendicular to $y + 2x = 9$?

A $y = ^-2x - 3$

B $x = 2y + 14$

C $2y + x = 0$

D $y + 2x = \frac{-1}{9}$

F5 (a) Give an equation for the line parallel to $y - 3x = 2$ through $(0, 3)$.

(b) Give an equation for the line perpendicular to $2y + 3x = 7$ through $(0, ^-5)$.

F6 Shape ABCD is a rectangle.

C is the point with coordinates $(0, ^-8)$.

The equation of the straight line through AB is $x + 2y = 4$.

Find the equations of the lines DC, AD and BC.

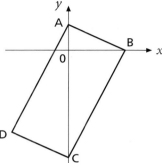

Sometimes the equation of a straight line is not in the form $y = mx + c$.

Examples

Find the gradient and y-intercept of the line with equation $y + 4x = 8$.

$$y + 4x = 8$$
$$y = 8 - 4x$$
$$y = ^-4x + 8$$

So the gradient is **$^-4$** and the y-intercept is **8**.

Find the gradient and y-intercept of the line with equation $3y + 1 = 2x$.

$$3y + 1 = 2x$$
$$3y = 2x - 1$$
$$y = \frac{2x - 1}{3}$$
$$y = \tfrac{2}{3}x - \tfrac{1}{3}$$

So the gradient is $\tfrac{2}{3}$ and the y-intercept is $^-\tfrac{1}{3}$.

G *Equation of a line through two given points*

Example

What is the equation of the straight line through $(5, 8)$ and $(7, 16)$?

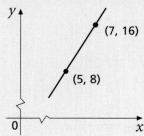

The gradient is $\dfrac{16 - 8}{7 - 5} = \dfrac{8}{2} = 4$

So the equation has the form $y = 4x + c$.

The line goes through $(5, 8)$ so $8 = 4 \times 5 + c$

so $c = 8 - 20 = ^-12$

So the equation of the line is $y = 4x - 12$.

G1 Without drawing …

    (a) find the gradient of the line through $(4, 8)$ and $(10, 11)$.

    (b) find the equation of the line in the form $y = mx + c$.

    (c) find the coordinates of the point where the line crosses the y-axis.

G2 Find, in the form $y = mx + c$, the equation of the line through

    (a) $(2, 12)$ and $(5, 21)$ (b) $(9, 15)$ and $(11, 23)$ (c) $(6, 5)$ and $(9, 6)$

G3 Find, in the form $y = mx + c$, the equation of the line through

    (a) $(3, 8)$ and $(5, 4)$ (b) $(8, 13)$ and $(12, 1)$ (c) $(^-3, 4)$ and $(6, 7)$

G4 The shape ABCD is a square.

The coordinates of three vertices are A $(18, 14)$, B $(20, 20)$ and C $(26, 18)$.
Find the equations of each of the four lines AB, BC, DC and AD.

G5 Jane walks from work to her evening class.

Distances from her home at various times are shown in the table below.

Time (minutes)	2	6	10
Distance from home (metres)	1360	1680	2000

Here, the relationship between distance and time is linear (this means
the graph is a straight line.)

    (a) Draw a sketch of distance from home (h) against time (t).

    (b) Find the equation that links h and t in the form $h = \ldots\ldots$

    (c) What does the gradient of this line represent?

    (d) Use the equation to find how far Jane is from home after 3.5 minutes.

    (e) How far was she from home at the start of her journey?

G6 The volume of oil in a tank at various times is shown in the table below.

Time (seconds)	10	100	200
Volume (litres)	1988	1880	1760

Here, the relationship between volume (v) and time (t) is linear.

    (a) Find the equation that links v and t in the form $v = \ldots\ldots$
         Write any non-integers as decimals in your equation.

    (b) What does the gradient of this line represent?

    (c) How much oil was in the tank to start with?

    (d) Use the equation to find how much oil is in the tank after 2 minutes.

*G7 A is the point $(8, 9)$ and B is $(12, 10)$. Without drawing …

    (a) find the equation of the line AB

    (b) find the equation of the line through A perpendicular to AB

H Lines of best fit

A pan of water was heated and the temperature measured at various intervals.

Time (minutes)	1	2	4	6	8
Temperature (°C)	32	40	53	75	85

We can investigate the relationship between time (t) and temperature (T).

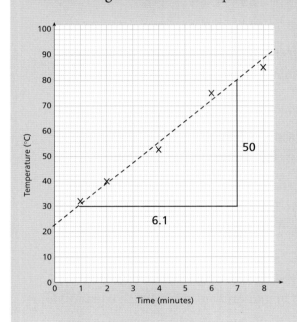

- Plot these pairs of values.
- They lie close to a straight line so draw the line of best fit.
- Choose two suitable points on the line and use them to estimate the gradient.
  $$\frac{50}{6.1} \approx 8.2$$
- Estimate where it cuts the vertical axis.
  The line cuts this axis at about 22.
- So the equation of the line of best fit is approximately
  $$T = 8.2t + 22$$

H1 Nigel lit a small candle. He put a beaker upside down over the candle and counted how many seconds it took for the flame to go out.

He repeated the experiment with larger containers.
Here are his results.

Volume (V ml)	200	250	500	750	1000
Time taken (t s)	12	17	32	48	60

(a) Plot the points and draw a line of best fit.
Estimate its gradient and y-intercept to one significant figure.
Hence find an approximate equation for your line of best fit.

(b) Use your equation to estimate how long it would take for the flame to go out in a beaker with volume 600 ml.

H2 For each of set of data …

- plot the pairs of values and draw a line of best fit
- estimate the gradient and vertical intercept correct to two significant figures
- hence, find an approximate equation for the line of best fit

(a) Here are the results of an experiment where different loads were added to a helical spring to see how much it stretched.

Load (L newtons)	1.7	3.2	5.0	6.7	8.3	10.0
Extension (E cm)	5.1	9.6	14.8	20.2	25.0	29.9

   (i) Use your equation to estimate the stretch in cm for a load of 7.5 newtons.

  (ii) Would you use this equation to estimate the stretch for a load of 1000 newtons? Explain your answer.

(b) Here are the some of the world records for women's indoor running events.

Distance run (s m)	200	400	800	1000
Time taken (t)	21.87	49.59	116.36	150.94

   (i) Use your equation to estimate what the world record would be for 600m.

  (ii) The world record for the 50 m race is 5.96 seconds. Does this fit your equation? Comment on your answer.

H3 (a) For each set of values …

- plot the pairs of values and decide if your graph shows linear correlation
- where appropriate, draw a line of best fit and find an approximate equation for it

   (i) This table shows the mean wing length of a group of ducks at various ages.

Age (d days)	2	4	6	8	10	13	16	19	22
Wing length (W mm)	26	29.2	33.9	37	41.6	57.4	73.7	81.5	105

  (ii) This table shows fat as a percentage of body weight for a group of adult men.

Age (y years)	23	39	41	49	50	53	54	56	57	60	61
Percentage of fat (F)	23	26	21	20	26	30	37	24	28	29	36

(iii) This table shows the results of aerial surveys carried out on each of 11 days in a particular part of Alaska. The average wind speed and the number of black bears sighted on each day are given.

Wind speed (w knots)	2.1	21.1	4.9	23.6	21.5	10.5	20.3	11.9	6.9	14	27.2
Number of bears (B)	99	30	82	43	49	79	54	69	87	72	23

(b) Decide, giving reasons, which of the following you can estimate with reasonable confidence.

   (i)   The wing length of a 100-day-old duck

   (ii) The percentage of a 40-year-old man's body that is fat

   (iii) The number of bears sighted on a day when the average wind speed is 15 knots

   (iv) The wing length of a 12-day-old duck

   (v) The number of bears sighted on a day when the average wind speed is 40 knots

Test yourself

T1 (a) What is the gradient and y-intercept of the line with equation $y = 4x - 7$?

   (b) What is the equation of the line parallel to $y = {}^-2x + 5$ going through $(0, 9)$?

T2 The line with equation $x + 2y = 6$ has been drawn on the grid.

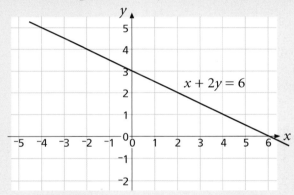

(a) Rearrange the equation $x + 2y = 6$ to make y the subject.

(b) Write down the gradient of the line with equation $x + 2y = 6$.

(c) Write down the equation of the line which is parallel to the line with equation $x + 2y = 6$ and passes through the point with coordinates $(0, 7)$. Edexcel

T3 What is the equation of the line perpendicular to $y = \frac{2}{3}x - 4$ through $(0, 6)$?

T4 Here are the world records for men's outdoor running events.

Distance run (s m)	200	400	800	1000
Time taken (t)	19.32	43.29	101.11	132.18

(a) Plot the pairs of values and find an approximate equation of the line of best fit.

(b) Use your equation to estimate what the world record would be for 600 m.

(c) Would you use this equation to estimate the world record for 100 metres? Explain your answer.

Review 3

1 Calculate the sides and angles marked with letters.

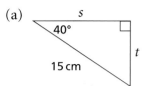 (a)

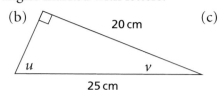 (b)

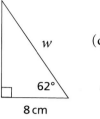 (c)

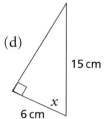 (d)

2 Calculate the volume of this skip.

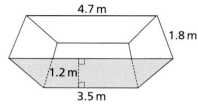

3 In Berlin in September 1923, a loaf of bread cost 1 512 000 marks.
By November the price had risen to 201 000 000 000 marks.
Write each of these prices in standard form.

4 Find the equation of
(a) the line parallel to $y = 3x - 5$ going through $(0, 7)$
(b) the line parallel to $2x + 3y = 6$ going through $(0, {}^-4)$
(c) the line perpendicular to $y = 2x + 3$, going through $(0, 0)$

5 Justin takes out a bank loan of £2000.
The bank charges interest at a rate of 2% per month on
the outstanding amount of the loan.
Justin pays back £500 after 1 month and another £500 after 2 months.
After 3 months he pays back the outstanding amount and ends the loan.

Complete this calculation to
find how much he has to pay
back at the end of the 3 months.

Starting amount	£2000.00
Interest for month 1	£ 40.00
Outstanding amount after 1 month	£2040.00
Repayment at end of month 1	- £ 500.00
Outstanding amount at start of month 2	£ 1540.00
Interest for month 2	......

6 ABCD is a trapezium, with AB parallel to DC.
AB = 15 cm, angle ADC = 20° and angle BCD = 35°.
The distance between AB and DC is 12 cm.

Calculate the length of DC.

7 (a) Write 0.000 0357 in standard form.

    (b) Work out $(3 \times 10^6) \times (5 \times 10^{-2})$, giving the result in standard form.

    (c) Work out $\dfrac{4 \times 10^3}{5 \times 10^{-2}}$, giving the result in standard form.

8 This box is in the shape of a prism whose cross-section is an equilateral triangle. Calculate

    (a) the height h of the equilateral triangle

    (b) the area of the equilateral triangle

    (c) the total surface area of the box

    (d) the volume of the box

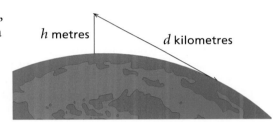

6.4 cm h 12.5 cm   60°

9 Find the value of n in each of these equations.

    (a) $2^5 \times 2^n = 2^{15}$ (b) $3^n \times 3^6 = 3^2$ (c) $2^2 \times 2^n = 4$ (d) $\dfrac{4^3}{4^n} = 4^{-5}$

10 From a point h m above the Earth's surface, the distance you can see on a clear day is d km, where d is approximately given by the formula
$$d = 3.6\sqrt{h}.$$

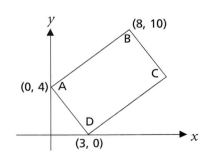

h metres d kilometres

    (a) The Eiffel tower is 300 m tall.
Use the formula to calculate roughly how far you can see from the top.

    (b) Rearrange the formula to make h the subject.

    (c) From what height would you be able to see a distance of 50 km?

11 This table shows the distribution of the weights of the pumpkins grown in a greenhouse.

    (a) Draw a frequency polygon for this data.

    (b) Calculate an estimate of the mean weight of the pumpkins.

Weight, w kg	Frequency
$0 < w \le 1$	5
$1 < w \le 2$	10
$2 < w \le 3$	7
$3 < w \le 4$	3

12 ABCD is a rectangle.
Find the equation of line

    (a) AB   (b) DC   (c) AD

    (d) BC   (e) BD   (f) AC

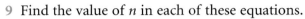

(8, 10)   B   C   (0, 4) A   D   (3, 0)

13 If 1 litre of water is poured into this plastic container, how deep will the water be (to the nearest 0.1 cm)?

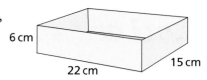

6 cm 22 cm 15 cm

14 In this diagram, the circle, whose radius is 5 cm, touches the sides of the regular pentagon.

O is the centre of the circle.

(a) Calculate angle a.

(b) Calculate the length of each side of the regular pentagon.

(c) Calculate the perimeter of the regular pentagon.

(d) Imagine that the regular pentagon is replaced by a regular polygon with 20 sides, all touching the circle.
Calculate the perimeter of the 20-sided regular polygon.

(e) Now imagine that the pentagon is replaced by a regular polygon with 180 sides, all touching the circle.
Calculate the perimeter of this polygon.

(f) Calculate the circumference of the circle and compare it with your answer to (e). Why should the two results be close?

15 Make the bold letter the subject of each of these formulas.

(a) $s = \sqrt{a\boldsymbol{r}}$

(b) $s = \sqrt{r + \boldsymbol{b}}$

(c) $s = \sqrt{a\boldsymbol{r} + b}$

(d) $s = \sqrt{ar + \boldsymbol{b}}$

(e) $s = \dfrac{\boldsymbol{a}\sqrt{r}}{b}$

(f) $s = \dfrac{a\sqrt{\boldsymbol{r}}}{b}$

(g) $s = \dfrac{\sqrt{a\boldsymbol{r}}}{b}$

(h) $s = \sqrt{\dfrac{a\boldsymbol{r}}{b}}$

*16 A train takes $4\frac{1}{2}$ hours to cover a distance of 540 km.
After improvements to the track, the journey time is reduced by 8%.
What is the percentage increase in the average speed of the train?

*17 The diagonals of a rectangle are each of length 24 cm.
The angle between the diagonals is 42°.
Calculate the area of the rectangle.

Challenge

An insect at A wants to crawl across the surface of the cuboid to the opposite vertex B.

The coloured lines show some possible routes.

- What is the length of the shortest route from A to B on the surface of the cuboid?

Explain why your route must be the shortest.

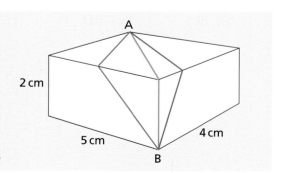

2 cm   5 cm   4 cm

14 Loci and constructions

You will revise
- the loci of points equidistant from two points or two lines
- constructions using a straight edge and compasses

This work will help you solve problems using loci.

A Standard loci and constructions

A1 A is a fixed point.
P, Q and R are points that are each 2 cm from A.

What is the locus consisting of all the points that
are 2 cm from A?

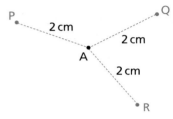

A2 *l* is a fixed (infinite) line.
X, Y and Z are points that are each 1.5 cm from *l*.

What is the locus consisting of all the points that
are 1.5 cm from *l*?

A3 A and B are fixed points.
P and Q are points that are each equidistant from A and B.

What is the locus of points that are equidistant from A and B?

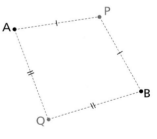

A4 *a* and *b* are fixed intersecting lines.
X and Y are points that are each equidistant
from *a* and *b*.

What is the locus of points that are equidistant
from *a* and *b*?

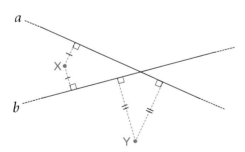

A5 Draw a circle and mark two points A, B on the circle.

Shade the locus of all the points inside the circle that are
nearer to A than to B.

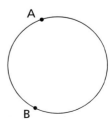

Constructions using straight edge and compasses

Arcs shown in the same colour have the same radius.

The perpendicular bisector of a line segment

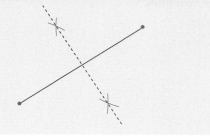

The bisector of an angle

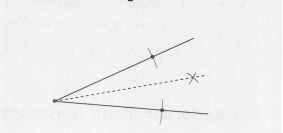

The line from a given point, perpendicular to a given line

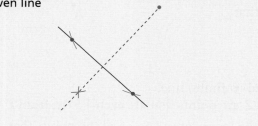

The line perpendicular to a given line, from a point on the given line

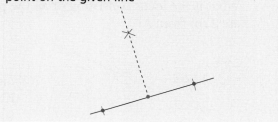

A6 Draw a triangle PQR.

Draw the perpendicular bisector of PQ and the perpendicular bisector of QR.
Label the point where the two perpendicular bisectors meet, O.

Construct a circle, centre O and radius OP.

The circle that you have drawn should pass through points P, Q and R. Explain this result.

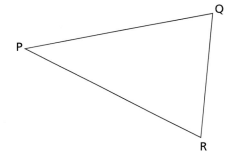

WJEC

A7 Draw this rectangle.

Using straight edge and compasses, construct the locus of points that are inside the rectangle and equidistant from A and C.

Shade the region consisting of all points inside the rectangle that are nearer to A than to C.

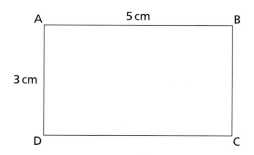

A8 Draw the rectangle in question A7 again.

Using straight edge and compasses, construct the locus of points that are inside the rectangle and equidistant from the sides AB and AD.

Shade the region consisting of all points inside the rectangle that are nearer to AB than to AD.

A9 Mark two points P and Q and draw a line *l* that does not go through either P or Q.

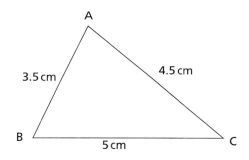

(a) Explain how to find, using only straight edge and compasses, the point on *l* which is equidistant from P and Q.

(b) If P, Q and *l* are arranged in a special way, then there is **no** point on *l* equidistant from P and Q. What is this special arrangement?

(c) What is special about the arrangement of P, Q and *l* if **every** point on *l* is equidistant from P and Q?

A10 Draw a triangle ABC.

Explain how to find, using only straight edge and compasses, the point on BC that is equidistant from lines AB and AC.

A11 Draw this triangle accurately.

Using straight edge and compasses, divide the inside of the triangle into three regions:

• points that are nearer to A than to either B or C

• points that are nearer to B than to either A or C

• points that are nearer to C than to either A or B

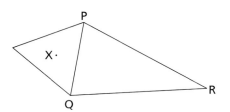

A12 Draw a triangle ABC.
Using straight edge and compasses, construct

• the line from A perpendicular to BC

• the line from B perpendicular to AC

• the line from C perpendicular to AB

What do you notice about the three lines?

*A13 Draw a scalene triangle PQR.

Using straight edge and compasses, construct an equilateral triangle on PQ (outside triangle PQR). Find by construction the centre of this triangle and label it X.

Now do the same for the other two sides of PQR. Label the centres of the equilateral triangles Y and Z.

What can you say about triangle XYZ?

B *Locus problems*

Example

A tree is to be planted in this garden.
It must be at least 5 m from corner C and
at least 7 m from the wall AD.

Shade the region where the tree may be planted.

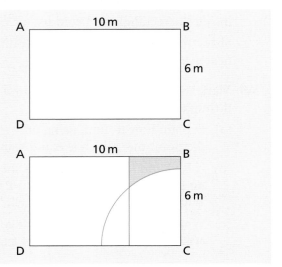

The green arc shows points that are 5 m from C.

The red line shows points that are 7 m from AD.

The tree may be planted anywhere inside the
shaded region.

B1 This is a triangular plot of land.
There is an electrical cable running along PQ
and a radio mast at the point R.

It is dangerous for a person to be within 10 m of the cable
or within 20 m of the mast.

Draw the triangle to scale and shade the region where
it is safe to be.

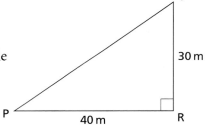

B2 A, B and C are three points situated at the vertices of an equilateral triangle
whose sides are each 40 km.

There is a transmitter at each of the points A, B and C.
Signals from transmitter A can be picked up at any point within 30 km of A.
30 km is the 'range' of transmitter A.

The range of transmitter B is 25 km and of C 30 km.

David is somewhere inside triangle ABC. He can pick up signals from A
and B but not C.

Draw triangle ABC to scale and shade the region where David could be.

B3 Draw the rectangle PQRS, where PQ = RS = 7 cm and PS = QR = 5 cm.

Draw the region consisting of all the points inside the rectangle that
are both nearer to P than to R and nearer to PQ than to PS.

B4 Draw the rhombus ABCD such that the sides are each 4 cm and the angle
between AB and AD is 60°.

Construct and shade the region consisting of all points inside the rhombus
that are nearer to B than to any other vertex.

B5 The diagram shows part of the boundary between two countries.
For security reasons, nobody is allowed to go within
1 km of the boundary on either side.

Copy the diagram and shade the prohibited region.

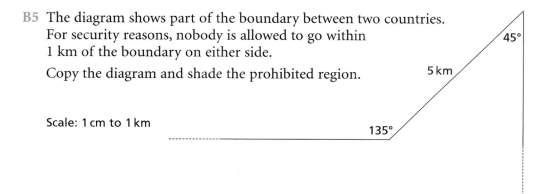

Scale: 1 cm to 1 km

B6 A field is in the shape of a right-angled triangle with
sides of length 30 m, 40 m and 50 m.
A hedge runs along each side of the field.

A goat is tethered to a post in the field.
The goat can just reach every hedge

Draw the field to scale and use straight edge and compasses to construct
the position of the post.

Shade the part of the field that the goat cannot reach.

Test yourself

The diagrams for these questions are on sheet G158.

T1 **In this question you should use ruler and compasses only for the constructions.**

In diagram 1, ABC is a triangular field.
A mobile phone company plans to erect a mast in the field.
The mast must be exactly the same distance from A as it is from B.
The mast must also be closer to AB than it is to BC.
Indicate on the diagram all possible positions of the mast. OCR

T2 Diagram 2 shows the wall of a house drawn to a scale of 2 cm to 1 m.
A dog is fastened by a lead 3 m long to a point X on the wall.
Copy the diagram and shade the area that the dog can reach. OCR

T3 The plan in diagram 3 shows the landing area, ABCD, for a javelin event.
AD is the throwing line.
The arc BC is drawn from centre X.
The plan has been drawn to a scale of 1 cm to 5 m.

The landing area is fenced off in front of the throwing line.
The position of the fence is always 10 m from the boundaries AB, BC and CD
of the landing area.

Draw accurately the position of the fence on the plan. AQA(SEG) 2000

15 Cumulative frequency

You should know how to calculate an estimate of the mean of a frequency distribution.

This work will help you
- ◆ make a cumulative frequency table and draw a cumulative frequency graph
- ◆ find and interpret the median, quartiles and interquartile range
- ◆ interpret and draw a box-and-whisker plot

A Hold your breath!

How long can you hold your breath?

As well as affecting your ability to swim underwater, there are many situations where your ability to hold your breath could be a lifesaver.

Measure how long people in your class can hold their breath for.

Suppose that, in a fire training exercise, advisers want to be able to evacuate a building in a time for which 90% of all people could hold their breath. How long do you think this would be?

B Cumulative frequency tables

This table shows the distribution of the birth weights of some babies.

Altogether there are 15 babies who weigh up to and including 2 kg.

Weight (w kg)	Frequency
$0 < w \leq 1$	2
$1 < w \leq 2$	13
$2 < w \leq 3$	20
$3 < w \leq 4$	16
$4 < w \leq 5$	9

} 15 babies weigh up to 2 kg.

This table is called a **cumulative frequency table**.

Weight (w kg)	Cumulative frequency
$w \leq 1$	2
$w \leq 2$	15
$w \leq 3$	35
$w \leq 4$	
$w \leq 5$	

This is another way of writing 'weights up to and including 3 kg'.

35 babies weigh up to 3 kg.

B1 (a) What are the two numbers missing from the cumulative frequency table above?

(b) What does the last number in the cumulative frequency column tell you?

B2 This table shows the distribution of the lengths of some snakes.

Length (*l* cm)	Frequency
0 < *l* ≤ 50	4
50 < *l* ≤ 100	10
100 < *l* ≤ 150	17
150 < *l* ≤ 200	14
200 < *l* ≤ 250	5
250 < *l* ≤ 300	2

(a) How many snakes have lengths up to 150 cm?

(b) How many have lengths up to 200 cm?

(c) How many have lengths up to 250 cm?

(d) How many were measured altogether?

B3 This table shows the distribution of the heights of children at a playgroup.

Height (*h* cm)	Frequency
60 < *h* ≤ 65	2
65 < *h* ≤ 70	9
70 < *h* ≤ 75	12
75 < *h* ≤ 80	6
80 < *h* ≤ 85	4

Copy and complete this cumulative frequency table.

Height (*h* cm)	Cumulative frequency
h ≤ 65	2
h ≤ 70	

B4 Make a cumulative frequency table for each of these.

(a)

Age (*a* years)	Frequency
10 < *a* ≤ 20	4
20 < *a* ≤ 30	5
30 < *a* ≤ 40	10
40 < *a* ≤ 50	7
50 < *a* ≤ 60	2

(b)

Height (*h* cm)	Frequency
140 < *h* ≤ 150	3
150 < *h* ≤ 160	7
160 < *h* ≤ 170	14
170 < *h* ≤ 180	20
180 < *h* ≤ 190	12

B5 This cumulative frequency table gives information about the lengths of a collection of fish.

Length (*l* cm)	*l* ≤ 30	*l* ≤ 35	*l* ≤ 40	*l* ≤ 45	*l* ≤ 50
Cumulative frequency	7	18	43	52	55

(a) How many fish are there in the collection?

(b) How many fish have lengths in the interval 30 < *l* ≤ 35?

(c) How many are there in the interval 40 < *l* ≤ 45?

(d) How many are there in the interval 35 < *l* ≤ 50?

B6 Calculate an estimate of the mean length of the fish in question B5.

C Cumulative frequency graphs

These tables give information about the breath-holding times of a group of people.

Time (t seconds)	Frequency
$30 < t \le 40$	4
$40 < t \le 50$	8
$50 < t \le 60$	12
$60 < t \le 70$	14
$70 < t \le 80$	2

Time (t seconds)	Cumulative frequency
$t \le 40$	4
$t \le 50$	12
$t \le 60$	24
$t \le 70$	38
$t \le 80$	40

We can use the numbers in the cumulative frequency table to draw a **cumulative frequency graph**.

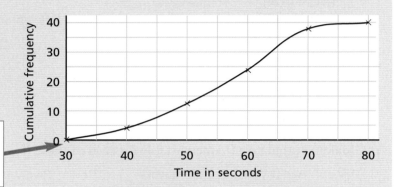

The graph starts here because the frequency table tells us that nobody holds breath for less than 30 seconds.

C1 Use the cumulative frequency graph above to estimate how many people can hold their breath for up to
(a) 45 seconds   (b) 55 seconds   (c) 65 seconds   (d) 75 seconds

C2 Estimate the percentage of people who can hold their breath for
(a) up to 45 seconds   (b) up to 55 seconds   (c) up to 65 seconds
(d) over 65 seconds   (e) over 45 seconds   (f) between 40 and 70 seconds

C3 Estimate the time for which 90% of the people can hold their breath.

C4 This table gives information about the waist measurements of a group of boys.

(a) Make a table of cumulative frequencies.

(b) Draw a cumulative frequency graph.

(c) Estimate the number of boys whose waists are between 67 cm and 77 cm.

(d) Estimate what percentage of the group have waists over 72 cm.

Waist (w cm)	Frequency
$60 < w \le 65$	12
$65 < w \le 70$	20
$70 < w \le 75$	28
$75 < w \le 80$	12
$80 < w \le 85$	8

C5 This table gives information about the weekly milk yields of a herd of cows.

Yield (y litres)	Cumulative frequency
$y \leq 260$	30
$y \leq 270$	80
$y \leq 280$	150
$y \leq 290$	220
$y \leq 300$	250

(a) How many cows are there in the herd?

(b) Draw a cumulative frequency graph. Start it at (250, 0).

(c) Use the graph to estimate how many cows have milk yields

    (i) up to 285 litres (ii) over 265 litres

(d) Use the graph to estimate the percentage of the herd which yielded

    (i) up to 275 litres (ii) up to 295 litres (iii) over 295 litres (iv) over 285 litres

C6 This table shows the distribution of marks in an examination.

Number of marks (m)	Frequency
$0 < m \leq 10$	4
$10 < m \leq 20$	7
$20 < m \leq 30$	12
$30 < m \leq 40$	23
$40 < m \leq 50$	38
$50 < m \leq 60$	46
$60 < m \leq 70$	20
$70 < m \leq 80$	16
$80 < m \leq 90$	11
$90 < m \leq 100$	3

(a) Make a cumulative frequency table.

(b) Draw a cumulative frequency graph.

(c) Estimate the number of people who scored 75 marks or more.

(d) The pass mark was 55. Estimate the number who passed.

D Median, quartiles and interquartile range

Here is the cumulative frequency graph of the heights of a group of boys.

- How can you use the graph to estimate the median height of the group?

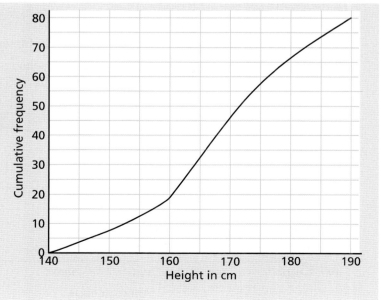

This is a cumulative frequency graph of the armspans of a group of 200 girls in year 10.

The **median** armspan is such that 50% of the group are above it and 50% below it.

The **lower quartile** is the armspan which is a quarter of the way up the group:
25% are below it and 75% above.

The **upper quartile** is the armspan which is three quarters of the way up the group:
75% are below it and 25% above.

The 'middle half' of the group has armspans between the lower and upper quartiles.

The difference between the two quartiles is called the **interquartile range**.
For this data its value is

$$180 - 163 = \mathbf{17\,cm}$$

The interquartile range is often used as a measure of how spread out the data is.

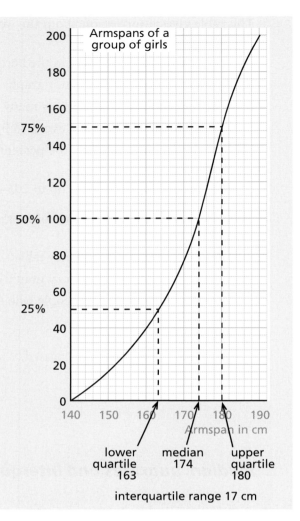

lower quartile 163 median 174 upper quartile 180

interquartile range 17 cm

D1 The graph below is based on the armspans of 400 fifteen-year-old boys.

Estimate from the graph (a) the median armspan (b) the lower quartile
(c) the upper quartile (d) the interquartile range

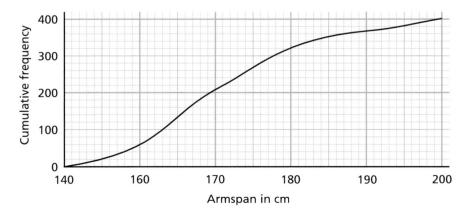

D2 Write a couple of sentences comparing the armspans of the girls and the boys shown on the opposite page.

D3 A firm making light bulbs tested 500 of them to see how long they lasted.
The results are summarised in this frequency table.

(a) Make a cumulative frequency table for the data.

(b) Draw a cumulative frequency graph.

(c) Estimate

    (i) the median lifetime

    (ii) the lower quartile

    (iii) the upper quartile

    (iv) the interquartile range

    (v) the percentage of bulbs which lasted longer than 2200 hours

Lifetime (x hours)	Frequency
$0 < x \le 500$	20
$500 < x \le 1000$	80
$1000 < x \le 1500$	140
$1500 < x \le 2000$	180
$2000 < x \le 2500$	50
$2500 < x \le 3000$	30

D4 This table shows the distribution of the birth weights of babies born in a hospital.

Draw a cumulative frequency graph for the data and use it to estimate the median, quartiles and interquartile range.

Weight (w kg)	Frequency
$0 < w \le 1$	6
$1 < w \le 2$	18
$2 < w \le 3$	48
$3 < w \le 4$	36
$4 < w \le 5$	12

D5 This table shows the distribution of the weights of the apples picked from a tree.

The frequencies have been given as percentages of the total number of apples.

(a) Make a cumulative percentage table (like a cumulative frequency table, but with percentages).

(b) Draw a cumulative percentage graph.

(c) Estimate the median weight, the upper and lower quartiles and the interquartile range.

Weight (w g)	Percentage
$50 < w \le 60$	8%
$60 < w \le 70$	13%
$70 < w \le 80$	24%
$80 < w \le 90$	27%
$90 < w \le 100$	16%
$100 < w \le 110$	9%
$110 < w \le 120$	3%

D6 Calculate an estimate of the mean weight of the apples in question D5.

*D7 This cumulative frequency table gives information about the weights of a group of babies.

Weight (w kg)	$w \le 1$	$w \le 2$	$w \le 3$	$w \le 4$	$w \le 5$
Cumulative frequency	4	11	25	36	38

Calculate an estimate of the mean weight of the babies, showing your method.

E Box-and-whisker plots

Here is some information about the distribution of the heights of all the
Y10 girls in a school.

Minimum	Lower quartile	Median	Upper quartile	Maximum
144 cm	159 cm	165 cm	169 cm	181 cm

A **box-and-whisker plot** is a way of displaying this information.

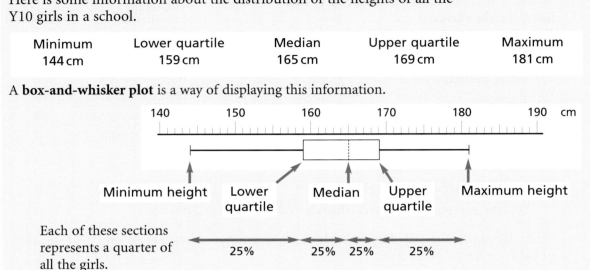

Each of these sections represents a quarter of all the girls.

E1 What feature of the box-and-whisker plot shows the interquartile range?

E2 These two box-and-whisker plots show the distributions of the ages
of the people living in two towns, A and B.

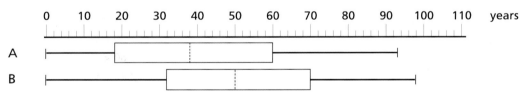

(a) What is the median age of the population of town A?

(b) What is the lower quartile of the ages of the population of town B?

(c) What percentage of the population of town A are over 60?

(d) What percentage of the population of town B are over 50?

(e) Find the interquartile range of the ages in each town.

(f) In which town are people older on the whole? Explain your answer.

(g) In which town are people's ages more spread out on the whole? Explain.

E3 Draw a box-and-whisker plot to show this information about the weights
of a group of men.

> A quarter of the men weigh 60 kg or less, the lightest being 52 kg.
> A quarter weigh 75 kg or more, the heaviest being 93 kg.
> The median weight is 66 kg.

E4 These box-and whisker plots show the distributions of the handspans of Y10 boys and girls.

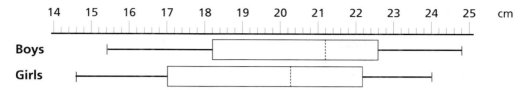

Write a couple of sentences comparing the boys' and girls' handspans.

E5 These box-and whisker plots show the distributions of the salaries in five companies.

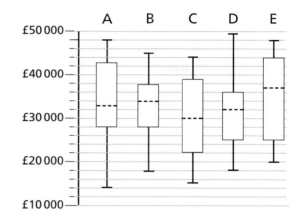

(a) In which company are the salaries highest on the whole?
Explain the reasons for your answer.

(b) In which company is there the greatest variation in salaries?
Explain your answer.

E6 These graphs show the distribution of the lengths of two groups of worms.

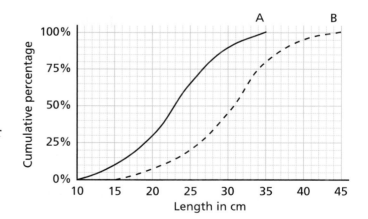

(a) Find the median and quartiles of the lengths of each group.

(b) Draw two box-and-whisker plots, one for each group.

(c) Write down the interquartile range for each group.

(d) Write a couple of sentences comparing the two groups.

Test yourself

T1 This table shows the distribution of the weights of a group of boys.

Weight (w kg)	Frequency
$40 < w \le 50$	15
$50 < w \le 60$	22
$60 < w \le 70$	48
$70 < w \le 80$	25
$80 < w \le 90$	10

(a) Make a cumulative frequency table.

(b) Draw a cumulative frequency graph.

(c) Use the graph to estimate the median weight of the group.

(d) Estimate the lower and upper quartiles, and the interquartile range.

(e) Draw a box-and-whisker plot for the data.

T2 Here are summaries of the data obtained from measuring the weights of two varieties of apple.

Variety A

Minimum	131 g
Lower quartile	142 g
Median	155 g
Upper quartile	171 g
Maximum	185 g

Variety B

Minimum	139 g
Lower quartile	150 g
Median	160 g
Upper quartile	173 g
Maximum	182 g

(a) What percentage of the apples of variety A weigh up to 142 g?

(b) What percentage of the apples of variety B weigh over 173 g?

(c) Use the information in the two tables to compare the weights of the two varieties.

T3 This was the money received by the charity Megalife each week for a year.

Money received (£x)	Frequency
$0 < x \le 50$	32
$50 < x \le 100$	12
$100 < x \le 150$	4
$150 < x \le 200$	2
$200 < x \le 250$	1
$250 < x \le 300$	0
$300 < x \le 350$	1

(a) Calculate an estimate of the mean amount received each week.

(b) Copy and complete the cumulative frequency table below and draw the cumulative frequency graph of the distribution.

Money received (£)	≤ 0	≤ 50	≤ 100	≤ 150	≤ 200	≤ 250	≤ 300	≤ 350
Cumulative frequency	0							

(c) Use your graph to find the median and quartiles of the distribution. Show your method clearly.

(d) A spokesperson for Megalife said 'The weekly average was nearly £60'. Comment on this statement.

OCR

Combining transformations

You will revise how to describe and carry out rotations, reflections, translations and enlargements on shapes.

This work will help you

◆ understand negative enlargements

◆ find a centre of enlargement

◆ recognise and carry out a one-way stretch

◆ recognise combinations of transformations

A *Transformations*

The pattern on the grid below is made up of trapeziums.

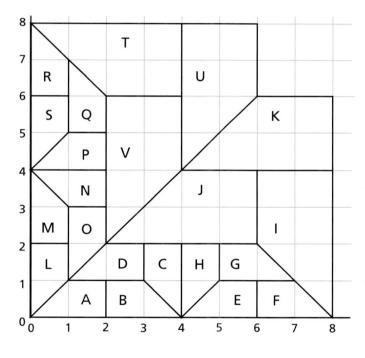

- To map shape A on to shape B needs a reflection in the line $x = 2$. What other reflections can be used to map shapes on to other images?

- To transform shape B on to shape H needs a 90° clockwise rotation with centre (4, 0). What other rotations can be used to map shapes on to other images?

- To transform shape L on to shape O needs a translation of $\begin{bmatrix} 1 \\ 1 \end{bmatrix}$.

  What other translations can be used in this pattern?

- Describe the enlargement which transforms shape D on to shape J. What other enlargements can be used in this pattern?

Congruent means having the same shape and the same size.
All the shapes in this diagram are congruent to one another.

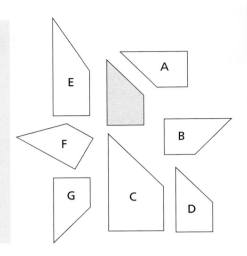

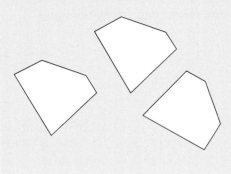

A1 Which of the trapeziums in the diagram on the right are congruent to the shaded one?

A2 Which transformations always produce shapes which are congruent to the original shape?

A translation is carried out by moving every point in the shape by the same column vector.

A3 Describe the translations needed to transform the shaded shape to each of the lettered shapes on this grid.

A4 Copy the grid and shaded shape. Show the image of the shaded shape after a translation using these vectors.

(a) $\begin{bmatrix} 2 \\ 3 \end{bmatrix}$ (b) $\begin{bmatrix} 2 \\ -4 \end{bmatrix}$

(c) $\begin{bmatrix} -6 \\ -3 \end{bmatrix}$

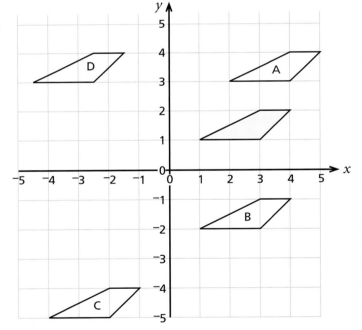

A5 (a) Where will these points move to after a translation of $\begin{bmatrix} 2 \\ -5 \end{bmatrix}$?

(i) (3, 3)   (ii) (−2, 0)
(iii) (12, 7)   (iv) (−5, −5)

(b) What is the image of the point (a, b) after a translation of $\begin{bmatrix} 2 \\ -5 \end{bmatrix}$?

(c) What is the image of the point (a, b) after a translation of $\begin{bmatrix} s \\ t \end{bmatrix}$?

A reflection is defined simply by describing the mirror line used.
If a coordinate grid is used this can be done by giving the equation of the mirror line.

A6 Give the equations of the lines of reflection used to reflect the shaded shape on to each of the other shapes on this diagram.

A7 Copy the grid and the shaded shape on to squared paper. Show the result of reflecting the shaded shape in each of these lines.

(a) $y = 0$ Label it J.

(b) $x = 1$ Label it K.

(c) $y = x + 1$ Label it L.

(d) $y = 4 - x$ Label it M.

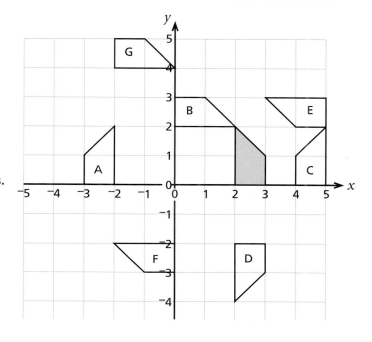

A8 On three separate sketches of this diagram show the shape and its reflection in

(a) $y = 1$ (b) $y = x$

(c) $y + x = 5$

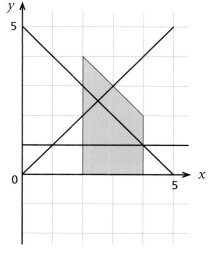

A9 (a) Copy and complete this table showing the images of points after reflection in $y = x$.

Point	Image
(3, 1)	
(2, 6)	
(3, -2)	
	(5, 5)
	(-4, -2)

(b) What is the image of the point (a, b) after reflection in the line $y = x$?

(c) What is the image of the point $(23, -17)$ after reflection in the line $y = x$?

A10 By testing with a range of coordinates give the image of the point (a, b) after reflecting in these lines.

(a) $x = 0$ (b) $y = 0$ (c) $x = 4$ (d) $y = -x$

To describe a rotation fully requires
- the centre of rotation
- the angle of rotation
- the direction of rotation

The convention used is that an anticlockwise rotation is a positive (+) rotation and a clockwise rotation is negative (−).

A11 This pattern has been drawn on triangular dotty paper by rotating about the centre of the pattern.

What is the angle of rotation from

(a) B to A   (b) E to C

(c) F to A

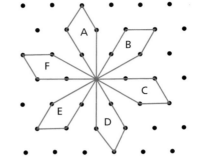

A12 Describe fully the rotations needed to map the green shape on to each of the lettered shapes.

A13 Copy the grid and green shape on squared paper.
Show the images of these rotations of the green shape.

  (a) A rotation of 180°, centre (0, 1)
       Label it H.

  (b) A $^-$90° rotation, centre (0, 0)
       Label it J.

  (c) A $^+$90° rotation, centre (3, 3)
       Label it K.

  (d) A 180° rotation, centre ($^-$1, $^-$1)
       Label it L.

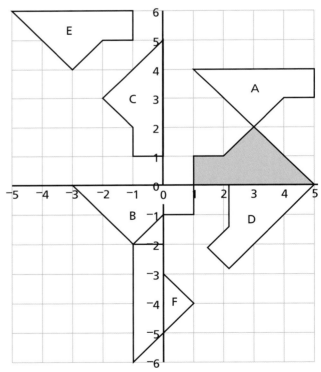

A14 (a) Where is the point (5, 3) mapped to after a rotation of 180°, centre (0, 0)?

   (b) What is the image of the point (*a*, *b*) after this rotation?

   (c) Where would the point ($^-$12, $^-$6) move to after this rotation?

A15 What is the image of the point (*a*, *b*) after a $^-$90° rotation, centre (0, 0)?

*A16** What is the image of the point (*a*, *b*) after a 180° rotation, centre (3, 2)?

B *Enlargements*

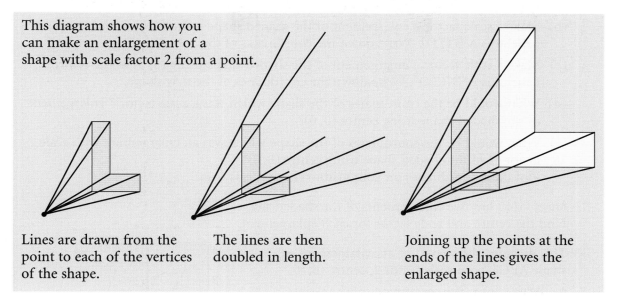

This diagram shows how you can make an enlargement of a shape with scale factor 2 from a point.

Lines are drawn from the point to each of the vertices of the shape.

The lines are then doubled in length.

Joining up the points at the ends of the lines gives the enlarged shape.

B1 Use this method on sheet G159 to make an enlargement with scale factor 2 of the shaded shape ABCDE using point P as the centre of enlargement.

B2 Now mark points halfway along the lines from P to the shaded shape ABCDE. Join these points together.
What is the scale factor of this enlargement of the shaded shape?

Here, each line from C to a vertex of the pink shape is extended backwards from C by its original length.

Joining the ends of these lines gives an enlargement with scale factor ⁻1.

Extending backwards by twice the original length gives an enlargement with scale factor ⁻2.

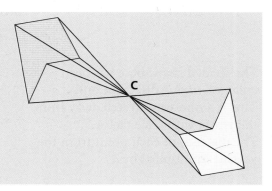

B3 Enlarge the shaded shape on G159 by a scale factor of ⁻1 using point P as the centre of enlargement.

B4 Describe how you would make enlargements of the shaded shape with these scale factors.
(a) 3   (b) ⁻2   (c) $-\frac{1}{2}$

B5 Draw lines between matching points on the shaded shape ABCDE and the other shape A′B′C′D′E′ on sheet G159. What do you notice?
What is the scale factor of the enlargement in this case?

B6 (a) Draw a grid going from ⁻8 to 8 on both axes.
Draw the quadrilateral with coordinates A (2, 1), B (4, 1), C (4, 3), D (3, 3).

(b) Make a scale factor 2 enlargement of the shaded shape using (0, 0) as the centre.
Label this A′B′C′D′. Write down the coordinates of the enlarged shape.

(c) Make a scale factor $\frac{1}{2}$ enlargement of the shaded shape using (0, 0) as the centre.
Label this A″B″C″D″. Write down the coordinates of the new shape.

(d) What would be the coordinates of the shape which was a scale factor 3 enlargement
of the shaded shape using centre (0, 0)?

(e) What would be the coordinates of the shape which was an enlargement with scale
factor ⁻2 of the shaded shape using centre (0, 0)?
Plot these coordinates on your grid to check this is true.

B7 Sheet G160 has five enlargements of the shaded shape.
Find the centre and scale factor for each enlargement.

B8 This diagram shows part of an enlargement of
shape ABCD with scale factor 2, centre (0, 0).

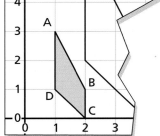

(a) Write down the coordinates of the points
A′, B′, C′ and D′ of the enlarged shape.

(b) What would be the image of the point (a, b)
under an enlargement with scale factor 2 and
centre (0, 0)?

(c) What is the image of the point (a, b) after
enlargements from centre (0, 0) using these scale factors?

(i) 4 (ii) $\frac{1}{2}$ (iii) ⁻1 (iv) ⁻3

The shaded shape here has been
enlarged in one direction only.
This is a **one-way stretch** away from
the y-axis with scale factor 3.
The distance of each point from the
y-axis is multiplied by 3.

A one-way stretch can be described by the
scale factor and the line it moves from.

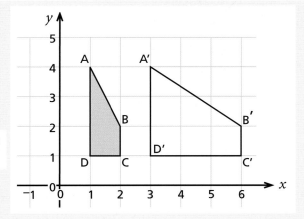

B9 The one-way stretches of the shaded shape in this diagram are all from the y-axis.
Give the scale factor for each stretch.

B10 What is the image of the point (a, b) after a one-way stretch from the y-axis of scale factor 2?

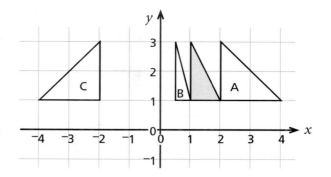

B11 This diagram shows a one-way stretch of the shaded shape from the x-axis.

(a) What is the scale factor of the stretch?

(b) What is the image of the point (a, b) after this transformation?

(c) Where would these points be mapped to under this transformation?

  (i) $(3, 2)$ (ii) $(5, 0)$ (iii) $(2, ^-4)$

(d) What is the image of the point (a, b) after a one-way stretch from the x-axis with scale factor f?

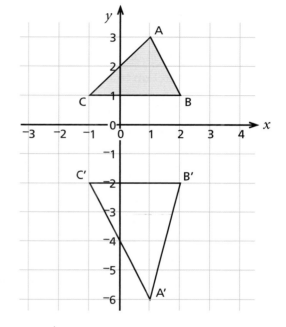

B12 Copy this shape on to a grid.
Show the image of this shape after a one-way stretch from the y-axis with scale factor 2.

Label the image M′N′P′Q′.

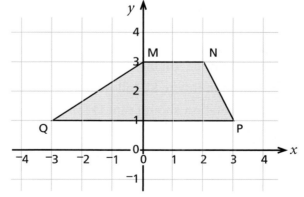

B13 (a) What points will not be moved by a one-way stretch from the y-axis?

(b) What points will not be moved by a one-way stretch from the line $y = 4$?

C *Describing transformations*

If the shaded shape ABOE is reflected in the
line BO, the result is shape CBOF.

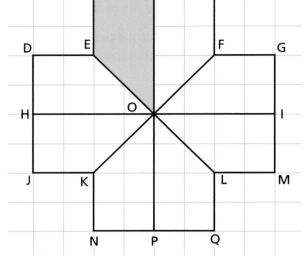

C1 What shape will be the result of

   (a) reflecting HOKJ in line BP

   (b) rotating FGIO $^+90°$ about point O

   (c) reflecting KOPN in line EL

C2 Describe the transformation
which maps

   (a) DEOH on to MLOI

   (b) FGIO on to EDHO

   (c) HOKJ on to POKN

C3 What shape is the image of ABOE after

   (a) a reflection in EL followed by a reflection in BP

   (b) a $^-90°$ rotation centre O followed by a reflection in HI

   (c) a reflection in FK followed by a 180° rotation centre O

   (d) a translation of $\begin{bmatrix} 0 \\ -4 \end{bmatrix}$ followed by a $^-90°$ rotation centre L

C4 (a) What transformation maps ABOE on to QPOL?

   (b) What transformation maps OLQP on to OKNP?

   (c) What **single** transformation maps ABOE on to NPOK?

C5 Reflecting ABOE in the line HI and then reflecting it in BP has the
same effect as rotating ABOE 180° about point O.

Describe the single transformation that is equivalent to the following pairs of
transformations of ABOE.

   (a) A reflection in line EL followed by a reflection in line HI

   (b) A $^-90°$ rotation about point O followed by a reflection in line OI

   (c) A translation of $\begin{bmatrix} 4 \\ 0 \end{bmatrix}$ followed by a $^-90°$ rotation about point F

   (d) A $^-90°$ rotation about point O followed by a $^+90°$ rotation about point I

   (e) A translation of $\begin{bmatrix} 0 \\ -4 \end{bmatrix}$ followed by a $^-90°$ rotation centre O

C6 (a) Describe the single transformation on this diagram that maps ABCD on to A′B′C′D′.

(b) Describe the single transformation that maps A′B′C′D′ on to A″B″C″D″.

(b) Describe the **single** transformation that maps ABCD on to A″B″C″D″.

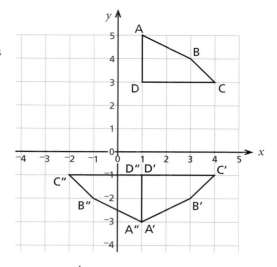

C7 Copy shape PQRS on squared paper using a grid going from ⁻5 to 8 on both axes.

(a) Show on your diagram the reflection of shape PQRS in the line $x = 4$. Label this shape P′Q′R′S′.

(b) Show on your diagram the reflection of shape P′Q′R′S′ in the line $x = 2$. Label this shape P″Q″R″S″.

(c) Describe the single transformation which maps PQRS on to P″Q″R″S″.

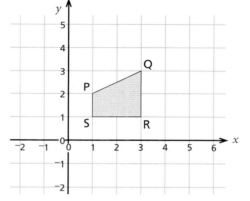

C8 Describe the single transformation that has the same effect as these pairs of transformations.

(a) A reflection in $x = ⁻1$ followed by a reflection in $x = 2$

(b) A reflection in $y = 0$ followed by a reflection in $y = 4$

*****C9** What single transformation has the same effect as reflecting in one line and in another parallel to it?
Describe the single transformation that has the same effect as reflecting in the line $x = m$ followed by a reflection in the line $x = n$.

C10 Copy shape KLMN on squared paper using a grid going from ⁻6 to 10 on both axes.

(a) Show on your diagram an enlargement of KLMN scale factor 3 using centre (0, 0). Label this shape K′L′M′N′.

(b) Show on your diagram an enlargement of K′L′M′N′ scale factor $\frac{1}{2}$ using centre (0, 0). Label this shape K″L″M″N″.

(c) Describe the single transformation which maps KLMN on to K″L″M″N″.

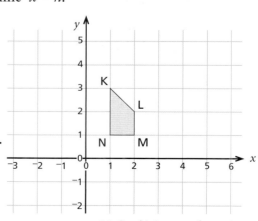

C11 (a) On the same grid as question C10 show the result of enlarging shape
KLMN by scale factor ⁻3 using centre (0, 3).
Label this shape K'''L'''M'''N'''.

(b) Describe the single transformation that would map K'''L'''M'''N''' on to K'L'M'N'.

C12 Describe the single transformation that would have the same effect as each of
these pairs of transformations.

(a) An enlargement with scale factor 4 and centre (0, 0) followed by
an enlargement with scale factor $\frac{1}{2}$ and centre (0, 0)

(b) An enlargement with scale factor ⁻3 and centre (3, 0) followed by
an enlargement with scale factor $\frac{1}{2}$ and centre (3, 0)

(c) A one-way stretch with scale factor 3 from the x-axis followed by
a one-way stretch with scale factor 3 from the y-axis.

C13 For each of these statements write 'always true', 'sometimes true' or 'never true'.
Give examples to justify your answers.

A *One translation followed by another translation
can be replaced by a single translation.*

B *A reflection in one line followed by a reflection in a different
line can be replaced by one reflection in a single line.*

C *A rotation about a point followed by another rotation about the
same point can be replaced by a single rotation about that point.*

D *A reflection followed by another reflection in a different line
can be replaced by a translation.*

Write some statements of your own that are always true or sometimes true.
Give examples to justify your answers.

Test yourself

T1 This diagram shows three quadrilaterals P, Q and R.

(a) Describe fully the transformation that maps P on to Q.

(b) Describe fully the transformation that maps P on to R.

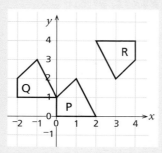

AQA(SEG) 1998

T2

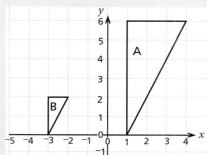

Describe fully the single transformation which maps triangle A to triangle B.

Edexcel

T3 Shape P has been drawn on a grid. Copy this diagram on squared paper.

(a) Reflect the shape P in the *y*-axis. Label the image Q.

(b) Rotate the shape Q through 180° about (0, 0). Label this image R.

(c) Describe fully the single transformation which maps shape P to shape R.

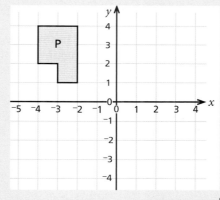

Edexcel

T4

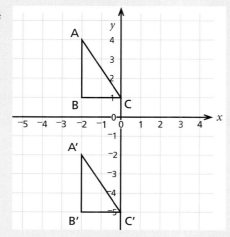

The grid here shows a triangle ABC and a triangle A′B′C′.

(a) On a copy of this grid, draw the triangle A″B″C″ which is an enlargement of ABC with a scale factor $-\frac{1}{2}$ with centre (2, 1).

(b) Describe fully the transformation that takes triangle A″B″C″ to triangle A′B′C′.

AQA(NEAB) 1998

17 Probability

You will revise how to
◆ calculate a probability using equally likely outcomes
◆ estimate a probability from data

This work will help you calculate probabilities in the case of
◆ mutually exclusive events
◆ independent events
◆ dependent events

A Crash!

People who go on holiday often pay for travel insurance.

They pay an amount, called the premium, to an insurance company. If they have an accident on holiday, or have something stolen, or fall ill and need medical treatment, the insurance company pays the bill.

In the following game, you have to imagine that you run a travel insurance company. People going on holiday insure against being involved in a car accident.

There are two kinds of car accident:

Minor – insurance company pays **£200**

Major – insurance company pays **£500**

When a person goes on holiday,

the probability of a **minor** accident is $\frac{1}{10}$,

the probability of a **major** accident is $\frac{1}{20}$.

You are in competition with other insurance companies.
The amount of business you get depends on the premium you charge.

Lowest premium gets **20** customers.   **Highest** gets **none**.

Second highest gets **5**.   All others get **10**.

Every company starts with £1000.

Stage 1 Decide on the premium you will charge.

Stage 2 You will find out how many customers you get.

Stage 3 Give each of your customers a random number from 000 to 999.

• If it is from 000 to 099, the customer has a minor accident.

• If it from 950 to 999, the customer has a major accident.

• Otherwise, the customer has no accident.

Stage 4 Work out your profit or loss. Then start again at stage 1.

B *Calculating and estimating probabilities*

The solid on the right is called a regular octahedron.

When it is rolled, each face is equally likely to be uppermost.
There are 8 faces, so the probability that a particular face is uppermost is $\frac{1}{8}$.

The solid on the left is called a cuboctahedron.
It has 14 faces: 6 squares and 8 triangles.

The shape is not regular.
If it is rolled, you can't tell just by looking at it what is the probability that
the uppermost face will be a square.

But if you roll it a large number of times, you can use **relative frequency** as
an estimate of probability.

$$\text{relative frequency of 'square on top'} = \frac{\text{number of times square was on top}}{\text{number of times solid was rolled}}$$

In the questions below, write probabilities as fractions (in their simplest form) or decimals.

B1 A cuboctahedron was rolled 400 times.
It landed with a square face uppermost 340 times.
Estimate the probability that the cuboctahedron will land with

(a) a square uppermost   (b) a triangle uppermost

B2 (a) Sharmila takes a cube and writes these numbers on its six faces: 1, 1, 2, 2, 2, 3.
Sharmila rolls the cube.
What is the probability that the number 1 is uppermost?

(b) David writes numbers on the faces of a cube.
He rolls the cube many times and makes this record of how it lands.

Number uppermost	1	2	3	4
Frequency	42	79	85	34

What numbers do you think he wrote on the six faces of the cube?

B3 Britalite tested 250 of their 100 watt bulbs, to see how long they lasted.
The results of the test are shown in this table.

Lifetime, L, of bulb (hours)	$0 \leq L < 1000$	$1000 \leq L < 2000$	$2000 \leq L < 3000$	$3000 \leq L$
Frequency	24	62	145	19

Estimate the probability that a Britalite 100 watt bulb will last

(a) less than 1000 hours   (b) less than 3000 hours   (c) 2000 hours or more

C Mutually exclusive events

Imagine that these cards are turned over and shuffled.
One card is picked at random.

| 2 | 3 | 4 | 5 | 6 | 7 | 8 | 9 | 10 | 11 |

An **event** is something that may or may not be true about the outcome.
Here are some examples of events, their **favourable outcomes** and probabilities.

Event	Favourable outcomes			Probability
A: The number picked is a multiple of 3.	3	6	9	$\frac{3}{10}$
B: The number picked is a factor of 20.	2	4 5	10	$\frac{4}{10}$

The lists of favourable outcomes for A and B are completely different. There is no number common to both lists. We say the events A and B are **mutually exclusive**.

In other words A and B can't both happen.

Suppose you win a prize if the number picked is **either** a multiple of 3 **or** a factor of 20.

The probability of winning is $\frac{3}{10} + \frac{4}{10} = \frac{7}{10}$.

You can add the probabilities because the two events are mutually exclusive.

Now look at these two events.

Event	Favourable outcomes					Probability
C: The number picked is even.	2	4	6	8	10	$\frac{5}{10}$
D: The number picked is a factor of 60.	2 3	4 5	6		10	$\frac{6}{10}$

Events C and D are not mutually exclusive.
They can both happen (if the outcome is 2, 4, 6 or 10).

Suppose you win a prize if the number picked is **either** even **or** a factor of 60.
You can't find the probability of winning by adding the probabilities of C and D, because the events are not mutually exclusive.

The favourable cases for winning the prize are 2, 3, 4, 5, 6, 8, 10. So the probability is $\frac{7}{10}$.

C1 (a) Are the events A and C mutually exclusive?

(b) Are the events A and D mutually exclusive?

(c) E is the event 'The number picked is a factor of 21'.
Are the following events mutually exclusive?

(i) A and E   (ii) B and E   (iii) C and E   (iv) D and E

C2 In a fairground game, players each choose a number on this board. An electronic device lights up and turns off the numbers in a random way. When it stops, one number is lit up.

1	2	3	4
5	6	7	8
9	10	11	12
13	14	15	16

What is the probability that the lit-up number is

(a) in the first row   (b) in the first column

(c) either in the first row or the first column

(d) on the edge of the board

(e) on the diagonal from top left to bottom right

(f) either on the edge or on the diagonal from top left to bottom right

(g) either a square number or an odd number

C3 These cards are turned over and shuffled. A card is picked at random.

20	21	22	23	24	25	26	27	28	29

The event A is 'The number picked is less than 24'.

The event B is 'The number picked is a multiple of 5'.

The event C is 'The number picked is prime'.

The event D is 'The number picked is a multiple of 3'.

(a) Are these pairs of events mutually exclusive?

(i) A, B   (ii) A, C   (iii) A, D   (iv) B, C   (v) B, D

(b) What is the probability that the number picked is either prime or a multiple of 3?

(c) Jake said:
'The probability of picking a number less than 24 is $\frac{4}{10}$.
The probability of picking an even number is $\frac{5}{10}$.
So the probability of picking either a number less than 24 or an even number is $\frac{9}{10}$.'

Is he right? If not, why not?

C4 A number is picked at random from this collection.

25	26	27	28	29	30	31	32	33	34	35	36

The event P is 'The number picked is odd'.

The event Q is 'The number picked is a multiple of 4'.

The event R is 'The number picked is prime'.

The event S is 'The number picked is a square number'.

(a) Are these pairs of events mutually exclusive?

(i) P, Q   (ii) Q, R   (iii) R, S   (iv) Q, S   (v) P, S

(b) What is the probability that the number picked is

(i) either odd or prime   (ii) either prime or square

(iii) either prime or a multiple of 4   (iv) either even or a multiple of 3

(v) either a multiple of 5 or a multiple of 6

D Independent events

For the rest of this unit, assume that coins, dice and spinners are fair, unless told otherwise.

Paul spins a coin and rolls a dice.

Dice	H(ead)	T(ail)
6	H, 6	T, 6
5	H, 5	T, 5
4	H, 4	T, 4
3	H, 3	T, 3
2	H, 2	T, 2
1	H, 1	T, 1

Coin

The coin and the dice do not affect each other, so their outcomes are **independent**.

There are 12 equally likely outcomes of the coin and dice, as shown in the diagram on the right.

The probability of each outcome (for example H, 5) can be found by **multiplying** the separate probabilities:

$$\underset{\substack{\text{probability}\\\text{of 'head'}}}{\frac{1}{2}} \times \underset{\substack{\text{probability}\\\text{of 5}}}{\frac{1}{6}} = \underset{\substack{\text{probability}\\\text{of 'head' and 5}}}{\frac{1}{12}}$$

The **multiplication rule for independent events** says:

> If two events are independent, the probability that they **both** happen is found by multiplying their probabilities.

D1 (a) What is the probability of getting an even number on the dice above?

 (b) Use the multiplication rule to find the probability of getting an even number on the dice and a 'head' on the coin.

 Check from the diagram of outcomes that your result is correct.

D2 Sam has two regular octahedral dice, both numbered 1 to 8. She rolls them both.

What is the probability that

 (a) both dice show 5

 (b) the first dice shows 5 and the second shows an even number

 (c) both dice show multiples of 3

 (d) both dice show prime numbers

D3 These two spinners are spun. What is the probability that

 (a) spinner A shows red

 (b) spinner B shows red

 (c) both spinners show red

 (d) A shows red and B shows blue (e) both show blue

 (f) both show white (g) neither shows white

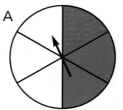

A   B

E Tree diagrams

The possible outcomes when these two spinners are spun can be shown in a **tree diagram**.

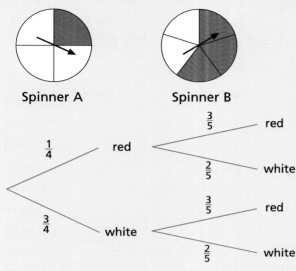

Spinner A   Spinner B

Probabilities along the branches are found by multiplying.

$\frac{3}{5}$ — red   probability of (A red, B red) $= \frac{1}{4} \times \frac{3}{5} = \frac{3}{20}$

$\frac{1}{4}$ — red

$\frac{2}{5}$ — white   probability of (A red, B white) $= \frac{1}{4} \times \frac{2}{5} = \frac{2}{20}$

$\frac{3}{5}$ — red   probability of (A white, B red) $= \frac{3}{4} \times \frac{3}{5} = \frac{9}{20}$

$\frac{3}{4}$ — white

$\frac{2}{5}$ — white   probability of (A white, B white) $= \frac{3}{4} \times \frac{2}{5} = \frac{6}{20}$

The probability that both spinners give the same colour is found by adding:

probability of same colour = probability of (A red, B red) + probability of (A white, B white)

$$= \frac{3}{20} + \frac{6}{20} = \frac{9}{20}$$

It is better **not** to simplify the fractions: it makes them easier to compare and add.

E1 (a) Copy and complete the tree diagram for these two spinners.

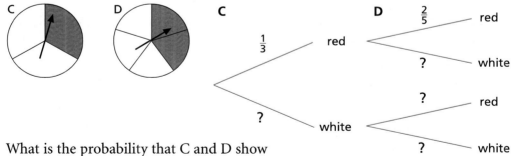

    (b) What is the probability that C and D show
        (i) the same colour   (ii) different colours

E2 (a) Draw a tree diagram for these two spinners.
    (b) Find the probability that the two spinners
        show the same colour.

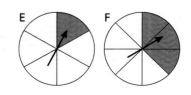

E3 Gill has a coin which is weighted so that the probability that it lands 'head' is $\frac{3}{5}$ and 'tail' $\frac{2}{5}$.

(a) Copy and complete the tree diagram for two throws of the coin, writing the probabilities on the branches.

(b) Find the probability of getting one 'head' and one 'tail'.

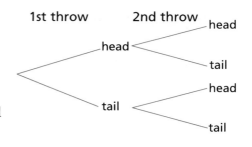

E4 Patrick has two cubes, A and B.

The faces of cube A are coloured red, red, blue, blue, green, green.

The faces of cube B are coloured red, blue, blue, blue, green, green.

Patrick rolls the two cubes to see which colours come on top.

(a) Copy and complete the tree diagram, writing the probabilities on the branches.

(b) Find the probability that both cubes land with the same colour on top.

(c) Find the probability that one of the cubes lands blue and the other green.

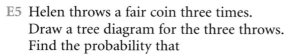

E5 Helen throws a fair coin three times. Draw a tree diagram for the three throws. Find the probability that

(a) all three throws give the same result

(b) two throws give 'head' and the other 'tail'

E6 Hitesh throws an ordinary dice three times.

(a) Copy and complete this tree diagram.

(b) Find the probability that Hitesh gets two or more sixes in his three throws.

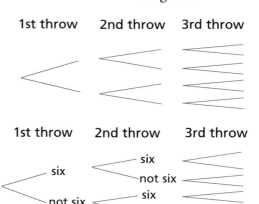

Chinese dice

A set of 'Chinese dice' consists of three dice numbered

    A: 6 6 2 2 2 2   **B:** 5 5 5 5 1 1   **C:** 4 4 4 3 3 3

One player chooses a dice. Then the second player chooses one of the remaining two dice. Both players roll their dice and the higher score wins.

It is said that it is better to be the player who chooses second. Is this true?

F Dependent events

A box contains 2 black counters
and 4 white counters.

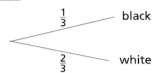

One counter is picked at random.
This tree diagram shows the probabilities.

Suppose the counter is **not** put back in the box.
The contents of the box will be different, depending on whether the counter
taken out was black or white.

If it was black, the box would now contain 1 black and 4 white counters.

If it was white, it would contain 2 black and 3 white counters.

If another counter is now taken out at random, the probability that it is black is
dependent on the colour of the first counter.

The complete tree diagram looks like this.

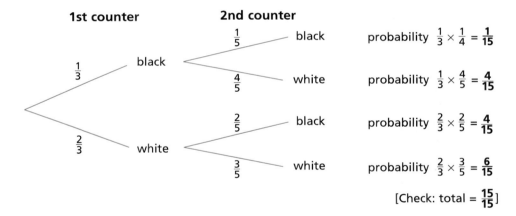

F1 From the tree diagram above, find the probability that the two counters are

(a) the same colour   (b) different colours

F2 A bag contains 3 red cubes and 2 blue cubes.
A cube is taken at random from the bag and not replaced.
Then a second cube is taken at random.

(a) Draw a tree diagram for this situation.

(b) Find the probability that both of the cubes taken out are the same colour.

(c) Find the probability that the two cubes are of different colours.

F3 Two counters are taken at random from a box containing 4 red and 6 blue counters.
Use a tree diagram to find the probability that the two counters are the same colour.

F4 The probability that it will be raining tomorrow morning is $\frac{1}{3}$.
If it is raining, the probability that Simon will be late for school is $\frac{1}{4}$.
If it is not raining, the probability that he will be late is $\frac{1}{5}$.

(a) Copy and complete this tree diagram.

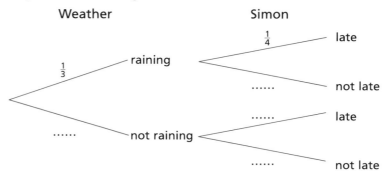

(b) Find the probability that Simon will be late for school tomorrow.

F5 In a TV game show, contestants are given two tasks.
In each task they either succeed or fail.

The probability of succeeding in the first task is 0.8.

If a contestant succeeds in the first task, the probability of succeeding in the second is 0.6.
If a contestant fails in the first task, the probability of succeeding in the second is 0.3.

What is the probability that a contestant

(a) succeeds in both tasks   (b) fails in at least one task

F6 A pack of cards consists of 3 'wolf' cards and 2 'goat' cards.
Here are the rules for a game.

A player takes a card at random from the pack.
If it is a wolf, the player keeps it. If it is a goat, it goes back in the pack.
The player then takes a card at random again.

If the player took a wolf both times, they win 5 points.
If they took a wolf once (either first or second time), they win 2 points.
If they took a goat both times, they do not win any points.

Find the probability that a player wins

(a) 5 points   (b) 2 points

F7 Jack takes a card at random from this pack.

| 1 | 2 | 3 | 4 |

He keeps the card and then takes a second card at random.

What is the probability that the second number he takes is higher than the first?

G Mixed questions

G1 These nine cards are placed face down and shuffled.

| 1 | 2 | 3 | 4 | 5 | 6 | 7 | 8 | 9 |

One card is picked at random.
Pat says:
  'The probability that the number on the card is even is $\frac{4}{9}$.
  The probability that the number is less than 6 is $\frac{5}{9}$.
  So the probability that the number is either even or less than 6 is $\frac{4}{9} + \frac{5}{9} = 1$.'

What is wrong with Pat's reasoning?

G2 What's wrong here?

  The probability that a person stopped at random is male is about 0.5.
  The probability that a person stopped at random is wearing a skirt is about 0.25.
  Therefore the probability that a person stopped at random is male and wearing a skirt is 0.5×0.25 which is 0.125 or 12.5%.

G3 This table gives information about the children in a primary school class.

	Left-handed	Right-handed
Girls	5	15
Boys	4	9

(a) One of the children is picked at random from the class.
    What is the probability that the child is a girl?

(b) One of the boys in the class is picked at random.
    What is the probability that he is left-handed?

(c) A boy in the class is picked at random, and a girl is picked at random.
    What is the probability that they are both left-handed?

(d) One of the right-handed children is picked at random.
    What is the probability that the child is a boy?

G4 Josh assembles alarms which consist of a lamp, a buzzer and a battery.
He takes each component from a separate bin and puts them together.
The probability that the lamp he takes is faulty is 0.4.
The probability that the buzzer is faulty is 0.5, and the battery 0.3.

An alarm fails to work if any of the components is faulty.
What is the probability that the next alarm Josh makes is faulty?

G5 Josie takes a card at random from this pack and keeps it.
Then she takes a second card at random.
Find the probability that she takes one odd number and one even number (in either order).

| 1 | 2 | 3 | 4 | 5 |

Test yourself

T1 A spinner is divided into three colours: red, white and blue.

When it is spun, the probability that it lands on red is $\frac{7}{20}$.

The probability that the spinner lands on white is $\frac{2}{5}$.

(a) If the spinner is spun 600 times, how many times would you expect it to land on blue?

(b) The spinner is spun twice.
Calculate the probability that one spin results in red and the other white, in either order.

AQA(NEAB) 1997

T2 The table shows information about a group of adults.

	Can drive	Cannot drive
Male	32	8
Female	38	12

(a) A man in the group is chosen at random.
What is the probability that he can drive?

(b) A man in the group is chosen at random and a woman in the group is chosen at random.
What is the probability that both the man and the woman **cannot** drive?

AQA(SEG) 2000

T3 On his way to work, Nick goes through a set of traffic lights and then passes over a level crossing.
Over a period of time, Nick has estimated the probability of stopping at each of these.

The probability that he has to stop at the traffic lights is $\frac{2}{3}$.
The probability that he has to stop at the level crossing is $\frac{1}{5}$.

These probabilities are independent.

(a) Construct a tree diagram to show this information.

(b) Calculate the probability that Nick will not have to stop at either the lights or the level crossing on his way to work.

OCR

T4 A packet contains stamps from three different countries.
The packet contains 4 Spanish stamps, 10 French stamps and 6 German stamps.

Two stamps are to be removed at random, without replacement.
Calculate the probability that both stamps will be from the same country.

Edexcel

18 Simultaneous equations

You should know how to

◆ draw the graph of a straight line, given its equation

◆ rearrange simple formulas

This work will help you

◆ form and solve a variety of linear simultaneous equations, using elimination and substitution

◆ interpret the solution of simultaneous equations as the point of intersection of two graphs

A Problems

> A bag contains a mixture of 5p coins and 1p coins.
> There are 17 coins in the bag and their total value is 61p.
> How many of each type of coin are there?

> The sum of two numbers is 40.
> The difference between them is 7.
> What are the numbers?

> In a pet shop there are some mice and some budgies.
> Altogether they have 27 heads and 70 legs.
> How many mice and how many budgies are there?

A1 Lorna has 5p stamps and 6p stamps.
She sticks 11 stamps on a parcel, with a total value of 63p.
How many of each type of stamp does she use?

A2 Jupiter chocolate bars are made in two sizes: regular and king size.
2 regular bars and 5 king size bars weigh 760 g altogether.
1 regular bar and 7 king size bars weigh 920 g altogether.

(a) How much does a regular bar weigh?

(b) How much does a king size bar weigh?

A3 At a fairground stall, Colin bought 5 burgers and 4 hot dogs and paid £6.20.
Judy bought 7 burgers and 1 hot dog and paid £5.69.

(a) How much does a burger cost?   (b) How much does a hot dog cost?

A4 Rory has some 2-litre bottles of lemonade.
Sarah has some 3-litre bottles of lemonade.
Altogether they have 27 litres.
Rory has 6 bottles more than Sarah.
How many bottles does each person have?

B Simple elimination

Examples

Solve this pair of simultaneous equations.

$$3x + 5y = 1$$
$$7x - 5y = 19$$

If we add the equations, the terms involving y on the left-hand side will 'disappear'.

$$
\begin{array}{r}
3x + 5y = 1 \\
7x - 5y = 19 \\
\hline
10x = 20
\end{array}
\quad \text{Add}
$$

so $\quad x = 2$

To find the value of y, substitute x in one equation. Choosing the first equation gives

$$3 \times 2 + 5y = 1$$
$$5y = {}^-5$$
$$y = {}^-1$$

So the solution is $x = 2$, $y = {}^-1$.
Substitute in the other equation to check:
$7x - 5y = 7 \times 2 - 5 \times {}^-1 = 19$ ✓

Solve this pair of simultaneous equations.

$$5a + 2b = 8$$
$$5a - b = 6$$

If we subtract the equations, the terms involving a on the left-hand side will 'disappear'.

$$
\begin{array}{r}
5a + 2b = 8 \\
5a - b = 6 \\
\hline
3b = 2
\end{array}
\quad \text{Subtract}
$$

so $\quad b = \frac{2}{3}$

To find the value of a, substitute b in one equation. Choosing the second equation gives $\quad 5a - \frac{2}{3} = 6$
$$5a = 6 + \frac{2}{3} = \frac{20}{3}$$
$$a = \frac{4}{3}$$

So the solution is $a = \frac{4}{3}$, $b = \frac{2}{3}$.
Substitute in the other equation to check:
$5a + 2b = 5 \times \frac{4}{3} + 2 \times \frac{2}{3} = \frac{24}{3} = 8$ ✓

B1 Solve each pair of simultaneous equations.

(a) $v - w = 12$
$v + w = 20$

(b) $3p - 4q = 10$
$5p + 4q = 6$

(c) $7n + 3m = 20$
$5m - 7n = {}^-4$

(d) $4k + 6h = 11$
$4k + 2h = 5$

(e) $2x + 3y = 4$
$2x - 5y = 20$

(f) $5a - 3b = 29$
$a - 3b = 5$

(g) $c - 5d = {}^-4$
$4c - 5d = 14$

(h) $r - 5s = {}^-11$
$r - 7s = {}^-13$

(i) $3x - 9y = {}^-7$
$3x + 9y = 8$

B2 (a) Rearrange the equation $3x = 4 - 2y$ into the form $ax + by = c$.

(b) Use your rearrangement to solve the simultaneous equations
$$3x = 4 - 2y$$
$$5x - 2y = 12$$

B3 Solve each pair of simultaneous equations.

(a) $x + y = 14$
$x = 6 + y$

(b) $9y = 22 - 5x$
$5x + 2y = 1$

(c) $2x - 3y = 19$
$5x = 3y + 49$

B4 (a) Where do the lines with equations $3x + 4y = 22$ and $x + 4y = 2$ intersect?

(b) Where do the lines with equations $x + 2y = 6$ and $y - x = 2$ intersect?

For questions B5 to B9, form a pair of simultaneous equations and then solve them.

B5 The sum of two numbers is 36 and the difference between them is 9.
If x is the smaller number and y is the larger, find x and y.

B6 If you take Poppy's age away from Daniel's age you get 7.
If you take Poppy's age away from three times Daniel's you get 40.

How old are Poppy and Daniel?

B7 On these two balances,
each apple weighs the same and
each strawberry weighs the same.

Find the weight of an apple and
the weight of a strawberry.

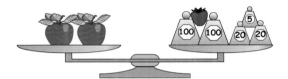

B8 On these two balances,
each pear weighs the same and
each cherry weighs the same.

Find the weight of a pear and
the weight of a cherry.

B9 One angle in a triangle is 50°.
The difference between the other two angles is 64°.

(a) What is the smallest angle in the triangle?

(b) What size are the other two angles?

B10 At what point do the lines with equations $3y + 2x = 7$ and $y = \dfrac{13 + 2x}{2}$ intersect?

C *Multiply to eliminate*

Examples

Solve this pair of simultaneous equations.

   (A) $x - y = 2$
   (B) $4x + 3y = 29$

If we multiply both sides of equation (A) by 3, the coefficients of y will be the same size.

$$\begin{array}{ll} \text{(A)} \times 3 & 3x - 3y = 6 \\ \text{(B)} & \underline{4x + 3y = 29} \\ & 7x = 35 \end{array} \quad \text{Add}$$

$$\text{so} \quad x = 5$$

Substitute in equation (A) to give
$$5 - y = 2$$
$$y = 3$$

So the solution is $x = 5$, $y = 3$.
Substitute in equation (B) to check:
$4x + 3y = 4 \times 5 + 3 \times 3 = 29$ ✓

Solve this pair of simultaneous equations.

   (A) $5x - 2y = 11$
   (B) $2x - 3y = 11$

If we multiply both sides of equation (A) by 2 and both sides of equation (B) by 5, the coefficients of x will be the same size.

$$\begin{array}{ll} \text{(A)} \times 2 & 10x - 4y = 22 \\ \text{(B)} \times 5 & \underline{10x - 15y = 55} \\ & 11y = {}^-33 \end{array} \quad \text{Subtract}$$

$$\text{so} \quad y = {}^-3$$

Substitute in equation (B) to give
$$2x + 9 = 11$$
$$2x = 2$$
$$\text{so } x = 1$$

So the solution is $x = 1$, $y = {}^-3$.
Substitute in equation (A) to check:
$5x - 2y = 5 \times 1 - 2 \times {}^-3 = 11$ ✓

C1 Solve each pair of simultaneous equations.

   (a) $3v - 2w = 10$
        $5v + 4w = 2$

   (b) $2p + 3q = 26$
        $6p + 5q = 30$

   (c) $4n + 5m = 13$
        $7m + 3n = 13$

   (d) $5h + 2k = 4$
        $7k - 2h = 1$

   (e) $4x - 3y = 14$
        $7x - 5y = 23$

   (f) $2a + 3b = 11$
        $6a - 5b = 5$

   (g) $2c + 4d = {}^-3$
        $2d + 16c = 1$

   (h) $7r - 3s = 2$
        $3r + 8s = 25$

   (i) $4x - y = 7$
        $5y - 2x = {}^-11$

C2 Solve each pair of simultaneous equations.
    (You may need to rearrange some of the equations first.)

   (a) $2v = 31 - 3w$
        $3v + w = 64$

   (b) $5p + 3q = 4$
        $8p = 5 + 5q$

   (c) $3n = 4m - 8$
        $5n - 16 = 7m$

   (d) $\dfrac{3a + b}{4} = 5$
        $6a - b = 25$

   (e) $\dfrac{h}{4} + 5g = 15$
        $3h + 2g = 35$

   (f) $\dfrac{3p - 2q}{5} = 3$
        $2p = 3(q + 5)$

C3 (a) Where do the lines with equations $2x + 3y = 2$ and $7y - 10x = 2$ intersect?

      (b) Where do the lines with equations $6x - 5y = 94$ and $5x = 3y + 55$ intersect?

For questions C4 to C11, form a pair of simultaneous equations and then solve them.

C4 A potter sells large and small mugs.
    Two small mugs and one large mug weigh 758 grams.
    Four small mugs and three large mugs weigh 1882 grams.
    How heavy is each mug?

C5 Two people were in front of me in the queue for ice cream.
    The man bought 5 cones and 7 tubs for £6.70.
    The woman bought 3 cones and 5 tubs for £4.50.
    How much will it cost me for my cone?

C6 Calculate the weight of a grape and the weight of an orange.

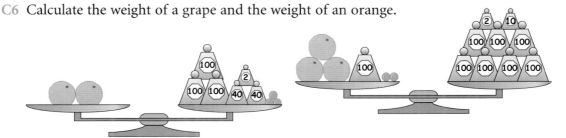

C7 Calculate the weight of a peach and the weight of a banana.

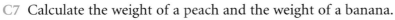

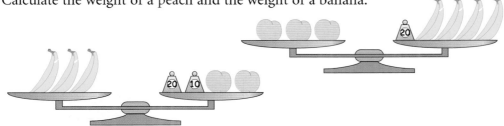

C8 Some hens and a herd of cows are in a field.
    Between them they have 50 heads and 180 legs.
    How many cows and how many hens are in the field?

C9 A drinks machine takes 20p and 50p coins.
    In the machine, there are 31 coins altogether.
    They are worth £12.20.
    How many of each type of coin are there?

C10 In 5 years time a dog will be as old as his owner was 3 years ago.
    Now, the sum of their ages is 17 years.
    How old is the dog now?

C11 A dolphin can swim at 10 m/s with the current and at 3 m/s against it.
    Find the speed of the current and the speed of the dolphin in still water.

D Substitution

Examples

The sum of two numbers is 220.
The result of dividing the larger number by the smaller is 19.

What are the numbers?

Form a pair of simultaneous equations.

(A) $p + q = 220$

(B) $\dfrac{p}{q} = 19$

We can rearrange equation (B) to give
$$p = 19q$$

Now substitute into equation (A) to give
$$19q + q = 220$$
$$20q = 220$$
$$q = 11$$

Substitute again in equation (A) to give
$$p + 11 = 220$$
$$p = 209$$

So the numbers are 11 and 209.
Substitute in equation (B) to check:
$$\frac{p}{q} = \frac{209}{11} = 19 \checkmark$$

At what point do the lines with equations $3y + 5x = 2$ and $y = x + 1$ intersect?

Substitute the expression for y into the first equation
$$3y + 5x = 2$$
$$3(x + 1) + 5x = 2$$
$$3x + 3 + 5x = 2$$
$$8x + 3 = 2$$
$$8x = {}^-1$$
$$x = \frac{{}^-1}{8}$$

Substitute in the second equation to give
$$y = {}^-\tfrac{1}{8} + 1$$
$$y = \tfrac{7}{8}$$

So the point of intersection is $({}^-\tfrac{1}{8}, \tfrac{7}{8})$.
Substitute in the first equation to check:
$$3y + 5x = 3 \times \tfrac{7}{8} + 5 \times \tfrac{{}^-1}{8}$$
$$= \tfrac{21}{8} - \tfrac{5}{8} = \tfrac{16}{8} = 2 \checkmark$$

D1 Solve each pair of simultaneous equations using substitution.

(a) $y = 2x$
$y + 8x = 20$

(b) $6w - 5v = 18$
$w = \tfrac{1}{3}v$

(c) $p = 5q$
$2p - 9q = 2$

(d) $n = m + 5$
$3n + 11m = 57$

(e) $4j - k = 0$
$j = \tfrac{1}{2}k - 3$

(f) $h = 2g + 1$
$5g - 3h = 2$

(g) $2f = 10e - 1$
$2f + 5e = 2$

(h) $d = 4c - 1$
$d = 11 - 5c$

(i) $5b - 3a = 8$
$20 - 9b = 3a$

D2 Solve each pair of simultaneous equations using substitution.
(You will need to rearrange one of the equations first.)

(a) $y - 2x = 0$

$3y = 10x - 14$

(b) $\dfrac{p}{q} = 10$

$5q + p = 3$

(c) $\dfrac{3n}{m} = 1$

$m = 4n - 7$

D3 (a) Where do the lines with equations $y = 2x + 3$ and $2x - 3y = 11$ intersect?

(b) Where do the lines with equations $y = \frac{1}{2}(x + 5)$ and $2y + 3x = 17$ intersect?

For D4 to D8, form a pair of simultaneous equations and solve them using substitution.

D4 Joy is three times as old as her brother Ken.
Ken is eleven years younger than Joy.
How old are they?

D5 A goose lays gold and silver eggs.
A gold egg weighs three times as much as a silver egg.

One day she lays 9 gold and 5 silver eggs.
Their total weight is 1200 grams.

How heavy is a gold egg?

D6 A wallet contains £120 in £5 notes and £10 notes.
There are twice as many £5 notes as £10 notes.
Find the number of each kind of note in the wallet.

D7 The difference between two numbers is 40.
The result of dividing the larger number by the smaller is 6.
What are the numbers?

D8 A fraction is equivalent to $\frac{1}{3}$.
The sum of the numerator and the denominator is 36.
Find the fraction.

E *Simultaneous equations and graphs*

Use the equations below to form pairs of simultaneous equations.

A $x + 2y = 9$

B $2x + 2y = 10$

C $x + y = 5$

D $y = 2 - x$

Can you solve them?

E1 (a) Use the graphs to solve each
pair of simultaneous equations.

(i) $y = 2x + 4$
$x + y = 1$

(ii) $y - 2x = 1$
$x + y = 1$

(b) How do you know there is no
solution to the following pair
of simultaneous equations?

$y = 2x + 4$

$y - 2x = 1$

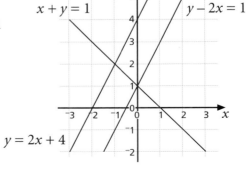

E2 Use the graphs to solve each pair of simultaneous equations.

(a) $10y + 7x = 64$
$x + y = 10$

(b) $y = 2x + 1$
$x = 6$

(c) $y - 2x = 1$
$7x + 10y = 64$

(d) $y = 10 - x$
$x - 6 = 0$

(e) $10y = 64 - 7x$
$x + 1 = 7$

(f) $2x = y - 1$
$x = 10 - y$

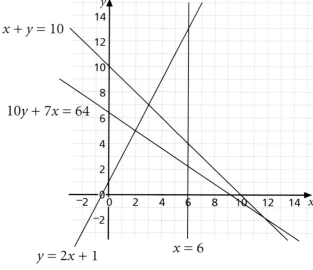

E3 Use graphs to solve each pair of simultaneous equations.

(a) $y = 5 - x$
$y = 2 + x$
Draw axes for x and y from 0 to 7.

(b) $2y - x = 0$
$y = x + 1$
Draw axes for x and y from ⁻4 to 4.

Check each solution by substituting into the equations.

E4 (a) Use the graphs to solve this pair of simultaneous equations.

$$y = \frac{17 + x}{6}$$

$$x + 4y = 19$$

(b) Check your solution by substituting into the equations.

(c) Comment on the accuracy of your solution.

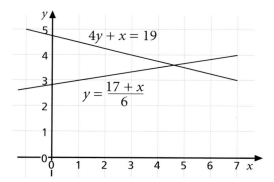

Simultaneous equations and graphs

A pair of simultaneous linear equations can have

one solution, …
$x + y = 6$
$2x + y = 10$

no solution, …
$2x + y = 6$
$4x + 2y = 15$

or **infinitely many** solutions.
$2x + y = 6$
$4x + 2y = 12$

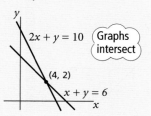

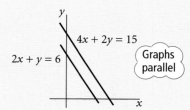

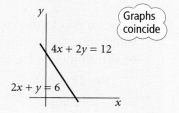

F Mixed problems

F1 Solve the following simultaneous equations by an algebraic (not graphical) method.
Show all your working.

$$6x + y = 37 \qquad 2x + 2y = 19$$

WJEC

F2 Solve each pair of simultaneous equations.

(a) $2a + 5b = 9$
   $3b + 2a = 3$

(b) $6c - 5d = {}^-3$
   $5d + 6c = 7$

(c) $e + 3f = 9$
   $5f + 4e = 1$

(d) $7g + 11h = 8$
   $2g = 11h - 26$

(e) $j - k = 3$
   $j + k = 2$

(f) $4n + 5m = 66$
   $9m = 60 - 3n$

(g) $2(p - q) = 6$
   $3p + 7q = 4$

(h) $2r - 3s = 0$
   $s = 2 - r$

(i) $u = 5t + 6$
   $u = 3t - 10$

F3 Explain how you can tell there is no solution to the simultaneous equations below.

$$x - 3y = 2 \qquad y = \frac{3x - 7}{9}$$

F4 A holiday for 3 adults and 5 children costs £3258.
The same holiday for 4 adults and 6 children costs £4098.
How much would the same holiday cost for 2 adults and 2 children?

F5 The shape ABCD is a parallelogram.
All lengths marked are in cm.
Find the lengths of AB and BC.

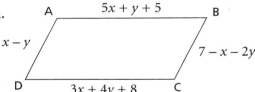

F6 A box contains £110 in £2 coins and £5 notes.
There are three times as many coins as there are notes.
How many coins are there?

F7 The shape ABC is an equilateral triangle.
All lengths marked are in cm.
Find the perimeter of the triangle.

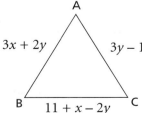

F8 A fraction is equivalent to $\frac{2}{7}$. If the numerator and denominator are both increased by 1, the fraction is equivalent to $\frac{3}{10}$. Find both fractions.

F9 A line has gradient 2 and cuts the y-axis at $(0, 1)$.
Another line has equation $9y = 17x - 9$.

(a) Without drawing them, show that the two lines are not parallel.

(b) What is the point of intersection of these two lines?

F10 (a) Explain how you can tell that there are an infinite number of solutions to the simultaneous equations $a + 2b = 5$ and $4b = 10 - 2a$.

(b) Find three of these solutions.

*F11 (a) Solve the simultaneous equations $5a - 2b = 16$ and $4a + 7b = 30$.

(b) **Use your answers** to solve the simultaneous equations

$$\frac{5}{c} - \frac{2}{d} = 16 \quad \text{and} \quad \frac{4}{c} + \frac{7}{d} = 30$$

*F12 Solve the simultaneous equations $\frac{4}{a} + 3b = 11$ and $\frac{2}{a} - 5b = {}^-14$

*F13 A motorist drives at one speed for $2\frac{1}{2}$ hours and at a different speed for $1\frac{1}{2}$ hours. She drives 190 kilometres in total.
If she had driven at the first speed for 3 hours and the second speed for 1 hour she would have travelled 180 kilometres.
What was the faster of the two speeds?

*F14 In seven years time a father will be three times as old as his son.
Three years ago he was five times as old as his son.
How old is each person now?

*F15

> A ship is twice as old as its boilers were when the ship was as old as its boilers are.

The ship is now 40 years old. How old are its boilers?

Test yourself

T1 Solve each pair of simultaneous equations.

(a) $6v + w = 1$
     $2w - 6v = 20$

(b) $2p - 3q = 4$
     $3p - 2q = 11$

(c) $n = 5m + 16$
     $3m + 2n = {}^-7$

T2 Some people and some octopuses are swimming in a tank.
Between them they have 11 heads and 52 legs.

How many people and how many octopuses are in the tank?

T3 Where do the lines with equations $4x + 7y = 5$ and $3x = 9y - 1$ intersect?

T4 A fraction is equivalent to $\frac{1}{5}$. If the numerator and denominator are both decreased by 1, the fraction becomes equivalent to $\frac{1}{6}$. Find both fractions.

Review 4

1 Draw the square ABCD whose sides are 6 cm.

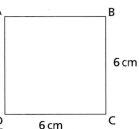

(a) Draw the locus of all the points inside the square that are nearer to AB than to AD, and more than 2 cm from BC.

(b) What fraction of the area of the square does this locus occupy?

2 This cumulative frequency graph shows the distribution of marks (out of 100) scored by 200 students in a test.

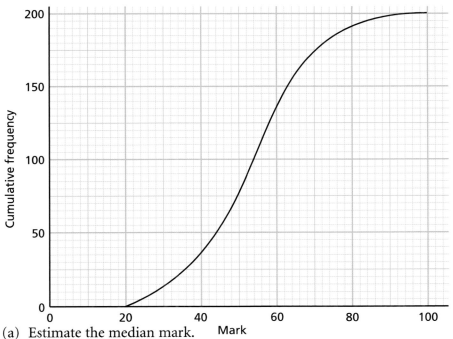

(a) Estimate the median mark.

(b) Estimate the lower and upper quartiles.

(c) Estimate the percentage of students who will pass the test if the pass mark is 35.

3 A fair 12-sided dice is numbered from 1 to 12.

(a) The dice is rolled once.
What is the probability that the score is either a multiple of 2 or a multiple of 3?

(b) The dice is rolled twice. What is the probability that

(i) both scores are multiples of 2 or 3   (ii) neither score is a multiple of 2 or 3

4 (a) Shape A can be mapped on to B by a reflection in the line $x = 7$ followed by a rotation.

What are the centre and angle of the rotation?

(b) A can also be mapped on to B by a rotation followed by a reflection in the line $x = 7$.

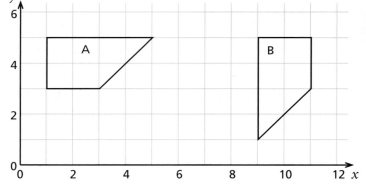

What are the centre and angle of this rotation?

5 A group of three adults and one child pay £12 altogether for concert tickets.
Another group of two adults and four children pay £13 altogether.

Form a pair of simultaneous equations.
Solve them to find the cost of an adult ticket and the cost of a child ticket.

6 The diagram shows the end wall of a greenhouse.
Calculate the area of the wall.

Give your answer to an appropriate degree of accuracy.

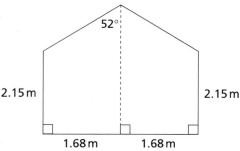

7 A is the point $(1, 7)$ and B is $(5, 1)$.

(a) M is the midpoint of AB.
What are the coordinates of M?

(b) Explain why the gradient of the line through M perpendicular to AB is $\frac{2}{3}$.

(c) The equation of this line is of the form
$$y = \tfrac{2}{3}x + c$$
Use the fact that the line goes through M to work out the value of c.

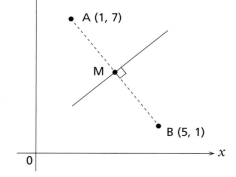

(d) Find the coordinates of the point on the y-axis which is equidistant from A and B.

(e) Use Pythagoras's theorem to calculate the distance of this point from A and from B.

8 Solve the simultaneous equations

$$6x + y = 28 \qquad\qquad x - 3y = 11$$

9 $\frac{2}{3}$ of the children in a primary class are boys and $\frac{1}{3}$ are girls.

$\frac{1}{4}$ of the boys are left-handed and $\frac{1}{6}$ of the girls are left-handed.

(a) What fraction of all the children are left-handed?

(b) What is the smallest possible value for the number of children in the class? Explain.

10 A circular crater on the Moon has a diameter of 1.8×10^3 metres.
Calculate its area in m^2, in standard form to two significant figures.

11 The quantities u, v, w and x are connected by the formula $w = v - ux$.

(a) Find the value of x when $u = 5$, $v = 4$ and $w = 6$.

(b) Rearrange the formula to make x the subject.

12 In a photographic competition, each entry was marked out of 50.
There were 127 entries altogether.
23 entries scored more than 40; 56 scored more than 30; 97 scored more than 20.
The minimum score was 11.

(a) Copy and complete this table.

Score	11–20	21–30	31–40	41–50
Frequency				

(b) Calculate an estimate of the mean score.

*(c) If the median score was 30, what was the minimum number of entries scoring 30?

*13 A bag contains four coins.
Two of the coins are Russian and two are Mexican.

Josie takes a coin at random from the bag.
Then, without returning the first coin, she again takes a coin at random.

Find the probability that

(a) both coins are Russian

(b) one is Russian and the other is Mexican

Challenge

The points K, L, M, N are the midpoints of the sides of the square ABCD.

- Explain why the area of the yellow square is one-fifth of the area of the square ABCD.

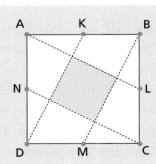

19 Solving inequalities

You will revise simple and combined inequalities, and how to
represent them on a number line.

This work will help you to solve inequalities.

A Review

A1 For each of these inequalities, say whether it is true or false.
 (a) $3 > {}^-5$ (b) $^-1 < {}^-3$ (c) $2 \leq 2$ (d) $^-1.9 \leq {}^-2$

A2 Write an inequality, using n, for each of these diagrams. For example, the first is $n < 2$.

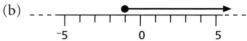

(a)

(b)

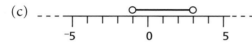
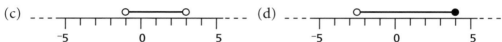

(c)

(d)

A3 Draw number lines to show these inequalities.
 (a) $n \geq {}^-3$ (b) $0 < n < 5$ (c) $^-1 < n$ (d) $^-4.5 < n \leq {}^-1$

The **integers** are the positive and negative whole numbers, including 0.

A4 List all the integers, x, such that
 (a) $5 < x \leq 10$ (b) $^-2.5 < x < 0$ (c) $3.5 > x > 0$ (d) $4 < 2x \leq 12$

A5 There are seven integers, p, such that $p^2 \leq 9$. What are they?

A6 Write down four integers, n, such that $n^2 < 6$.

A7 List all the integers, x, such that $0 \leq x^3 \leq 20$.

A8 List all the prime numbers, p, such that $0 < p^2 < 100$.

A9 If w stands for the weight of my suitcase in kg, then
 $w \leq 20$ means that my suitcase weighs at most 20 kg.

Write inequalities for each of these.
Choose your own letters, but say what they stand for.
 (a) Maximum number of passengers 65 (b) Children must be at least 12 years old
 (c) Baggage must weigh less than 40 kg (d) Free travel for the over 60s
 (e) No-one over 21 admitted (f) Lose at least 30 pounds guaranteed!

B *Manipulating inequalities*

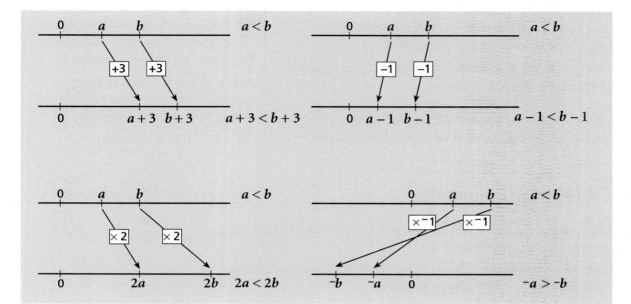

B1 Which of the following are equivalent to $n \leq 3$?

   A $n + 1 \leq 4$ B $3n \leq 9$ C $3 \leq n$ D $\frac{n}{2} \leq 1\frac{1}{2}$ E $2n \leq 6$

B2 Which of the following are equivalent to $p < 10$?

   A $p - 10 < 0$ B $2p < 10$ C $2p < 20$ D $\frac{p}{2} < 5$ E $p + 10 < 20$

B3 Which of the following inequalities are equivalent to $a \geq 10$?

   A $a - 5 \geq 5$ B $2a \geq 20$ C $a + 5 \geq 15$ D $\frac{1}{2}a \geq 5$ E $a + \frac{1}{2} \geq 10\frac{1}{2}$

B4 Which of the following are equivalent to $x \leq 4$?

   A $^-x \leq {}^-4$ B $^-x \geq {}^-4$ C $2x \leq 8$ D $^-2x \geq {}^-8$ E $^-2x \leq {}^-8$

B5 Find the four equivalent pairs in these eight inequalities.

  A $\quad p \leq 8$ **B** $\quad p + 4 \leq 14$ **C** $\quad 2p \leq 20$ **D** $\quad 4p \leq 24$

  E $\quad 2p \leq 18$ **F** $\quad p - 2 \leq 6$ **G** $\quad p \leq 9$ **H** $\quad 3p \leq 18$

B6 Find the four equivalent pairs in these eight inequalities.

  A $\quad 2p \leq 8$ **B** $\quad p \leq {}^-4$ **C** $\quad p \geq 4$ **D** $\quad 4p \geq {}^-16$

  E $\quad ^-2p \geq {}^-8$ **F** $\quad ^-2p \leq {}^-8$ **G** $\quad ^-p \geq 4$ **H** $\quad ^-p \leq 4$

Summary

If we are given an inequality, we can add or subtract
any number on both sides. The number we add
or subtract can be positive or negative.

$$x - 5 \leq 7 \Rightarrow x \leq 12 \quad \text{(add 5)}$$
$$n + 6 > 2 \Rightarrow n > {}^-4 \quad \text{(subtract 6)}$$

We can multiply or divide both sides of an inequality
by the same **positive** number.

$$\frac{x}{2} \leq 12 \Rightarrow x \leq 24 \quad \text{(multiply by 2)}$$
$$2n > 8 \Rightarrow n > 4 \quad \text{(divide by 2)}$$

If we multiply or divide both sides of an inequality
by a **negative** number, the inequality 'turns round'.

$${}^-x \leq 12 \Rightarrow x \geq {}^-12 \quad \text{(multiply by }{}^-1)$$
$${}^-2n > 8 \Rightarrow n < {}^-4 \quad \text{(divide by }{}^-2)$$

C *Solving inequalities*

You can solve inequalities by doing the same thing to both sides,
just as you can with equations.

But if you multiply or divide both sides by a negative number,
you must 'turn round' the inequality.

'Solving' means getting just a single letter on one side of the inequality
(usually the left), and a number on the other side of the inequality.

Solve $2x + 17 \geq 25$.
Draw the solution on a number line.

$2x + 17 \geq 25$
$2x \geq 8$ *(take 17 off both sides)*
$x \geq 4$ *(divide both sides by 2)*

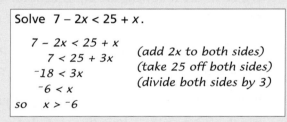

Solve $7 - 2x < 25 + x$.

$7 - 2x < 25 + x$
$7 < 25 + 3x$ *(add 2x to both sides)*
${}^-18 < 3x$ *(take 25 off both sides)*
${}^-6 < x$ *(divide both sides by 3)*
so $x > {}^-6$

C1 Solve each of these inequalities.
Check by finding two numbers that fit each of your solutions and
confirming that they fit the original inequality.

   (a) $x + 5 \leq 11$ (b) $x - 5 \geq 11$ (c) $2x \leq 12$ (d) $2x \leq 0$

C2 Solve the following inequalities.

   (a) $4r - 3 \geq 9$ (b) $t - 1 \leq {}^-6$ (c) $1 + 2f > 11$

   (d) $10 \leq 3d - 2$ (e) $x + 9 < 5$ (f) $17 \leq 3 + 2z$

   (g) $\frac{a}{2} + 4 < 8$ (h) $\frac{s}{3} + 2 \geq 5$ (i) $6k - 5 \leq {}^-5$

C3 Solve the inequality $4r - 3 \leq 3$, and hence write down
three integers that satisfy the inequality.

C4 Solve each of these.

(a) $5d \le 4d - 2$ (b) $8w < 6w + 14$ (c) $6z \le 18 - 3z$

(d) $3a < a - 20$ (e) $s \ge \frac{s}{2} + 3$ (f) $\frac{w}{2} - 10 \le w$

C5 Solve $4 - x > 6$ by first adding x to both sides.

C6 Solve $3x + 19 > 5x + 7$ by first subtracting $3x$ from both sides.

C7 Solve $x - 5 \ge 3x - 8$ and sketch the solution on a number line.

C8 Solve each of these inequalities.

(a) $4s + 2 < 3s + 12$ (b) $5z - 5 \ge 3z + 7$ (c) $4p - 1 > p + 5$

(d) $3 - 5t > 7t + 1$ (e) $2 - w \le w - 4$ (f) $2g - 3 \ge 6 - g$

(g) $1 - 5p \le 9 - 3p$ (h) $2d - 6 < 4 + 4d$ (i) $1 - k \le k - 1$

(j) $\frac{v}{2} + 1 > v + 2$ (k) $\frac{e}{2} + 1 \le e + 1$ (l) $4 + \frac{m}{2} \ge 1 + 2m$

C9 Multiply the brackets out in this inequality: $2(x - 3) \le 15$.
Then solve the inequality.

C10 Solve these.

(a) $2(n + 1) > n + 15$ (b) $2(k + 2) \le 3(k - 3)$ (c) $5(x - 5) \ge 3(x - 3)$

(d) $4(p - 1) \le 3(1 - p)$ (e) $3(2 - 3h) > 2(2 - 3h)$ (f) $4 - 5f \ge 2(2 - 4f)$

C11 By first multiplying both sides by 3, solve the inequality $\frac{n+2}{3} > n$.

C12 Copy and complete this working
to solve $\frac{x-3}{5} \le x + 1$.

$$\frac{x - 3}{5} \le x + 1$$
$$x - 3 \le 5(x + 1)$$

C13 Solve these.

(a) $\frac{x+5}{3} \le x$ (b) $\frac{x+5}{3} \le x - 1$ (c) $\frac{x+5}{3} \le 1 - x$ (d) $\frac{5-x}{3} \le 1 - x$

C14 By first multiplying both sides by 15, solve $\frac{x+2}{3} > \frac{x+3}{5}$.

C15 Solve these.

(a) $\frac{n-2}{2} > \frac{n-1}{3}$ (b) $\frac{2-n}{4} < \frac{1-n}{8}$ (c) $\frac{2n-3}{3} \ge \frac{2n-5}{4}$

C16 Solve the inequality $\frac{r+1}{3} \le r - 2$.   Edexcel

D *Further inequalities*

Solve $5 \le 2x - 1 < 11$. Represent the solution on a number line.

To solve this inequality we must do the same thing to each part.

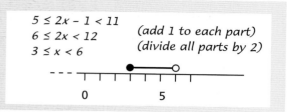

D1 Solve $^-2 < 3x + 1 \le 4$ and sketch your solution on a number line.

D2 List the values of n, where n is an integer, such that $1 \le 2n - 3 < 5$.

AQA 1999

D3 Solve these inequalities.

(a) $2 \le 3x - 4 < 8$ (b) $^-8 \le 4x < 10$ (c) $^-8 \le 2x + 4 < 6$

D4 Solve the inequality $^-1 \le 3x + 2 < 5$.

AQA(SEG) 2000

D5 (a) Find the integer values of n which satisfy the inequality $7 < 5n < 34$.

(b) Solve the inequality $5x - 2 < 18$.

OCR

D6 Solve these inequalities.

(a) $80 < 4(2x + 1) \le 100$ (b) $^-2 < 2(x + 3) < 7$ (c) $0 \le \dfrac{5x - 1}{2} \le 7$

D7 Find the integer values of n which satisfy each of these inequalities.

(a) $^-8 \le 2(3n - 1) < 9$ (b) $^-1 < \frac{1}{2}(4 + n) < 1$

D8 Find the largest possible integer value of x if

(a) $5x - 2 < 39$ (b) $2(4x - 1) < 11$ (c) $3(2x + 1) < 20$

Quadratic equations

When solving quadratic inequalities, a sketch graph may help.

Solve $x^2 < 16$.

From the sketch graph you can see immediately that if $x^2 < 16$, then $^-4 < x < 4$.

What if $x^2 > 16$?

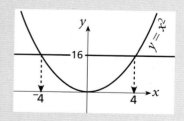

D9 (a) Which of the following numbers satisfy the inequality $x^2 < 25$?

0, 3, 4, 5, 6, 7, $^-4$, $^-5$, $^-6$, $^-7$, $^-4.99$, $^-5.01$

(b) Mark on a number line the numbers in part (a) which satisfy the inequality.

(c) Write down the solution to the inequality $x^2 < 25$.

D10 (a) Which of the following numbers satisfy the inequality $x^2 > 25$?
  0, 3, 4, 5, 6, 7, $^-4$, $^-5$, $^-6$, $^-7$, $^-4.99$, $^-5.01$

(b) Mark on a number line the numbers in part (a) which satisfy the inequality.

(c) Write down the solution to the inequality $x^2 > 25$, in the form '$x > \ldots$ or $x < \ldots$'

D11 Solve the inequality $x^2 < 9$. AQA(NEAB) 1998

☒ D12 (a) Solve the inequality $3x - 7 > x - 3$.   (b) Solve the inequality $x^2 > 16$.   AQA 2000

D13 Solve these inequalities. Represent each solution on a number line.
  (a) $x^2 + 10 \le 35$ (b) $2m^2 \ge 98$ (c) $2n^2 - 5 < 67$

D14 Solve the inequality $(x + 3)^2 > 25$. OCR(MEG)

Test yourself

T1 Draw a number line for each of these inequalities.
  (a) $n < {}^-1$ (b) $^-3 \le n < 4$ (c) $^-2 \ge n$

T2 List all the integers, n, such that
  (a) $^-2 < n < 2$ (b) $n^2 \le 1$ (c) $0 \le 2n \le 11$

T3 n is a whole number such that $6 < 2n < 13$. List all the possible values of n. Edexcel

T4 (a) Solve the inequality $3n > {}^-8$.

(b) Write down the smallest integer which satisfies the inequality $3n > {}^-8$. Edexcel

T5 Solve each of these inequalities.
  (a) $5 - 2h \le 3h$ (b) $2y + 1 < 4y - 2$ (c) $1 - 3u \le 3 - u$
  (d) $2(x + 3) > 3(1 - x)$ (e) $\frac{s-3}{2} \le 4$ (f) $\frac{s+1}{4} < 1 - s$

☒ T6 Solve the inequality $x + 20 < 12 - 3x$. AQA 2000

T7 Solve the inequality $9n + 1 < 14n - 2$. OCR

T8 Solve the following inequalities.
  (a) $2 > x - 4$ (b) $2(x + 3) > 3(2 - x)$ AQA(NEAB) 1997

T9 Solve each of these.
  (a) $4 < 3x + 1 \le 10$ (b) $2 \le 5x - 8 < 32$ (c) $^-1 < \frac{w}{2} - 1 < 1$

T10 Solve $x^2 < 49$. Represent your solution on a number line.

☒ T11 (a) List all the solutions of the inequality $^-6 < 3n \le 8$ where n is an integer.

(b) Solve the inequality $y^2 > 4$. AQA(SEG) 2000

20 Brackets and quadratic equations 1

You should know how to
- multiply powers by adding indices, for example $a^2 \times a^3 = a^5$
- multiply simple expressions such as $3m \times 2n = 6mn$
- add, subtract and multiply simple fractions

This work will help you
- multiply out brackets from expressions such as $a(ab + 5b)$, $(a + 2)(a - 3)$
- factorise expressions such as $a^2b + 5ab$, $a^2 - a - 6$
- solve quadratic equations such as $x^2 + 5x + 6 = 0$ by factorising
- solve problems by forming and solving quadratic equations

A Simplifying

Here are some important things to remember when simplifying algebraic expressions.

Examples

- Only **like** terms can be grouped together.

$$a^2 + 3ab + 4a^2 - 5ab + 10$$
$$= a^2 + 4a^2 + 3ab - 5ab + 10$$
$$= 5a^2 - 2ab + 10$$
This cannot be simplified further.

- Multiplication signs are usually missed out in a simplified expression.

$$2q^2 \times 3q$$
$$= 2 \times q^2 \times 3 \times q$$
$$= 2 \times 3 \times q^2 \times q$$
$$= 6q^3$$

- To multiply out brackets, multiply each term inside the brackets.

$$3a(\tfrac{1}{3}a + b)$$
$$= 3a \times \tfrac{1}{3}a + 3a \times b$$
$$= a^2 + 3ab$$

- To subtract an expression in brackets, change the signs inside the brackets.

$$4xy - x(3x - 5y)$$
$$= 4xy - (3x^2 - 5xy)$$
$$= 4xy - 3x^2 + 5xy$$
$$= 9xy - 3x^2$$

A1 Simplify these.

(a) $2a + 5b + 3a - 4b$ (b) $p^2 + 4p - 6p + 12$ (c) $4xy - 5y - xy + y$

(d) $2a \times 5b$ (e) $3p^2 \times p \times 2p^3$ (f) $4mn \times 7m^2$

A2 Multiply these out.

(a) $7a(3a + 5)$ (b) $k(3k^4 - h)$ (c) $2pq(3q + 5p)$

A3 Multiply out and simplify these.

(a) $4(a - 2) + 3(a - 1)$ (b) $4p - 3(p - 5)$ (c) $x(x + 2) - 3(x + 2)$

A4 Find four pairs of equivalent expressions.

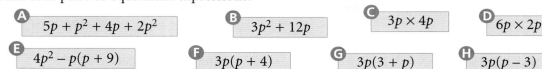

A $5p + p^2 + 4p + 2p^2$ **B** $3p^2 + 12p$ **C** $3p \times 4p$ **D** $6p \times 2p$

E $4p^2 - p(p + 9)$ **F** $3p(p + 4)$ **G** $3p(3 + p)$ **H** $3p(p - 3)$

A5 Find pairs of expressions from
the bubble that multiply to give

(a) $10xy^2$ (b) $2xy + 8y^2$

(c) $10y^2$ (d) $5x^2y - 5x$

(e) $2xy - 2x^2y^2$ (f) $5y - 5xy^2$

$2xy$ $5y$ $1 - xy$

$x + 4y$ $2y$ ^-5x

A6 Simplify these.

(a) $8x \times \frac{1}{4}y$ (b) $\frac{1}{2}h + 5h^2 + \frac{1}{4}h$ (c) $\frac{1}{2}m \times \frac{1}{4}mn$

A7 Multiply out and simplify these.

(a) $\frac{1}{4}(8x + 4) + 2(x - y)$ (b) $\dfrac{b(9b - 6)}{3}$ (c) $2x(\frac{1}{2}y + \frac{1}{3}x)$ (d) $\dfrac{3c(8 - c)}{2}$

A8 Find the missing expressions to make these statements correct.

(a) $2x(3x - \blacksquare) = 6x^2 - 10x$ (b) $4y(2x + \blacksquare) = \blacksquare + 12y^2$

(c) $2a(\blacksquare - b) = 4a^2 - \blacksquare$ (d) $\blacksquare(a - b) = 3a^2b - \blacksquare$

A9 Find an expression for the area of each shaded shape.
Multiply out any brackets and give each expression in its simplest form.

(a)

$4n - 3$

$3n$

(b)

p

q

2

$p - 5$

(c)

a

$a - 5$

b $2b$

(d)

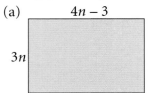

y

$y - 1$

x

$2x$

(e)

$2h - 3$

$\frac{1}{2}k$

h

k

B Factorising

A $6 = 2 \times 3$

B $18 = 2 \times 3 \times 3$

C $20 = ?$

D $4x + 20 = 4(x + 5)$

E $12y - 6 = 2(6y - 3)$

F $18k + 12 = ?$

G $a^2 - 4a = ?$

H $4x + 8x^2y = 4(x + 2x^2y)$

I $8h^2 + 16h = ?$

J $3p^2 - pq^2 = ?$

B1 Factorise these completely.

(a) $2p + 16$ (b) $4q - 4p$ (c) $6r + 3s$ (d) $6t - 3$

(e) $5x + 10y - 15$ (f) $4a - 8b$ (g) $10v - 15w^2$ (h) $24c^2 + 36d^2$

B2 Factorise these completely.

(a) $x^2 + 6x$ (b) $p^2 - pq$ (c) $4d^2 + 5d$ (d) $2b^2 - 6b$

(e) $6a^2 - 3a$ (f) $10m^2 + 25m$ (g) $8b^2 + 2bc$ (h) $2m + m^2 - mn$

B3 Factorise these completely.

(a) $y^2z + 4y^2$ (b) $2k^3 - k^2$ (c) $6a^2b - 3a^2$ (d) $2c^2 + 7cd$

(e) $12gh^3 - 18gh^2$ (f) $2mn^5 + 3n^7$ (g) $8gh^2 - 8g^2h$ (h) $5a^2b + 15ab^3$

B4 (a) Factorise $3n + 6$.

(b) Explain how the factorisation tells you that $3n + 6$ will be a multiple of 3 for any integer n.

B5 (a) Factorise $n^2 + n$.

(b) Use the factorisation to explain how you can tell that $n^2 + n$ will always be even for any integer n.

B6

A	B	D	E	G	H	L	O	R	S	U	W
$x - 6$	$y - 1$	$y - x$	$2x$	xy	$y^2 - 2$	x	$xy + 1$	$6y$	$x^2 + 3y$	5	$2x + 1$

Fully factorise each expression below as the product of two factors.
Use the code above to find the letter for each factor.

Rearrange each set of letters to spell an animal.

(a) $x^2 - 6x$ $12xy + 6y$ $5x^2 + 15y$

(b) $2x^2 + x$ $2x^3 + 6xy$ $2x^2 - 12x$

(c) $2xy - 2x$ $x^2y - 6xy$ $6y^2 - 6xy$

(d) $2xy^2 - 4x$ $2xy - 2x^2$ $xy^3 - 2xy$ $x^2y^2 + xy$

When factorising an expression, remember to factorise completely.

Examples

$3a - 6b^2 = 3(a - 2b^2)$

$8k + 6k^2 = 2(4k + 3k^2)$
$= 2k(4 + 3k)$

$4xy^2 - 12x^2y = 4xy(y - 3x)$

This is unfinished because $4k + 3k^2 = k(4 + 3k)$.

C *Two brackets*

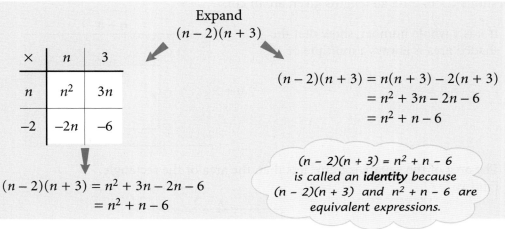

Expand
$(n - 2)(n + 3)$

×	n	3
n	n^2	$3n$
-2	$-2n$	-6

$(n - 2)(n + 3) = n(n + 3) - 2(n + 3)$
$= n^2 + 3n - 2n - 6$
$= n^2 + n - 6$

$(n - 2)(n + 3) = n^2 + 3n - 2n - 6$
$= n^2 + n - 6$

*$(n - 2)(n + 3) = n^2 + n - 6$ is called an **identity** because $(n - 2)(n + 3)$ and $n^2 + n - 6$ are equivalent expressions.*

C1 Multiply out and simplify these.
- (a) $(x + 2)(x + 3)$
- (b) $(x + 4)(x + 6)$
- (c) $(x + 1)(x + 3)$
- (d) $(n - 2)(n + 5)$
- (e) $(n + 6)(n - 5)$
- (f) $(n + 2)(n - 3)$
- (g) $(p - 7)(p + 1)$
- (h) $(p - 2)(p - 5)$
- (i) $(p - 1)(p - 8)$

C2 Solve these equations.
- (a) $(n + 1)(n + 6) = n(n + 8)$
- (b) $(x - 1)(x + 4) = 3x$
- (c) $(n - 8)(n + 7) = n^2 - 2n + 1$
- (d) $(p - 2)(p - 4) = (p + 1)(p + 3)$

C3 For each statement below, decide if it is an identity or an equation.
Solve each equation.
Show each identity is true.
- (a) $(n + 2)(n + 4) = n^2 + 8n + 6$
- (b) $(n - 6)(n - 1) = n^2 - 7n + 6$
- (c) $(n + 4)(n - 3) = n(n + 1) - 12$
- (d) $(n + 5)(n - 3) = n(n + 1)$

C4 (a) Multiply out and simplify these.
- (i) $(n + 7)(n - 7)$
- (ii) $(n + 5)(n - 5)$
- (iii) $(n - 3)(n + 3)$

Comment on your results.

(b) Write down the expansion of $(n + 4)(n - 4)$.

C5 (a) Expand and simplify these squares.
      (i) $(x + 5)^2$ (ii) $(x + 6)^2$ (iii) $(x - 3)^2$ (iv) $(x - 1)^2$
  (b) Expand and simplify these.
      (i) $(x + a)^2$ (ii) $(x - b)^2$

C6 Copy and complete these.
  (a) $(n + 1)(\ \ldots\ldots\ ) = n^2 + 6n + 5$ (b) $(\ \ldots\ldots\ )(n - 4) = n^2 + 2n - 24$
  (c) $(n - 6)(\ \ldots\ldots\ ) = n^2 - 8n + 12$ (d) $(n + 2)(\ \ldots\ldots\ ) = n^2 \ldots\ldots + 14$

In questions C7 to C10, all lengths given are in cm.

C7 If n is a whole number, show that the shaded area is always a multiple of 10.

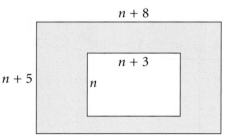

C8 The area of the square is $5\,\text{cm}^2$ less than the area of the rectangle. Find x.

C9 The area of the square is $5\,\text{cm}^2$ more than the area of the rectangle. What is the area of each shape?

C10 Find the length of each side for these right-angled triangles.
  (a)   (b)   (c)

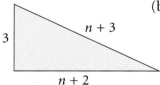
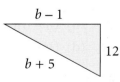
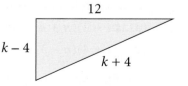

*C11 Copy and complete these.
  (a) $(\ \ldots\ldots\ )(n + 1) = n^2 - 9n \ldots\ldots$ (b) $(n - 5)(\ \ldots\ldots\ ) = n^2 - 11n \ldots\ldots$
  (c) $(n + 6)(\ \ldots\ldots\ ) = n^2 - 36$ (d) $(\ \ldots\ldots\ )^2 = n^2 + 8n \ldots\ldots$

D *Further factorising*

An expression that can be written in the form

$$ax^2 + bx + c$$

is called a **quadratic** expression.

Can you factorise these quadratic expressions?

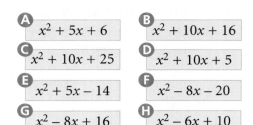

A $x^2 + 5x + 6$

B $x^2 + 10x + 16$

C $x^2 + 10x + 25$

D $x^2 + 10x + 5$

E $x^2 + 5x - 14$

F $x^2 - 8x - 20$

G $x^2 - 8x + 16$

H $x^2 - 6x + 10$

D1 Factorise these.

(a) $x^2 + 8x + 7$ (b) $x^2 + 8x + 15$ (c) $x^2 + 10x + 21$

(d) $x^2 + 8x + 16$ (e) $x^2 + 3x - 4$ (f) $x^2 - 4x - 5$

(g) $x^2 - 6x + 5$ (h) $x^2 - 15x + 14$ (i) $x^2 - 6x + 9$

D2 Which two of these expressions cannot be factorised using integers only?

A $x^2 + 13x + 40$

B $y^2 + 10y + 12$

C $z^2 + 5z - 50$

D $k^2 - 2k - 35$

E $m^2 - 3m + 2$

F $n^2 - 4n + 2$

D3 The expression $n^2 - 10n + 25$ can be factorised to $(n - 5)(n - 5)$ or $(n - 5)^2$.
So we can call $n^2 - 10n + 25$ a **perfect square**.

Which of the expressions below are perfect squares?

A $x^2 + 6x + 9$

B $y^2 + 8y + 8$

C $z^2 - 10z - 25$

D $k^2 + 18k + 80$

E $m^2 - 4m + 4$

F $n^2 - 20n + 100$

D4 The nth term of a sequence is $n^2 + 2n + 1$.

(a) Work out the first five terms of this sequence.

(b) Show that **every** term in the infinite sequence must be a square number.

D5 Factorise $n^2 + 3n + 2$ and show that $n^2 + 3n + 2$ must be even for any integer n.

Difference of two squares

These expressions are examples of the **difference of two squares**: $x^2 - 25$ $b^2 - c^2$
Both terms in the expression are perfect squares.

D6 Paula is trying to factorise $a^2 - 9$.

(a) Expand $(a - 3)^2$ to show that she is wrong.

(b) Factorise $a^2 - 9$ correctly.

$a^2 - 9 = (a - 3)^2$ ✗
Try again, Paula.

D7 Factorise these.

(a) $x^2 - 16$ (b) $h^2 - 25$ (c) $n^2 - 1$ (d) $b^2 - 100$

D8 (a) Expand and simplify $(x + y)(x - y)$.

(b) Use your result from (a) to calculate $99^2 - 1^2$ without a calculator. Hence calculate 99^2.

D9 Without using a calculator, evaluate

(a) $999^2 - 1^2$ (b) 999^2 (c) $98^2 - 2^2$ (d) 98^2 (e) 998^2

- To factorise expressions of the form $x^2 + ax + b$ look for two numbers that **multiply** to make b and **add** to make a.

  Example

  Factorise $x^2 - 2x - 8$.

  > Look for two numbers that multiply to give $^-8$ and add to give $^-2$.
  > The numbers are $^-4$ and 2.
  > So $x^2 - 2x - 8 = (x - 4)(x + 2)$

- To factorise the **difference of two squares** remember that
  $$a^2 - b^2 = (a + b)(a - b)$$

E *Solving quadratic equations*

What values can x and y have if:

$$xy = 18 \qquad xy = {^-36} \qquad xy = 0$$

Factorising can be useful when solving quadratic equations.

You will also need to use the fact that when any two expressions multiply to give zero then one or other of the expressions must be zero.

Examples

Solve $n^2 + 3n - 4 = 0$

> $n^2 + 3n - 4 = 0$
> $(n - 1)(n + 4) = 0$
> Either $(n - 1) = 0$ or $(n + 4) = 0$
> So $n = 1$ or $n = {^-4}$

> You can check your solutions by substituting them in the original equation.

Solve $k^2 - 14k + 40 = 16$

> $k^2 - 14k + 40 = 16$
> $k^2 - 14k + 24 = 0$
> $(k - 2)(k - 12) = 0$
> Either $(k - 2) = 0$ or $(k - 12) = 0$
> So $k = 2$ or $k = 12$

E1 Solve these quadratic equations.

(a) $x^2 + 2x - 8 = 0$ (b) $y^2 + 12y + 20 = 0$ (c) $z^2 - z - 42 = 0$

(d) $p^2 - 5p + 6 = 0$ (e) $q^2 + 4q + 4 = 0$ (f) $h^2 - 12h + 36 = 0$

(g) $k^2 + 2k = 0$ (h) $m^2 + m = 0$ (i) $n^2 - 6n = 0$

E2 Rearrange and solve these equations.

(a) $g^2 + 2g + 1 = 4$ (b) $h^2 - 9h + 30 = 10$ (c) $k^2 - k = 12$

(d) $w^2 + 9w = {}^-8$ (e) $v^2 + 5v + 1 = v - 2$ (f) $m^2 - 4m + 9 = 2m + 1$

(g) $n(n + 1) = 12$ (h) $b(b - 1) = 6$ (i) $a(a + 5) = 29a - 144$

Some problems can be solved by forming and solving a quadratic equation.

Example

The length of a rectangle is 4 cm longer than its width.
The area of the rectangle is 77 cm^2.

What is the perimeter of the rectangle?

Let w be the width of the rectangle in cm.
The length of the rectangle is then $w + 4$.

The area of the rectangle is 77 cm^2 so

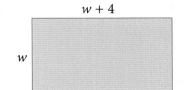

$$w(w + 4) = 77$$
$$w^2 + 4w = 77$$
$$w^2 + 4w - 77 = 0$$
$$(w - 7)(w + 11) = 0$$
$$w = 7 \text{ or } {}^-11$$

So the width of the rectangle is 7 cm (the width cannot be negative)
and the length is $7 + 4 = 11$ cm.

$2(7 + 11) = 36$ **So the perimeter is 36 cm.**

E3 The width of a rectangle is 3 cm less than its length.
The area of the rectangle is 54 cm^2.
What is the length of this rectangle?

E4 The area of this shape is 91 cm^2.
Find the value of x.

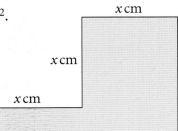

E5 Find the length of each side of this right-angled triangle.

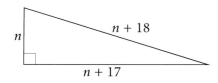

E6 The expression for the nth term of a sequence is $n^2 + n - 11$.

    (a) What is the 10th term of the sequence?

    (b) Which term of the sequence is 261?

Test yourself

T1 (a) Expand and simplify $5(2x + 3) - 2(x - 1)$.

    (b) (i) Factorise $4a + 6$ (ii) Factorise completely $6p^2 - 9pq$.   Edexcel

T2 (a) Expand and simplify $2(x - 1) + 3(2x + 1)$.

    (b) Factorise completely $6a^3 - 9a^2$.   Edexcel

T3 Factorise   (a) $4x^2y - 6xy^2$ (b) $x^2 - 4x - 12$ WJEC

T4 Factorise   (a) $x^2 - 2x - 15$ (b) $6x^2 - 8xy$ WJEC

T5 (a) Factorise $x^2 - 6x + 8$.

    (b) Solve the equation $x^2 - 6x + 8 = 0$.   Edexcel

T6 (a) Factorise $xy - y^2$.

    (b) Solve the equation $x^2 - 7x + 10 = 0$.   AQA 2001

T7 Solve the equation $x^2 - 5x = 0$.   AQA 2001

T8 Solve the equation $x^2 + 5x - 6 = 0$.   AQA 2002

T9 The dimensions of a rectangle are shown.

The rectangle has an area of $32\,\text{cm}^2$.

Form an equation, in terms of x, for the area of the rectangle.

$(x + 3)$ cm

$(x - 1)$ cm

Not to scale

Show that it can be written in the form $x^2 + 2x - 35 = 0$.   AQA 2001

T10 Solve the equation $x^2 + 2x - 35 = 0$ and hence write down the dimensions of the rectangle in question T9.

21 Sampling

You will revise how to carry out surveys.

This work will help you use

◆ a range of sampling methods to collect data

◆ random numbers to select a sample

A Truly representative

You need sheets G163 and G164.

Sheet G163 is taken from an aerial photograph of a large herd of African elephants.
There are 100 elephants on this diagram.

Researchers are interested in

the mean height of these elephants

the mean age of the elephants

the percentage that are bulls (males)

They do not have the resources to catch and examine all the elephants. They are able to catch 25 elephants.

- How would you choose the 25 elephants?
- Choose 25 elephants using your method.
  Use the database on sheet G164 to record the height, age and sex of each elephant in your sample.
- For your sample, find
  (a) the mean height
  (b) the mean age
  (c) the percentage that are bulls

Your teacher will tell you the mean height, mean age and percentage of bulls for all the elephants in the herd.

- How do your figures compare with the overall figures?
- Do you think your method of sampling was a good one? Was it fair?
- Were there any problems with your method?

If you want a sample to give reliable information about a population, the sample must be representative of the population.

An unrepresentative sample is called 'biased'.

Two common methods of selecting a sample are **random sampling** and **systematic sampling**.

Random sampling	**Systematic sampling**
Every member of the population has an equal chance of being chosen.	Individuals are chosen at regular intervals (for example, every 5th house in a street).

Sampling opinions for discussion

A local council wants to know what people want to do with a reclaimed rubbish tip.
The council decides to find out by asking a sample of people.

Below are some suggestions for how the sample should be chosen. For each method say, with reasons,

- whether it is random sampling, systematic sampling, or neither of these

- whether it would give a representative or a biased sample

- whether it would be easy to carry out in practice

A Ask the next 300 people who come into the Town Hall to fill in a questionnaire.

B Open 300 pages of the local residential telephone directory at random. Ring the first number on each page in the early evening.

C Stand in the High Street next Saturday morning and ask every 10th person who comes along.

D Put a page on the Internet and ask those in the area who read it to e-mail their opinion.

E Go through the local electoral register (a list of all those entitled to vote) and choose every 50th person to send a questionnaire to.

F Send researchers to every area in the town with instructions to call at every 10th house in every street and ask the person who answers the door.

B *Using random numbers*

To choose a random sample from a list, the list must first be numbered.
Individuals can then be chosen by using **random numbers**.

Random numbers can be obtained from the random-number generator
on a calculator or computer, or from random-number tables.

Random-number generator

This produces a sequence of digits.
Each digit in the sequence is equally likely to be 0, 1, 2, ... or 9.
The digits are usually grouped, for example in threes: 783 310 805 442 712 ...

Random-number table

This is a list printed from a random-number generator.
It is set out in columns and rows so that you can start at different places and
move through the table in different ways (across, down, etc).
There is a random-number table on sheet G162.

Choosing a random sample

Suppose you want to choose a random sample of 30 people from a numbered list of 430.

- Generate random numbers in three-digit groups.
- Ignore those outside the range 1 to 430 and any number that comes up twice.
- Continue until you have 30 numbers.

Example

  232 (854) 034 (551) (988) 365 007 273 419 (622) ... (ignoring those in brackets)

B1 Sheet G165 gives the weights in tonnes of 250 female elephants from the
Amboselli reserve.

   (a) Use random numbers to choose a sample of five elephants from the list.

   (b) Calculate the mean weight of the five elephants in your sample.
   (Keep a record of this – you will need it later.)

B2 (a) Describe how random numbers could be used to choose

   (i) 50 students from a school of 650, to take part in an experiment

   (ii) 150 people in a town of 25 000 residents, to be sent a questionnaire
   about local sports facilities

   (b) What practical problems might arise once the sample has been chosen
   in each case above?

C Sample size

TG

Collect together the mean weights of the samples of 5 elephants in question B1 for the whole class.

Display the results in a dot plot.
You will probably need a scale from 0.9 to 1.5.

- Suppose you want to estimate the mean weight of the 250 elephants. How reliable a result would you get by taking a random sample of 5 elephants?

Now investigate what happens when the size of the sample is increased.
Everyone selects 5 more elephants and calculates the mean of the sample of 10 selected so far.

Display the class results on another dot plot using the same scale as before.

- Does the result from a sample of 10 appear to be more reliable than from a sample of 5?

Finally everyone selects a further 15 elephants, making the samples of size 25.
Plot the sample means as before.

- Does a sample of 25 appear to give a reliable estimate of the mean weight of the population?
- Compare the sample means with the actual mean weight (which your teacher will have).

D Stratified sampling

The organisers of a sports club want to find out members' opinions about a proposed change to the rules.

The club members are 75% men and 25% women, and the organisers think that men and women could have different views about the proposed change.

If they choose a random sample, the proportions of men and women in the sample might not reflect the proportions in the membership as a whole.
So the sample might not be representative of members' opinions.

The organisers can get round this problem by using a **stratified sample**.
They split the population (the members) into two strata (men, women), and sample each stratum separately, in proportion to its size.

So if they want a sample of overall size 40, they would select 30 men and 10 women. (The individual men and women could be chosen by random or by systematic sampling.)

Example

There are 280 girls and 160 boys in a school.
A stratified sample of 30 pupils is to be selected for a survey.

The proportion of girls is $\frac{280}{440} = 0.636...$, so the sample must contain $0.636... \times 30 = \mathbf{19}$ girls

D1 Maddy is carrying out a survey on school uniform policy at her school in year 11.
There are 97 males and 115 females in year 11.
Maddy decides to take a stratified sample of 50 students.
How many males and females should she include in her sample?

D2 The table shows the number of students in years 7, 8 and 9 of a school.
A sample of 100 of these students were
asked some questions about homework.
The students were part of a stratified random sample.

How many students in each year group were
included in the sample?

Year	Number of students
7	118
8	165
9	142

AQA(SEG) 1999

D3 A vet wants to check a sample of 25 elephants from Amboselli for a disease.
The ages of the elephants in the sample of 100 in section A can be summarised as follows.

Age a (years)	$0 \leq a \leq 9$	$10 \leq a \leq 19$	$20 \leq a \leq 29$	$30 \leq a \leq 49$	$a \geq 50$
Frequency	39	26	19	12	4

(a) If the sample is to be stratified by age, how many elephants in each of the above age
groups would need to be included in the sample of 25?

(b) The overall gender split of 44% bulls and 56% females is also to be representative.
Copy and complete this table showing the number in each category that
should be sampled.

Age a (years)	$0 \leq a \leq 9$	$10 \leq a \leq 19$	$20 \leq a \leq 29$	$30 \leq a \leq 49$	$a \geq 50$
Bulls					
Females					

*D4 The adult population of a town, broken down by age and gender, is shown in this table.

Age group	18–40	41–60	61+	Total
Men	3420	4210	2230	9860
Women	3150	4620	3070	10840
Total	6570	8830	5300	20700

A stratified sample of 150 people is to be selected, in which both age group
and gender reflect their proportions in the population.
Make a table similar to the one above showing how many of each group should
be included in the sample (how many men aged 18–40, etc).

Test yourself

T1 After plans for a bypass to a large town were announced, the local newspaper received twelve letters on the subject. Eleven were opposed to it.
The newspaper claimed

OVER 90% ARE AGAINST NEW BYPASS

(a) Give two reasons why the newspaper could be criticised for making this claim.

(b) The local council is to carry out a survey to find the true nature of local opinion. Give two factors that should be taken into account when selecting the sample.

AQA(NEAB) 1997

T2 A random sample of size 25 is to be selected from the 180 boys in year 11 of a school.
Describe how you would use random-number tables to select a sample.

T3 In a school there are 420 pupils in the lower school, 310 pupils in the middle school, and 130 pupils in the upper school.

(a) How many pupils from each part of the school should be included in a stratified random sample of size 100?

(b) Explain briefly in what circumstances a stratified random sample might be taken rather than a simple random sample.

AQA(SEG) 1998

T4 There are 1000 students in Nigel and Sonia's school.
Nigel is carrying out a survey of the types of food eaten at lunchtime.

(a) Explain how Nigel could take a random sample of students to carry out this survey.

This table shows the gender and the number of students in each year group.

Year group	Number of boys	Number of girls	Total
7	100	100	200
8	90	80	170
9	120	110	230
10	80	120	200
11	100	100	200

Sonia is carrying out a survey about how much homework students are given.

She decides to take a stratified sample of 100 students from the whole school.

(b) Calculate how many in the stratified sample should be

    (i) students from year 9   (ii) boys from year 10.

Edexcel

22 Direct and inverse proportion

You will revise ratio.

This work will help you

◆ recognise direct and inverse proportion
◆ do calculations involving direct and inverse proportion
◆ solve problems involving other types of proportionality

A Ratio review

Adult lizard

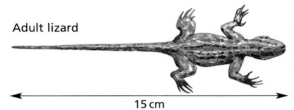

15 cm

Young lizard

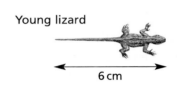

6 cm

The ratio of the adult's length to the young lizard's length can be written in different ways.

length of adult : length of young = 15 : 6
                                       = 5 : 2

$$\frac{\text{length of adult}}{\text{length of young}} = \frac{15}{6} = 2.5$$

A1 Write each of these ratios in its simplest form.
  (a) 10 : 5   (b) 12 : 20   (c) 9 : 15   (d) 200 : 25   (e) 144 : 360

A2 The 'aspect ratio' of a rectangular picture is the ratio $\frac{\text{height}}{\text{width}}$.

Find the aspect ratio of each of these pictures.

150 cm   (a)   125 cm

200 cm   (b)   80 cm

125 cm   (c)   200 cm

A3 The ratio of the number of rainy days to the number of dry days over a period of time is 2 : 5.
If there were 16 rainy days, how many dry days were there?

A4 Split £60 in the following ratios. (a) 2 : 3   (b) 4 : 1   (c) 2 : 3 : 7

A5 Write each of these ratios in the form $k : 1$. (a) 12 : 3 (b) 7 : 2 (c) 4 : 5

*A6 The ratio of sheep to goats on a farm is 5 : 4.
The ratio of goats to pigs is 3 : 2.
What is the ratio of sheep to pigs?

B *Related quantities*

Ribbon can be bought by length from a roll.
50 cm of ribbon costs 20p.

- How much ribbon can you buy for 1p?
- How much can you buy for 12p?
- How much does 1 cm of ribbon cost?
- What numbers are missing from this table?

Length (*L* cm)	0	1			20		50	80
Cost (*C* pence)	0		1	6		12	20	

The graph of *C* against *L* is a straight line which
goes through the origin (0, 0).

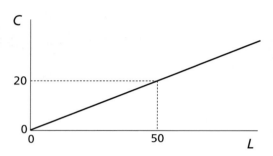

- What is the gradient of the graph?
- What is the equation connecting *C* and *L*?
- Use the equation to find *C* when *L* = 130.

B1 Stephanie works in a factory and is paid by the hour.
She works for 6 hours and is paid £45.

(a) How much does she earn in 1 hour?

(b) How much will she earn in a 40 hour week?

(c) How long does it take her to earn £1? (Give your answer as a fraction of an hour.)

(d) Copy and complete this table of values.
T is the time in hours, *W* the wages in £.

T	6	1	40			30
W	45			1	90	

(e) Draw a graph of *W* against *T*.

(f) What is the equation connecting *W* and *T*?

(g) Use the equation to find

(i) Stephanie's wages when she works 15 hours

(ii) the time Stephanie has to work to earn £60

B2 (a) Using suitable axes, plot values of *S* and *T*
as given in this table.

S	0	2	4	5	7	10
T	0	9	18	22.5	31.5	45

(b) Draw the graph of *T* against *S*
and write down its equation.

(c) Use the equation to find

(i) *T* when *S* = 30 (ii) *S* when *T* = 54

C *Direct proportion*

The cost of ribbon, C pence, is **directly proportional** to the length, L cm.
This means

- If L is multiplied by a number, then C is multiplied by the same number.
  (For example, if you buy 3 times as much, you pay 3 times as much.)

- The ratio $\dfrac{C}{L}$ is the same for every pair of values of L and C.

- The graph of C against L is **a straight line going through $(0, 0)$**.

- C is connected to L by a formula of the type
  $$C = kL$$
  k is called the **constant of proportionality**.
  (In the ribbon example, $k = 0.4$.)

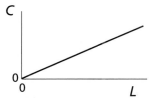

The symbol $\propto$ is used to mean 'is directly proportional to'. So we write $C \propto L$.

You can think of $C \propto L$ as another way of writing $C = kL$.

C1 In which of these graphs is Q directly proportional to P? Give reasons.

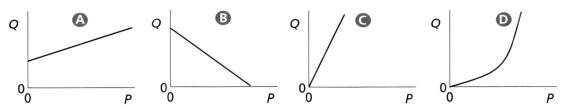

C2 Are these statements true or false?
  (a) The cost of petrol is directly proportional to the quantity purchased.
  (b) The height of a person is directly proportional to their age.
  (c) The area of a square is directly proportional to the length of the side.
  (d) The distance travelled in a certain time by a car travelling at constant speed
      is directly proportional to the speed of the car.
  (e) The time taken for a journey is directly proportional to the speed of travel.

C3 In each of these tables, is Q directly proportional to P?
  If so, give the equation connecting Q and P.

(a)

P	0	3	4	10
Q	0	12	16	40

(b)

P	0	2	4	6
Q	4	10	16	22

(c)

P	0	3	4	6
Q	0	18	32	72

(d)

P	4	6	7	10
Q	8	12	14	20

(e)

P	1	2	3	4
Q	3	7	11	15

(f)

P	2	10	12	20
Q	7	35	42	70

D *Calculating with direct proportion*

Example

When a spring is stretched, the extension, E cm, of the spring
is directly proportional to the stretching force, F newtons.

Given that $E = 12$ when $F = 5$, find   (a) E when $F = 8$ (b) F when $E = 15$

The information and the unknowns can be shown in a table.

F	5	8	?
E	12	?	15

E and F are connected by an equation of the form $E = kF$.
Use the known pair of values to find k:

$$12 = k \times 5, \text{ so } k = \frac{12}{5} = 2.4$$

Now use the equation $E = 2.4F$ to find the unknowns.

(a) When $F = 8$, $E = 2.4 \times 8 = $ **19.2**   (b) When $E = 15$, $15 = 2.4F$

$$\text{So } F = \frac{15}{2.4} = \textbf{6.25}$$

D1 The volume, V litres, of water that comes out of a tap is directly proportional to
the time, T minutes, for which the tap is turned on.

Given that $V = 125$ when $T = 2.5$, find

(a) the equation connecting V and T (b) the value of V when $T = 4.5$

(c) the value of V when $T = 10.5$ (d) the value of T when $V = 475$

D2 The quantity, P litres, of paint needed to paint a floor is directly proportional
to the area, A m², of the floor.

Given that $P = 3.0$ when $A = 40$, find

(a) the equation connecting P and A (b) the value of P when $A = 140$

D3 The length, L m, of a shadow is directly proportional
to the height, H m, of the object (at a given time of day).
Find

(a) the equation connecting L and H

(b) the length of the shadow of flagpole B

(c) the height of flagpole C

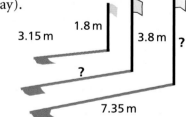

D4 The electric current I (measured in amps) in a lamp
is directly proportional to the voltage V (measured in volts).

When $V = 2.5$, $I = 1.5$. Find

(a) the equation for I in terms of V (b) the value of I when $V = 4.5$

(c) the value of V when $I = 6.9$

D5 The quantity Q is directly proportional to P.
When $P = 2.5$, $Q = 20$.

    (a) If $Q = kP$, find the value of k.   (b) What is Q when $P = 28$?

    (c) What is P when $Q = 4.5$?

D6 In this table, $Y \propto X$.

    (a) If $Y = kX$, find the value of k.

    (b) Copy and complete the table.

X	5	1		11
Y	18		12	

D7 On Mid Western Trains, the fare for a journey of 40 km is £7.20.
Fares are directly proportional to the distance travelled.

    (a) What is the fare for a journey of 70 km?   (b) How far could you travel for £18?

E $Q \propto P^2$, $Q \propto P^3$, ...

This table shows the mass, m grams, of discs
of diameter d cm cut from a sheet of plastic.

d (diameter in cm)	0	5	10	15	20
m (mass in grams)	0	50	200	450	800

m is **not** directly proportional to d.
(If you double d, then m is not doubled.)

The mass is directly proportional to the **square** of the diameter: $m \propto d^2$.

We can see this if we put an extra row in the table, showing d^2.

$\times 2 \Bigg($

d	0	5	10	15	20
d^2	0	25	100	225	400
m	0	50	200	450	800

The equation connecting m and d is of the form $m = kd^2$.

The constant of proportionality is 2, so $m = 2d^2$.

E1 The length of the skid mark made by a car after braking hard on a
particular type of road surface is proportional to the square of the speed.

This table shows some values of the speed, S km/h,
and the skid length L cm.

S	0	20	40	60	80
S^2					
L	0		800		

    (a) Copy the table and complete the values of S^2.

    (b) Find the value of k in the equation $L = kS^2$.

    (c) Use the equation to fill in the missing values of L.

    (d) Find the speed of a car whose skid mark is 450 cm long.

E2 The distance, s m, fallen by a stone dropped from a high point is proportional to the square of the time, t seconds, since it was dropped.

This table shows some values of t, t^2 and s.

(a) Use the known values to find the equation connecting s and t.

(b) Copy the table and use the equation to fill in the missing values of s and t.

t	0	4	7	
t^2	0			
s	0	80		980

E3 The force of air resistance, R newtons, on a car is directly proportional to the square of the speed, S km/h.

When the speed is 30 km/h, the air resistance is 1350 N.

(a) Find the equation connecting R and S in the form $R = kS^2$.

(b) Copy the table and fill in the missing values.

S	10	30	40	
S^2				
R		1350		2904

E4 A quantity Q is proportional to the square of a quantity P. When $P = 14$, $Q = 1078$.

(a) Find the equation connecting Q and P.

(b) Use the equation to find

    (i) the value of Q when $P = 8$ (ii) the value of P when $Q = 1782$

E5 Given that $Y \propto X^2$ and that $Y = 38.4$ when $X = 8$, find

(a) the equation connecting Y and X (b) the value of Y when $X = 13$

(c) the value of Y when $X = 22$ (d) the value of X when $Y = 173.4$

E6 The time taken for one complete swing (to and fro) of a simple pendulum is proportional to the **square root** of its length.

$T \propto \sqrt{L}$, where T is the time in seconds and L is the length in metres.

This table shows some values of L, $\sqrt{L}$ and T.

(a) Use the known values to find the equation connecting T and L.

(b) Copy the table and use the equation to fill in the missing values of T and L.

L	4	6.25	9	
$\sqrt{L}$	2			
T			6	50

E7 A weight is hung at the end of a beam of length L. This causes the end of the beam to drop a distance d. d is directly proportional to the cube of L. $d = 20$ when $L = 150$.

(a) Find a formula for d in terms of L.

(b) Calculate the value of L when $d = 15$.

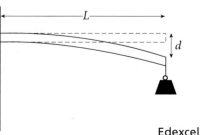

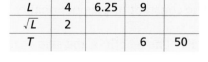

Edexcel

F Inverse proportion

A club is arranging a coach trip. The coach costs £200 to hire.
The cost will be shared equally between the people who go on the trip.

If n people go, the amount A (in £) that each person will have to pay
is given by the formula

$$A = \frac{200}{n}$$

This table shows some pairs of values of n and A.

n	8	10	20	25	40	50
A	25	20	10	8	5	4

Notice that

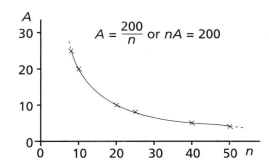

- If n is multiplied by a number, then
  A is **divided** by the same number.

  (For example, if twice as many people go,
  the amount per person is divided by 2.)

- The product nA is constant.
  (In this example it is always 200.)

- The graph of A against n is a curve
  (called a 'rectangular hyperbola').

A is **inversely proportional** to n. This can be written $A \propto \frac{1}{n}$.

When you see $A \propto \frac{1}{n}$ you can replace it by $A = \frac{k}{n}$ or $nA = k$.

F1 This table shows some pairs of values of
two quantities P and Q.

P	4	5	6	8	20	30
Q	15	12	10	7.5	3	2

(a) By working out PQ for each pair, check that Q is inversely proportional to P.

(b) Check that when P is multiplied by a number, then Q is divided by the
same number.

(c) What is the value of Q when P is   (i) 2   (ii) 3   (iii) 1   (iv) 0.5

F2 The time taken, T hours, for a journey is inversely
proportional to the speed of travel, S km/h.

S	4	5	8	20	40
T	10	8			

The table shows some values of S and T.

(a) Find the value of k in the equation $ST = k$.

(b) Use the equation to find the three missing values in the table.

(c) What is
   (i) the value of T when $S = 25$ (ii) the value of S when $T = 1.25$

Direct and inverse proportion compared

If Q is directly proportional to P, the **ratio** $\frac{Q}{P}$ is constant.

P	2	4	10
Q	5	10	25

$$\frac{Q}{P} = 2.5$$

If Q is inversely proportional to P, the **product** PQ is constant.

P	2	3	20
Q	15	10	1.5

$$PQ = 30$$

F3 For each table below, say whether it shows direct proportion, inverse proportion or neither.
If it shows direct or inverse proportion, write down the equation connecting Q and P.

(a)

P	4	5	8	20	40
Q	6	7.5	12	30	60

(b)

P	1	5	7	10	20
Q	20	5	4	3	1

(c)

P	2	4	8	10	40
Q	50	25	12.5	10	2.5

(d)

P	0.1	0.4	0.6	1.2	1.8
Q	3.6	0.9	0.6	0.3	0.2

F4 The frequency of a musical note is measured in hertz.
The frequency, F hertz, of the note produced by a guitar string is inversely proportional to the vibrating length, L cm, of the string.
Given that $F = 200$ when $L = 40$, find

(a) the equation connecting F and L (b) the value of F when $L = 16$

(c) the value of L when $F = 125$ (d) the value of L when $F = 1000$

F5 If $Y \propto \dfrac{1}{X}$ and $Y = 30$ when $X = 2$, find

(a) the value of Y when $X = 8$ (b) the value of X when $Y = 0.5$

F6 If $Q \propto \dfrac{1}{P}$ and $Q = 0.75$ when $P = 1.6$, find

(a) the value of Q when $P = 0.3$ (b) the value of P when $Q = 0.25$

F7 The quantity B is inversely proportional to the **square** of another quantity A.
(This is written $B \propto \dfrac{1}{A^2}$.)

This table shows some values of A, A^2 and B.

(a) Check that A^2B is constant for the first three sets of values in the table.

(b) Write the equation in the form $A^2B = k$.

(c) Find the missing values in the table.

A	2	3	4	5	
A^2	4	9	16		
B	90	40	22.5		10

Example

$Q \propto \dfrac{1}{\sqrt{P}}$, and $Q = 15$ when $P = 16$.

   (a) Find the equation connecting Q and P.

   (b) Find Q when $P = 25$

   (c) Find P when $Q = 20$

Make a table showing P, $\sqrt{P}$ and Q.
Fill in the known values.

P	16	25	
$\sqrt{P}$	4	5	
Q	15		20

P and Q are connected by an equation of the form $Q\sqrt{P} = k$.
Use the known pair of values of $\sqrt{P}$ and Q to find k

$$\sqrt{16} \times 15 = 4 \times 15 = k, \quad \text{so } k = 60$$

Now use the equation $Q\sqrt{P} = 60$ to find the unknowns.

(a) When $P = 25$, $Q \times 5 = 60$. So $Q = \dfrac{60}{5} = 12$

(b) When $Q = 20$, $20 \times \sqrt{P} = 60$. So $\sqrt{P} = \dfrac{60}{20} = 3$ So $P = 9$

F8 D is inversely proportional to the square root of M.

   (a) Given that $D = 10$ when $M = 4$, find the equation connecting D and M.

   (b) Copy this table and complete it using your equation.

M	4		16	
$\sqrt{M}$				
D	10	4		40

F9 You are given that y is inversely proportional to the square of x.
When $y = 10$, $x = 5$.

   (a) Find an expression for y in terms of x.   (b) Calculate y when $x = 10$.

AQA(SEG) 1999

F10 $V \propto \dfrac{1}{\sqrt{U}}$ and $V = 8$ when $U = 100$.

   (a) Find the equation connecting V and U.

   (b) Find (i) the value of V when $U = 16$ (ii) the value of U when $V = 5$

F11 $V \propto \dfrac{1}{U^2}$ and $V = 8$ when $U = 3$.

   (a) Find the equation connecting V and U.

   (b) Find (i) the value of V when $U = 20$ (ii) the value of U when $V = 2$

F12 The force, F newtons, between two magnets is inversely proportional to
the square of the distance, d metres, between them.
When the magnets are 2 metres apart, the force between them is 36 newtons.
Find (a) F when $d = 5$ (b) d when $F = 49$ OCR

*F13 Q is inversely proportional to P.
If P is increased by 50%, what is the percentage change in Q?

G Searching for proportionality

The table gives values of two quantities P and Q.

If P is multiplied by 2, Q is multiplied by 4, or 2^2.
If P is multiplied by 3, Q is multiplied by 9, or 3^2.

This tells us that $Q \propto P^2$.

The equation connecting Q and P is $Q = 3P^2$.

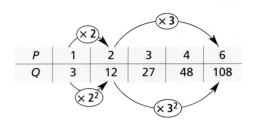

P	1	2	3	4	6
Q	3	12	27	48	108

*G1 This table gives values of two quantities, U and V.

U	2	4	6	8	10
V	4	32	108	256	500

(a) What is V multiplied by when U is multiplied by 2?

(b) What is V multiplied by when U is multiplied by 3?

(c) What is V proportional to?   (d) Find the equation connecting V and U.

*G2 This table gives values of two quantities, S and T.

S	2	4	6	10	20
T	180	45	20	7.2	1.8

(a) Check that when S is multiplied by 2, T is **divided** by 4, or 2^2.

(b) Check that when S is multiplied by 3, T is divided by 9, or 3^2.

(c) From (a) and (b) it follows that $T \propto \dfrac{1}{S^2}$. Find the equation connecting T and S.

*G3 Match each table to a type of proportionality and find each missing value.

(a)

P	2	6	12	24
Q	5	45	180	

(b)

P	2	8	18	72
Q	7	14	21	

$$Q \propto \frac{1}{P} \qquad Q \propto P^2$$

(c)

P	1	16	25	100
Q	100	25	20	

(d)

P	2	4	6	8
Q	5	40	135	

$$Q \propto \frac{1}{P^2} \qquad Q \propto P^3$$

(e)

P	2	3	4	6
Q	18	4.5	2	

(f)

P	6	8	10	15
Q	40	30	24	

$$Q \propto \sqrt{P} \qquad Q \propto \frac{1}{\sqrt{P}}$$

*G4 In each of the tables below, Q is either directly proportional to P^2, P^3 or $\sqrt{P}$, or inversely proportional to P^2, P^3 or $\sqrt{P}$.

Find the type of proportionality for each table and the equation connecting Q and P.

(a)

P	1	2	3	6
Q	5	20	45	180

(b)

P	1	4	16	36
Q	3	6	12	18

(c)

P	2	3	4	6
Q	18	8	4.5	2

(d)

P	1	2	3	4
Q	5	40	135	320

(e)

P	4	9	25	36
Q	30	20	12	10

(f)

P	3	5	6	15
Q	54	150	216	1350

Test yourself

☒ **T1** A company sells circular badges of different sizes.
The price, P pence, of a badge is proportional to the square of its radius, r cm.
The price of a badge of radius 3 cm is 180 pence.

    (a) Find an equation expressing P in terms of r.

    (b) Calculate the price of a badge of radius 4 cm.   AQA(SEG) 2000

T2 The wavelength, w metres, of radio waves is inversely proportional to
the frequency, f kHz, of the waves.

    (a) A radio wavelength of 1000 metres has a frequency of 300 kHz.
The frequency is doubled to 600 kHz. What is the new wavelength?

    (b) Calculate the frequency when the wavelength is 842 metres.

    (c) Radio NEAB has a frequency in kHz which is numerically equal to its
wavelength in metres. Calculate the wavelength of Radio NEAB.   AQA(NEAB) 1998

T3 C is inversely proportional to t^2.
When $C = 16$, $t = 3$.
Find t when $C = 9$.   AQA(NEAB) 2000

T4 y is inversely proportional to the square root of x. When $y = 6$ then $x = 4$.

    (a) What is the value of y when $x = 9$?   (b) What is the value of x when $y = 10$?

                                                                   AQA(NEAB) 1999

T5 Given that y is inversely proportional to x^3, and that $y = 5$ when $x = 4$,

    (a) find an expression for y in terms of x

    (b) calculate   (i) the value of y when $x = 2$ (ii) the value of x when $y = 0.32$ WJEC

☒ **T6** The intensity of light L, measured in lumens, varies inversely as
the square of the distance d m from the light.

    When the distance is 2 m, the light intensity is 250 lumens.

    (a) Calculate the value of L when d is 2.5 m.

    (b) Calculate the value of d when L is 90 lumens   Edexcel

T7 The mass of a solid sphere is directly proportional to the cube of its radius.
A sphere with radius 1.2 cm has mass 21.6 g.
Calculate the radius of a sphere of the same material with mass 34.3 g.   OCR

T8 y is inversely proportional to x^2. $y = 3$ when $x = 4$.

    (a) Write y in terms of x.   (b) Calculate the value of y when $x = 5$.   Edexcel

Review 5

1 Solve the following inequalities.

(a) $4z \geq 2z - 12$ (b) $3(p - 2) < 2(1 + p)$ (c) $\dfrac{x+3}{2} < \dfrac{4-x}{4}$

2 The intensity of illumination I (measured in lumens) on a surface is inversely proportional to the square of the distance, d m, between the light source and the surface.

Given that $I = 40$ when $d = 1.5$, calculate I when $d = 2.5$.

3 The staff of a public library want to find out the opinions of people who use the library. Each day for a week they interview 20 of the people using the library between 12 noon and 2 p.m.

Is the sample of people interviewed likely to be representative of library users? Explain your answer.

4 Factorise these.

(a) $x^2 + 7x + 12$ (b) $x^2 - 8x + 12$ (c) $x^2 + 4x - 12$ (d) $x^2 - x - 12$

5 Copy this diagram.

(a) Draw the image of triangle A after reflection in the y-axis. Label it A′.

(b) Draw the image of A′ after reflection in the line $y = x$. Label it A″.

(c) Describe the single transformation that maps A on to A″.

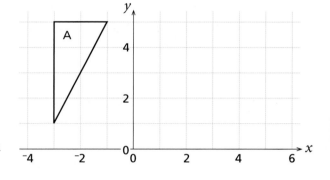

6 (a) Multiply out the brackets and simplify the expression $(x - 3)(x - 6) - 4$.

(b) Factorise your answer to (a).

(c) Hence write down the solutions of the equation $(x - 3)(x - 6) - 4 = 0$.

(d) Solve the equation $(x + 2)(x - 6) + 7 = 0$.

7 The diagram shows the circular cross-section of a tunnel of diameter 8.6 m.

The horizontal roadway is 6.4 m wide.

Calculate, to the nearest 0.1 m, the height of the top of the tunnel above the roadway.

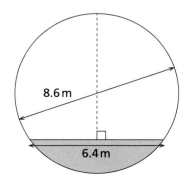

8 Find all of the integer values of n which satisfy the inequality $^-4 \le 2n + 5 < 3$.

9 A quantity V is inversely proportional to the square root of another quantity U.
   When U is 20, V is 60.
   (a) Calculate V when $U = 180$.   (b) Calculate U when $V = 120$.

10 Factorise these.
   (a) $x^2 - 1$ (b) $x^2 - 81$ (c) $x^2 - 6x + 9$

11 Rearrange the formula $q = a - \sqrt{bp}$ to make p the subject.

12 The owners of Heartbreak House have a business plan in which they expect
   the number of visitors to increase by 5000 each year.
   This year they have had 37 500 visitors.
   (a) What percentage increase (to the nearest 0.1%) do the owners expect between this
       year and next year?
   (b) What percentage increase do they expect between next year and the year after that?

13 (a) Write each of these expressions in the form a^n, where n is a positive or
       negative integer.
       (i) $a^2 \times a^3$ (ii) $(a^2)^3$ (iii) $a^{-2} \times a^3$ (iv) $\frac{a^2}{a^{-3}}$ (v) $(a^{-2})^{-3}$
   (b) Write 25^3 as a power of 5.
   (c) Solve the equation $3^n = 27^4$.

14 Copy this diagram.
   Draw an enlargement of triangle ABC
   with centre $(0, 4)$ and scale factor $^-2$.

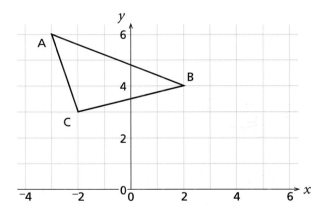

15 The mass of the Earth is about 6×10^{25} kg.
   The mass of the Moon is about $\frac{1}{80}$ of the mass of the Earth.
   Write the approximate mass of the Moon in kg in standard form.

16 Here is a simple rule for working out the squares of $1\frac{1}{2}, 2\frac{1}{2}, 3\frac{1}{2}, 4\frac{1}{2}, \ldots$
   Example: $(5\frac{1}{2})^2$ Work out $5 \times$ the next number after 5, and then add $\frac{1}{4}$.
             So $(5\frac{1}{2})^2 = (5 \times 6) + \frac{1}{4} = 30\frac{1}{4}$.
   Multiply out $(n + \frac{1}{2})^2$ and prove that the rule works.

17 Find the equations of the lines AB and CD and calculate the coordinates of the point where the two lines meet.

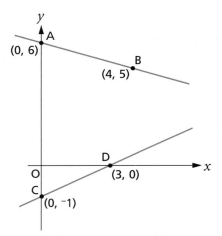

18 These cumulative percentage frequency graphs show the distribution of weekly income in two companies, A and B.

(a) What is the median weekly income in each company?

(b) What are the upper and lower quartiles of the weekly income in each company?

(c) What is the interquartile range of the weekly income in each company?

(d) Write a couple of sentences comparing the two companies.

Challenge

Three circles touch each other.
Their centres form a triangle whose sides are of length 8 cm, 9 cm and 10 cm.

• Find the radius of each circle.

23 Graphing inequalities

This work will help you graph and interpret inequalities in two variables.

A More or less

The line $x = 2$ has been drawn on this diagram.

The points that are on the line $x = 2$ or to the left of the line have x-coordinate 2 or less.

This is **the region $x \leq 2$**. (It includes the line itself.)

The points that are on the line or to the right of it have x-coordinate 2 or more.

This is **the region $x \geq 2$**. (It also includes the line itself.)

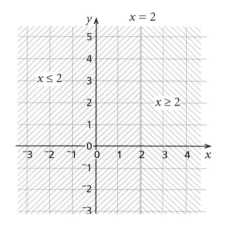

A1 Write an inequality for each region that is shaded.

(a)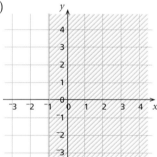

(b)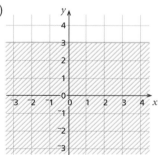

(c)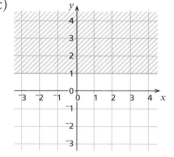

A2 (a) On squared paper draw axes with $^-3 \leq x \leq 5$ and $^-3 \leq y \leq 5$.

(b) Shade the region $x \geq 2$.

(c) Shade the region $y \leq 1$.

(d) Label clearly the region described by both $x \geq 2$ and $y \leq 1$.

A3 (a) Which of the four regions A, B, C or D
satisfies **both** these inequalities?

$$x \leq 2 \qquad y \geq 1$$

(You can assume each region includes its boundaries.)

(b) For each of the other three regions, write down
a pair of inequalities that all the points
in that region satisfy.

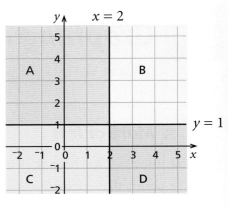

A4 The shaded region on the right
can be defined by four inequalities.
One of these is $x \geq 1$.

Write down the other three inequalities.

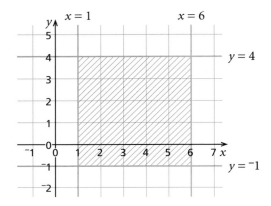

A5 The line $y = x + 2$ is shown on this diagram.

(a) ($^-1$, 5) is a point above the line $y = x + 2$.
Which of these two inequalities does this
point satisfy, $y > x + 2$ or $y < x + 2$?

(b) Check that other points above the line
all satisfy the same inequality.

(c) Check that points below the line satisfy
the other inequality.

(d) Sketch the diagram and shade the region
$y \geq x + 2$, which includes the line itself.

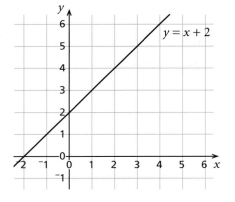

A6 The line $x + y = 2$ is shown on this diagram.

(a) Choose a point above the line.
Which of these two inequalities does this
point satisfy, $x + y < 2$ or $x + y > 2$?

(b) Check that points below the line satisfy
the other inequality.

(c) Sketch the diagram and shade the region
$x + y \leq 2$, which includes the line itself.

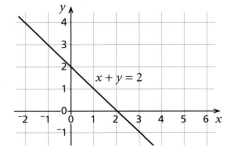

B *Further lines*

Show the region satisfied by the inequality $y \le 2x + 1$.

- First draw the boundary line $y = 2x + 1$.
- Choose a point on one side of the line.
  For example (3, 2)
- Check the inequality for your point.
  $y = 2$ $2x + 1 = 2 \times 3 + 1 = 7$
  As $2 \le 7$, then $y \le 2x + 1$ is true for this point.
- Shade the correct region.
  The inequality is true for (3, 2)
  so shade the region that includes this point.

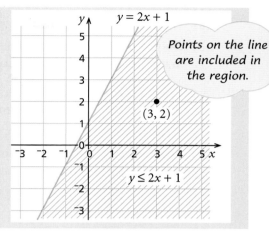

Points on the line are included in the region.

B1 The diagram shows the line $2x + 3y = 12$,
and a region that includes the line.

Which of these is the shaded region?

A $2x + 3y \le 12$

B $2x + 3y < 12$

C $2x + 3y \ge 12$

D $2x + 3y > 12$

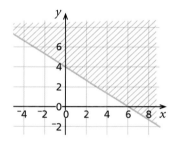

B2 For each diagram below

(i) write down the equation of the line

(ii) write down the inequality that the shaded region (including the line) satisfies

(a)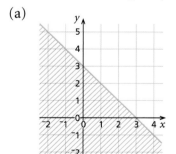

(b)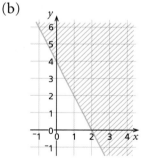

(c)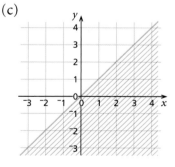

B3 (a) On squared paper sketch the line $3x + 4y = 12$.
Number each axis where the line crosses it.

(b) On your sketch, shade one side of the line to show the region $3x + 4y \le 12$.

B4 On squared paper, draw sketches to show clearly

(a) $3x + 2y \le 6$ (b) $y \le 2x + 1$ (c) $y \ge x - 2$

B5 A lorry carries a tonnes of sand together with b tonnes of gravel.
The total weight carried must not be greater than 30 tonnes.

(a) Write down an inequality in a and b that represents this.

(b) On graph paper draw suitable axes and show the region
that represents your inequality.

B6 A teacher is buying some prizes for a maths competition.
Calculators cost £3 each and geometry sets £5 each.

(a) Write down an expression for the total cost of n calculators and m geometry sets.

(b) The teacher cannot spend more than £60.
Use this fact to write down an inequality in n and m.

(c) On graph paper show the region that represents this inequality.

C Overlapping regions

Shade the region satisfied by all the inequalities $x \geq 1$ $y \geq 0$ $x + y \leq 4$

- Shade the three different
regions on the same set of axes.

- Where the shading overlaps
defines the region where **all**
the inequalities are satisfied.

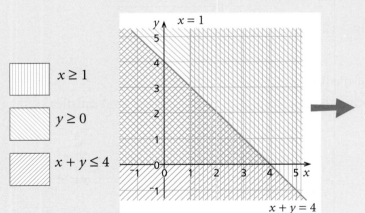

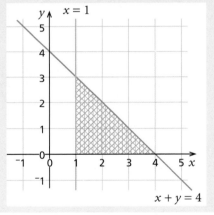

C1 (a) Which region, A, B, C or D, is
satisfied by the three inequalities

$x \leq 6$ $x + y \geq 7$ $y \leq x + 1$

(You can assume each region
includes its boundaries.)

Write down the coordinates of three
points that satisfy **all** these inequalities.

(b) For each other region, write down the
three inequalities that fully describe it.

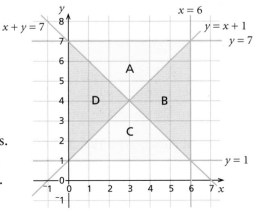

C2 Write down the three inequalities that define each shaded region.

(a)

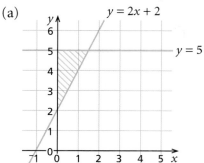

(b)

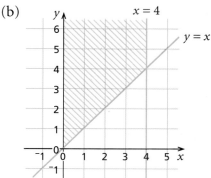

C3 Write down the three inequalities that define each shaded region.

(a)

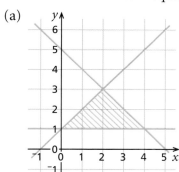

(b)

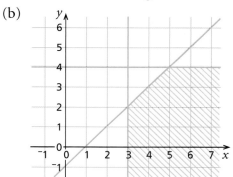

C4 Draw a set of axes, each numbered from 0 to 8.
Show clearly the single region that is satisfied by all of these inequalities

$x \geq 3$ $\qquad$ $y \leq 6$ $\qquad$ $y \leq x$

C5 On graph paper, draw and label axes with $^-3 \leq x \leq 4$ and $^-4 \leq y \leq 7$.
On the graph paper, draw the region which satisfies all of the following inequalities.

$x \geq ^-1$ $\qquad$ $y \leq 2$ $\qquad$ $x + y \leq 4$

Make sure that you clearly indicate the region that is your answer. WJEC

C6 On graph paper, draw and label axes with $^-2 \leq x \leq 6$ and $^-4 \leq y \leq 8$.
On the graph paper, shade the single region which satisfies both the inequalities

$y \geq 3x$ and $3x + 2y \leq 12$ OCR

C7 Do question A on sheet G167. AQA 2003 Specimen

D Boundaries

An inequality such as $y < 2x + 1$ does not include values for which $y = 2x + 1$.

So points on the boundary line are **not** included in the region defined by $y < 2x + 1$.

One way to show this is to use a **dotted** boundary line, as shown in the diagram.

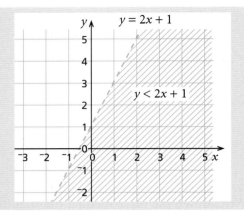

D1 Write an inequality for each shaded region.

(a)

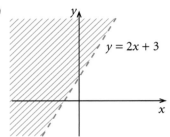

$y = 2x + 3$

(b)

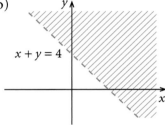

$x + y = 4$

D2 (a) Draw a set of axes, numbered from ⁻2 to 7 on both axes.
Draw the line $y = x + 1$ on these axes.

(b) On the same diagram shade the region that satisfies all the inequalities
$$x > 0, \quad y > 0 \quad \text{and} \quad y < x + 1.$$

D3 (a) Draw a set of axes, numbered from ⁻3 to 6 on both axes.

(b) Shade the part of your diagram that is in the region $4x + 3y > 12$.

(c) Similarly, shade the areas in   (i) $y - x > 2$ (ii) $y < \frac{1}{2}x + 1$.

(d) What inequalities are true for the region that is left unshaded in your diagram?

D4 (a) Draw axes on graph paper, numbered from ⁻2 to 6 on both axes.
On the graph draw the lines $x + y = 5$ and $y = 2x - 1$.

(b) Shade your graph to leave **unshaded** the region satisfying all the inequalities
$$x + y < 5, \quad y < 2x - 1 \quad \text{and} \quad y > 0.$$

(c) State all the points, with integer coordinates, that satisfy **all** the inequalities
in part (b).

OCR(MEG)

Test yourself

T1 (a) Factorise $5a - 10$.

(b) Solve the inequality $3x - 5 \leq 16$.

(c) On squared paper, draw axes with both x and y going from 0 to 6.
On the grid, shade the region where $x + y \leq 4$. AQA 2003 Specimen

T2 Use grid B on sheet G167 for this question.

Young Talent runs clubs for children on Saturdays.

Young Talent

Saturday clubs

(a) There are m children attending the music club.
There are d children attending the dance club.
The building cannot safely hold more than 35 children.

Write down an inequality which represents this information.

(b) On the grid, shade and label the region which shows this inequality. OCR

T3 Write down the three inequalities that define the shaded region.

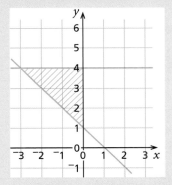

T4

The shaded region satisfies three inequalities.
One inequality is $y \leq 5$.

Write down the other two.

OCR

T5 Do question C on sheet G167. OCR

24 *Handling secondary data*

This work will help you interpret tabulated secondary data.

A *Drawing conclusions from data*

Many organisations and governments carry out surveys and collect data.
It helps them to plan new products, housing, medical care and so on.

Some of this data is published and may be used by other people.
For them it becomes secondary data, because they did not collect it themselves.

For example, this table gives the distances travelled by different
modes of transport in Great Britain.

Distance travelled: by mode

Billion passenger kilometres

	1961	1971	1981	1991	1995
Road					
Car and van[1]	157	313	394	584	594
Bus and coach	76	60	49	44	43
Pedal cycle	11	4	5	5	5
Motorcycle	11	4	10	6	4
All road	**255**	**381**	**458**	**639**	**646**
Rail[2]	39	36	34	38	37
Air[3]	1	2	3	5	6
All modes	**295**	**419**	**495**	**682**	**688**

Source: *Department of Transport*

[1] Includes taxis   [2] Data relates to financial years   [3] Includes Northern Ireland and Channel Islands

Units

A 'passenger kilometre' is one passenger travelling one kilometre.
So if a family of 3 travelled 20 km, this would be 60 passenger kilometres.

A1 What was the total distance travelled by road in 1961?

A2 Which number below gives the distance travelled by motorcycle in 1991?

6 000 000 km		6 000 km	
	6 km		6 000 000 000 km

Comparing

A3 Which modes of transport were used less in 1995 than in 1961?

A4 Can you say for certain that more people were flying in 1995 than in 1991?
Explain your answer carefully.

A5 (a) There were about 50 million people in Great Britain in 1961.
Calculate, approximately, the average distance travelled by a person in that year.
(A billion is 1000 million.)

(b) Repeat for 1995, assuming the population was about 60 million.

A6 Between 1961 and 1995, the distance travelled by rail didn't change much.
But travel as a whole increased.

(a) What percentage of all travel was by rail in

(i) 1961   (ii) 1995

(b) What happened to the percentage between 1961 and 1995?

A7 Between 1961 and 1995, what happened to the percentage of travel that was by road?

A8 Between 1961 and 1995, what happened to the percentage of
road travel that was by bus or coach?

A9 The data in the table came from
a survey by the United Nations.

'Brazil is better off for doctors
than either Egypt or Kenya.'

Country	Population (millions)	Number of doctors
Brazil	153.3	13 030
Egypt	54.7	10 010
Kenya	25.9	2330

Would you agree with this?
If not, how would you compare the three countries?

A10 Here is part of the British Crime Survey for England and Wales, giving
estimates about vandalism.

Vandalism, British Crime Survey estimates: England and Wales

per 10 000 households

Year	1981	1993	1995
Cases of criminal damage (vandalism)	1481	1638	1614

(a) What do these figures tell you about vandalism between 1981 and 1995?

(b) What other information would you need in order to find the number of cases of
vandalism in 1995?

B *Percentages from two-way tables*

This two-way table shows gender and age breakdowns for the population of a small town.

Many different percentages can be found using this data. Examples are given below.

	Age 0–59	Age 60+	Total
Males	358	82	440
Females	309	112	421
Total	667	194	861

What percentage of the whole population are male?

The whole population is 861, of whom 440 are male.

$\frac{440}{861} = 0.511$ (to 3 d.p.), so the percentage who are males is 51.1% (to 1 d.p.).

What percentage of the females are aged 60+?

The number of females is 421, of whom 112 are aged 60+.

$\frac{112}{421} = 0.266$ (to 3 d.p.), so the percentage of the females who are aged 60+ is 26.6% (to 1 d.p.).

What percentage of the people aged 60+ are male?

The number of people aged 60+ is 194, of whom 82 are male.

$\frac{82}{194} = 0.423$ (to 3 d.p.), so the percentage of the 60+ group who are male is 42.3% (to 1 d.p.).

B1 What percentage of the whole population are females aged 60+?

B2 What percentage of the people aged 0–59 are female?

B3 What percentage of the males are aged 0–59?

B4 What percentage of the whole population are aged 60+?

B5 This table gives information about the people who work for a company.

	Full-time	Part-time
Men	42	28
Women	33	57

Use the data to compare the proportions of part-time workers among the men and among the women.

B6 A researcher examined data relating to car accidents.
She found that in only 25% of accidents the driver had been drinking and in 75% the driver had not been drinking.

She drew the conclusion that driving was more dangerous if the driver had not been drinking.
Was she correct? If not, why not?

C *Getting the story right*

Here is some data published by the Home Office.

Selected offences recorded by the police: England and Wales

			thousands
	1971	1982	1983
Fraud and forgery	99.8	123.1	121.8
Criminal damage (vandalism)	27.0	417.8[a]	443.3
Other offences	5.6	3.8[b]	8.7[b]

Source: *Criminal Statistics, Home Office*

[a] Before 1982 vandalism causing less than £20 of damage was not recorded.

[b] The offence of 'abstracting electricity', of which there were 5688 in 1983, was included among 'other offences' in 1971 and 'theft' in 1982 and 1983.

This data was used for a newspaper article. Here is part of the article.

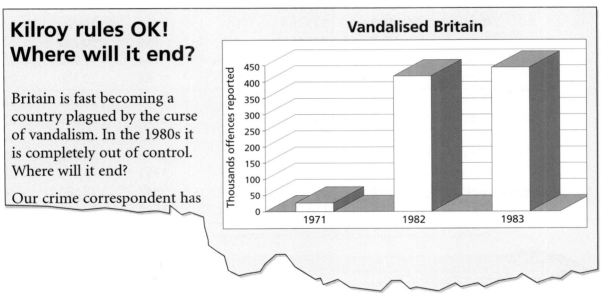

Kilroy rules OK! Where will it end?

Britain is fast becoming a country plagued by the curse of vandalism. In the 1980s it is completely out of control. Where will it end?

Our crime correspondent has

Vandalised Britain

C1 Does the article give an accurate account of the Home Office data?

C2 'When writing a report you should give the source of the data.' Why do you think this is?

D Investigating a large data set

Data is collected in order to answer questions.
But some large collections of data suggest interesting questions to ask.

This data about countries was collected by the United Nations.

	Population (million)	Area (square km)	Literacy[1] (%)	Doctors per million[2]	Main roads[3] (km)	Calories per person[4]
Information on twenty-one countries of the World (from South						
Argentina	33.8	2 780 400	94	2506	36 930	3075
Brazil	146.9	8 511 965	80	85	15 050	2723
Chile	15.2	756 626	94	760	10 300	2481
Columbia	34.0	1 138 914	88	778	25 600	2492
Ecuador	11.0	283 561	87	1082	6400	2410
Peru	22.5	1 285 216	81	856	7460	1890
Venezuela	20.8	912 050	90	1282	35 110	2741
Algeria	26.7	2 381 741	50	380	23 700	2989
Egypt	56.5	1 001 449	46	183	17 400	3318
Ethiopia	51.9	1 133 886	24	11	18 490	1694
Nigeria	105.3	923 768	51	108	29 700	2147
South Africa	39.7	1 221 037	76	656	50 730	3158
Tanzania	28.1	883 749	90	44	17 740	2181
Zaire	41.2	2 344 858	72	64	20 700	2094
France	57.4	551 500	99	2485	28 500	3618
Germany	81.2	356 733	100	3356	53 200	3500
Italy	57.1	301 268	97	4267	45 800	3484
Netherlands	15.3	40 844	100	2517	2060	3024
Poland	38.6	323 250	99	1993	45 300	3351
Spain	39.2	505 992	97	3547	18 500	3494
UK	58.3	244 100	100	1615	12 500	3282

Sources: United Nations 38/39/40th Year Books, 1998 World Almanac

[1] Percentage of adults who can read and write several simple sentences   [2] Number of doctors for one million people
[4] Mean number of calories eaten daily by each person   [5] Gross Domestic Product per person – the
[7] Annual deaths of under one year old per 1000 live births   total value of goods produced and
services provided per person in a year

America, Africa and Europe)

GDP per person[5] ($)	Life expectancy[6] men	women	Infant deaths[7]	Primary schools (thousands)
6912	65	73	25	21.7
2238	62	68	63	206.5
3030	68	75	17	8.6
1300	63	69	40	41.0
1142	63	68	63	15.0
1991	63	67	110	26.3
2994	67	73	23	15.8
1743	62	63	74	13.1
746	58	60	43	15.9
52	42	46	137	8.4
256	49	52	105	35.4
2882	58	64	72	–
98	51	55	106	10.4
93	50	54	83	10.8
23 149	73	81	6	42.2
24 157	73	79	6	19.8
21 177	73	80	7	22.9
21 130	74	80	5	8.5
2356	67	75	12	17.9
14 697	73	80	6	19.3
18 182	72	78	6	24.1

[3] Total length of good (A or B) roads   – No data available
[6] Mean life-time in years

D1 How could you compare countries by how crowded they are?
Give examples of countries which are relatively uncrowded and examples of those which are relatively crowded.

D2 'Brazil is better off for main roads than the UK.'
Is this hypothesis true? Explain your answer.

*D3 Imagine that Ecuador and Peru were combined into a single country.
Are these statements correct? If not, say why not.
(a) There would be 13 860 km of main roads.
(b) There would be 1938 doctors per million people.
(c) Literacy would be 84%.

D4 'Larger countries have larger populations.'
Draw a scatter diagram to decide whether or not this hypothesis is true.

D5 Investigate one or more of these hypotheses.
A   Smaller countries have less roads.
B   People in rich countries live longer.
C   People live longer in countries with a greater proportion of doctors.
D   Richer nations have higher literacy rates.
E   Countries where people have less to eat have high infant death rates.

E *Using more than one table*

Sometimes you need to use more than one table to answer a question.

For example, these tables give you some information about the population of the UK from 1901 onwards.

UK population 1901–91: numbers in millions

Year	Males	Females	Total
1901	18.49	19.75	38.24
1921	21.03	22.99	44.03
1941	23.22	24.95	48.22
1961	25.48	27.23	52.81
1981	27.10	28.74	56.35
1991	27.34	29.12	57.80

Source: *Pocket Britain in figures 1997*

UK population 1910–94: percentage in each age group

Year	0–14	15–34	35–64	65+
1910	32.3	35.6	27.2	4.9
1921	28.0	32.8	33.2	6.1
1941	21.0	32.4	37.5	9.2
1961	23.5	25.9	38.9	11.8
1981	20.6	30.2	34.2	15.0
1991	19.2	29.9	35.1	15.7
1994	19.5	29.0	35.8	17.7

Source: *Pocket Britain in figures 1997*

E1 These questions can be answered using one or both of the tables above.

> A What percentage of the UK population in 1991 were female?
>
> B How many people aged 15–34 were there in the UK in 1921?
>
> C What percentage of the UK population in 1941 were under 65?
>
> D How many people aged 65 or over were there in the UK in 1991?

(a) Answer each question.

(b) Which questions did you need both tables to answer?

E2 How many people aged over 34 were there in the UK in 1921?

E3 Can you use these tables to find out the number of males aged 65 and above in the UK in 1961? If not, why not?

Titanic

On 5 April 1912, the ocean liner *Titanic* hit an iceberg and many people died.

The number of people who died that night has never been known exactly but figures were published in a British Board of Trade enquiry report. These tables are based on data available on a website.

Males and females

Class	Numbers on board			Number of deaths		
	Male	Female	Both	Male	Female	Both
First class	180	145	325	118	4	122
Second class	179	106	285	154	13	167
Steerage	510	196	706	422	106	528
Crew	862	23	885	670	3	673
Total	1731	470	2201	1364	126	1490

Adults and children

Class	Numbers on board			Number of deaths		
	Adult	Child	Both	Adult	Child	Both
First class	319	6	325	122	0	122
Second class	261	24	285	167	0	167
Steerage	627	79	706	476	52	528
Crew	885	0	885	673	0	673
Total	2092	109	2201	1438	52	1490

E4 Use the data to answer these questions.

(a) (i) How many people were on board?

(ii) How many died?

(b) (i) How many children were on board?

(ii) Roughly, what percentage of the children died?

(c) Roughly what fraction of the men on board were crew?

(d) What percentage of the crew died?

(e) What percentage of children in steerage class died?

E5 Use the data to investigate one or more of these hypotheses.

A 'Women and children were placed in lifeboats first so were more likely to survive.'

B 'First class passengers were more likely to survive than other people on board.'

C 'Loss of life was greatest amongst the crew.'

F *Taking A-level mathematics – girls and boys*

The data on these two pages is from a secondary school in England.

All the students in a year 11 group are shown.
The table gives their sex, their GCSE mathematics grade and whether
they chose AS/A-level mathematics in the sixth form (Y = yes).

To be qualified to do AS/A-level mathematics, a student needs to have achieved
at least a grade C at GCSE.

- What conclusions can you draw from this data?

Reference number	Sex	GCSE mathematics	A level mathematics
1	M	A	
2	M	E	
3	F	C	
4	F	E	
5	M	A	Y
6	M	B	
7	M	C	
8	F	B	
9	F	B	
10	F	B	
11	M	B	
12	F	A	Y
13	F	B	
14	F	A*	
15	M	A	Y
16	M	C	
17	M	C	
18	M	C	Y
19	M	G	
20	M	A	Y
21	M	D	
22	M	D	
23	M	A*	Y
24	M	E	
25	M	D	

Reference number	Sex	GCSE mathematics	A level mathematics
26	M	F	
27	F	B	
28	F	B	
29	F	B	
30	F	A*	Y
31	F	A	
32	F	E	
33	M	D	
34	M	B	
35	M	E	
36	F	A*	Y
37	F	B	Y
38	F	F	
39	M	B	Y
40	M	D	
41	M	B	
42	M	C	
43	F	A	
44	F	B	
45	M	C	
46	M		
47	M	C	
48	F	B	
49	M	E	
50	M	C	

Reference number	Sex	GCSE mathematics	A level mathematics
51	M	C	
52	M	C	
53	F	C	
54	F	C	
55	F	A	
56	F	C	
57	F	B	
58	M	D	
59	M	C	
60	M	A*	Y
61	M	F	
62	M	F	
63	M	D	
64	M	A	Y
65	M	B	
66	F	D	
67	M	D	
68	F	E	
69	F	B	
70	F	D	
71	M	F	
72	M	C	
73	M	E	
74	M	E	
75	F	B	

Reference number	Sex	GCSE mathematics	A level mathematics
76	F	A	
77	F	C	
78	F	B	
79	F	B	
80	F	C	
81	F	B	
82	F	B	
83	F	D	
84	F	C	
85	F	A	
86	F	C	
87	F	A	
88	F	A	
89	M	B	
90	F	A	Y
91	M	A*	Y
92	F	D	
93	F	D	
94	F	B	
95	F	E	
96	F	E	
97	F	B	
98	F	B	
99	F	B	
100	F	C	

No.	Sex	Code	Y		No.	Sex	Code	Y		No.	Sex	Code	Y		No.	Sex	Code	Y
101	F	C			147	F	A			193	M	B	Y		239	F	D	
102	F	C			148	F	E			194	M	A			240	M	C	
103	F	B			149	F	B			195	F	E			241	F	A	Y
104	F	A	Y		150	F	B			196	M	G			242	F	B	
105	F	E			151	F	B			197	F	C			243	F	E	
106	M	B			152	F	E			198	F	C			244	F	A	Y
107	M	A	Y		153	M	E			199	M	A			245	F	A	
108	M	C			154	F	E			200	M	B			246	M	B	
109	F	E			155	F	C			201	F	C			247	F	B	
110	M	E			156	F	D			202	F	A	Y		248	F	C	
111	M	B			157	F	C			203	M	C			249	F	D	
112	M	C			158	F	G			204	M	A			250	F	E	
113	M	E			159	F	A*	Y		205	M	B			251	F	C	
114	M	B	Y		160	F	D			206	M	E			252	M	D	
115	M	A*	Y		161	F	D			207	M	A			253	M	D	
116	M	E			162	F	A	Y		208	M	C			254	F	F	
117	M	E			163	F	D			209	M	A	Y		255	M	F	
118	M	D			164	M	D			210	M				256	F	B	Y
119	M	B	Y		165	F	B			211	F	B			257	M	D	
120	M	B	Y		166	F	D			212	F	B			258	F	D	
121	M	F			167	M	F			213	F	E			259	F	B	
122	M	C			168	M	E			214	M	B	Y		260	F	E	
123	M	F			169	M	A	Y		215	F	A			261	M		
124	F	D			170	F	G			216	F	B			262	M	U	
125	F	C			171	F	A	Y		217	F	A			263	M	D	
126	F	A*			172	F	C			218	F	C			264	M	D	
127	F	E			173	F	B			219	F	B	Y		265	M		
128	M	D			174	F	F			220	F	D			266	M	D	
129	F	B			175	F	B			221	F	B			267	M	G	
130	F	B			176	F	C			222	F	D			268	M	D	
131	F	D			177	F	C			223	F	C			269	F	B	
132	F	B			178	F	A	Y		224	F	C			270	M	D	
133	F	A*			179	F	F			225	F	C			271	M	A	Y
134	F	E			180	M	C			226	M	C			272	M	B	Y
135	F	A			181	M	C			227	M	B	Y		273	F	E	
136	M	D			182	M	B			228	M	A	Y		274	M	B	
137	M	A	Y		183	M	D			229	M	D			275	M	D	
138	M	A	Y		184	M	C			230	M	C			276	M	E	
139	M	A*			185	M	D			231	F	E			277	M	F	
140	M	A			186	M	A*	Y		232	M	E			278	F	A	
141	F	C			187	M	B			233	F	A	Y		279	F	A	
142	M	D			188	F	D			234	F	F			280	F	A	Y
143	F	E			189	M	D			235	M	C			281	F	D	
144	M	A	Y		190	M	D			236	M	E			282	M	A	Y
145	M	D			191	M	B			237	F	A	Y		283	F	A	Y
146	M	B	Y		192	M	B			238	M	E						

25 Length, area and volume

You will need to use Pythagoras and trigonometry.
This work will help you calculate

◆ the length of an arc

◆ the area of a sector and a segment of a circle

◆ the surface area of a cylinder, a cone and a sphere

◆ the volume of a cylinder, a pyramid, a cone and a sphere

A Sector of a circle

Reminder When r is the radius of a circle,

$$\text{circumference } C = 2\pi r \qquad \text{area } A = \pi r^2$$

Part of a circle is called an **arc**.

A shape cut like this from a circle is called a **sector**.

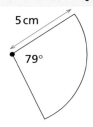

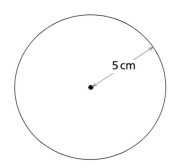

5 cm

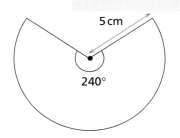

5 cm

240°

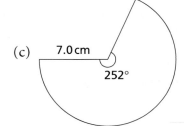

5 cm

79°

What is the circumference?
What is the area?

What fraction of the circle is this?
What is the length of the arc?
What is the area of the sector?

What fraction of the circle is this?
What is the length of the arc?
What is the area of the sector?

A1 For each of these, calculate

(i) the length of the arc to the nearest 0.1 cm

(ii) the area of the sector to the nearest 0.1 cm²

(a) 4.0 cm

28°

(b)

115° 6.0 cm

(c) 7.0 cm

252°

(d)

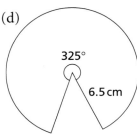

325°

6.5 cm

(e)

6.5 cm

85°

(f)

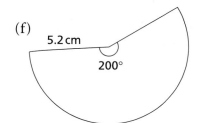

5.2 cm

200°

A2 Calculate

    (a) the area of a sector with radius 4.1 m and angle 48° (to the nearest 0.1 m²)

    (b) the length of its arc (to the nearest 0.1 m)

A3 Calculate

    (a) the area of a sector with radius 2.5 m and angle 220°

    (b) the length of its arc

A4 Calculate

    (a) the area of this sector

    (b) its total perimeter

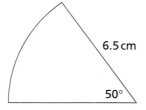

A5 These are the dimensions of the fabric part of this fan.

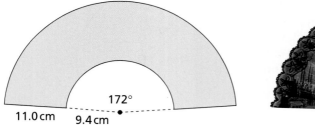 

    What is the area of the fabric part to the nearest cm²?

*A6 Find the radius of a circle that has area 100 cm².

*A7 Find the radius of a sector that has area 50 cm² and angle 125°.

*A8 Find, to the nearest degree, the angle of a sector that has area 35 cm² and radius 5 cm.

*A9 A fitter bends a 50 cm length of curtain track into an arc of a circle.
The track has turned through 48°.
What is the radius of the arc to the nearest cm?

B *Segment of a circle*

A **segment** of a circle is part of the circle cut off
by one straight line.

   • How can you find the area
of this segment?

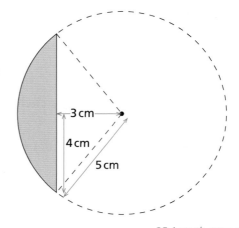

B1 Take the measurements you need from this diagram to find these to the nearest 0.1 cm².

(a) The area of sector OABC

(b) The area of triangle OAC

(c) The area of segment ABC

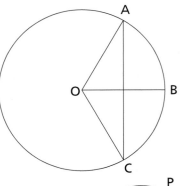

B2 (a) Calculate the area of sector OPQR.

(b) Use trigonometry to calculate lengths PX, XR and OX.

(c) Calculate the area of triangle OPR.

(d) Calculate the area of segment PQR.

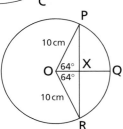

B3 Calculate the area of the shaded segment. You will first need to use trigonometry and Pythagoras to find some extra measurements. Mark them on a copy of the diagram and show all your working.

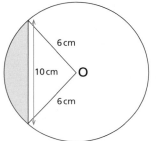

B4 This earring design is based on an equilateral triangle. The centres of the arcs are at the vertices. Find the shaded area to the nearest 0.1 mm².

B5 A segment can be more than half of a circle, like these. Devise methods to find their areas.

(a)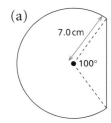

(b)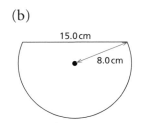

(c)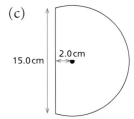

*B6 (a) This doorway has an arc of a circle at the top.

O is the centre of the circle.

What is the total area of the doorway, to the nearest 0.1 m²?

(b) What is the height of the middle of the doorway?

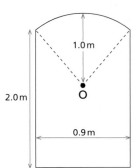

C Cylinder

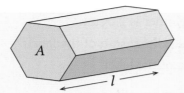

You can think of a cylinder as a special kind of prism.

C1 A cylinder has a cross-section with radius 4.0 cm, and length 10.0 cm.
   (a) Find the area of its cross-section to the nearest $0.1\,cm^2$.
   (b) Find its volume to the nearest cm^3, treating it as a prism.

C2 A cylinder has radius r and length l.
   (a) Write an expression for the area of its cross-section.
   (b) Write an expression for its volume, treating it as a prism.

C3 Check the expression you wrote for C2(b) then use it as
   a formula to find the volume of each of these cylinders.
   (a) Radius 3.0 cm   (b) Radius 4.2 cm   (c) Radius 3.0 m
        Length 8.5 cm   Length 9.0 cm   Length 4.7 m

C4 A tank to hold petroleum has these dimensions.
   How many litres of petroleum will it hold?
   (There are 1000 litres in a cubic metre.)

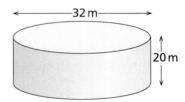

C5 A cylinder has volume $210\,cm^3$ and length 6.0 cm. Find its radius.

C6 A cylinder has volume $150\,cm^3$ and radius 2.8 cm. Find its length.

C7 Assuming that this slice of cake is a sector of a cylinder
   (like a sector of a circle), find its volume.

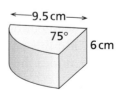

C8 A measuring cylinder for a science lab is being made.
   The radius on the inside is 1.50 cm.
   It needs to have a graduation mark for each cm^3 of liquid poured in.
   How far apart should these graduation marks be?

C9 A piece of copper wire has a **diameter** of 2 mm and is 1 m long.
   (a) Treating it as a cylinder, calculate its volume.
   (b) Given that copper has a density of 8.94 grams per cm^3, find its mass.

C10 This is a label that just fits round a cylinder.
The length of the cylinder is 8.0 cm.
The radius of the cylinder is 2.0 cm.

    (a) What is the other dimension of the label?

    (b) What is the area of the label?

C11 This label just fits round a cylinder.
The length of the cylinder is l.
The radius of the cylinder is r.

    (a) Write an expression (using π and r) for
        the other dimension of the label.

    (b) Write an expression for the area of the label.

C12 Check the expression you wrote for C11(b) then use it as a formula
to find the area of the curved surface of each of these cylinders.

    (a) Radius 3.0 cm   (b) Radius 7.6 cm   (c) Radius 18 m
        Length 5.0 cm   Length 12.3 cm   Length 5.2 m

C13 A cylinder has length 6.0 cm and curved surface area of 130 cm^2. Find its radius.

C14 A cylinder has radius 9.0 cm and curved surface area of 420 cm^2. Find its length.

C15 A cylinder has length 7.0 cm and radius 4.5 cm.
Find its total surface area (including the area of the circular ends).

C16 A cylinder has length l and radius r.
Write an expression for its total surface area (including the area of the ends).

C17 Check the expression you wrote for C16 then use it as
a formula to find the total surface area of each of these cylinders.

    (a) Radius 2.0 cm   (b) Radius 3.6 cm   (c) Radius 42 m
        Length 5.0 cm   Length 8.2 cm   Length 27 m

*C18 The largest gates on the Thames barrier have these dimensions.
Each gate is a segment of a cylinder.
(It is rotated into this position to keep back a flood tide.
Normally it is under the water with its curved surface
downwards.)

    (a) Work out the area of the curved surface.

    (b) Work out the areas of the three flat faces.

    (c) The curved surface is made from sheet steel
        weighing 314 kg per m^2.
        The flat faces are made from thinner sheet steel
        weighing 275 kg per m^2.

        Find the total mass of the surface steel in tonnes.

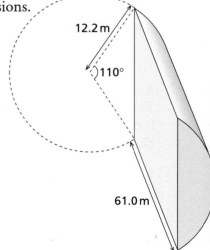

D *Pyramid and cone*

Draw this net accurately.

Cut, fold and glue the net to make
a **square-based pyramid**.

This is a **skew pyramid**: it does not have
a vertex over the centre of its base.

Work in groups of four.

- Can some of these pyramids be
  fitted together to make a cube?

- How many make a cube?

- What is the volume of the cube?

- What is the volume of the pyramid?

Now make other pyramids by putting together your models.

Record the volume, base area and perpendicular height in a table like the one below.

Pyramid	Volume	Base area	Perpendicular height
One pyramid			
Pyramid (with a rectangular base) made with two of the models			
Pyramid made with four of the models			

What do your results suggest about a formula for the volume of a pyramid?

D1 Here is a cube of side a cm. Its centre is joined to each vertex.
Then the cube is split into six identical pyramids (only four shown).

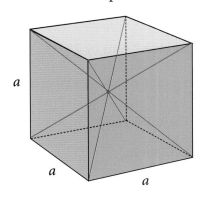
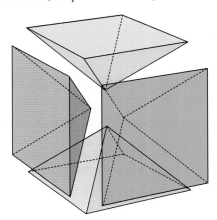

Write an expression for (a) the volume of one pyramid, (b) its base area, (c) its height.
Do these confirm the formula you found from the activity above?

The volume V of a pyramid is $\frac{1}{3}$ base area × height = $\frac{1}{3}Ah$

where A is the area of the base and h is the perpendicular height.

You have seen that the formula works for some pyramids with
a square or rectangular base. In fact it works for any shape base.

It works for a **skew pyramid** ...

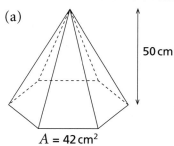

... and for one with a vertex above
the centre of its base (a **right pyramid**).

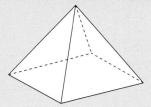

D2 Calculate the volume of each of these pyramids (to the nearest cm³).

(a)

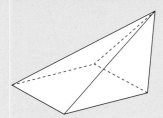

50 cm

$A = 42\,cm^2$

(b)

6.0 cm   6.0 cm

6.5 cm

(c)

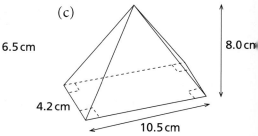

4.2 cm

10.5 cm

8.0 cm

D3 A pyramid-shaped hopper
needs to hold 50 m³ of grain.
It must be 6.0 m high.
What must this rectangular area be?

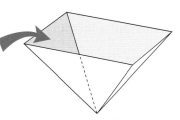

D4 When built, the Great Pyramid of Cheops had a square base 229 m by 229 m.
It was 146 m high.
Assume 95% of the pyramid to be stone (the rest being tunnels and chambers)
and that the density of the stone is 2.3 tonnes per m³.

Calculate an estimate of the mass of the stone in the Great Pyramid.

D5 An architect tries out different designs for a spire on a building.
She wants it to be a square-based pyramid.

Design P has a volume of 22.5 m³.

Design Q has the same base as P but twice the height.

(a) What is the volume of design Q?

Design R has the same height as P but the edges of the base are twice as long.

(b) What is the volume of design R?

D6 A pyramid has a base that is a regular nonagon with sides 6 cm.

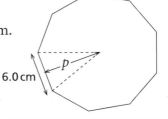

(a) Use trigonometry to find length p.

(b) Find the total area of the nonagon-shaped base.

(c) Given that the pyramid's height is 11 cm, find its volume.

6.0 cm

A regular polygon with very many sides starts to look like a circle …

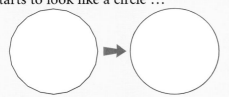

… so you can think of a cone as approximately a pyramid with a many-sided polygon for its base.

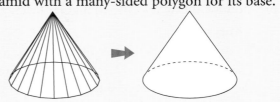

D7 A cone has base radius 4.0 cm and perpendicular height 10.0 cm.

(a) What is the area of its base?

(b) Treating this cone as you would a pyramid, what is its volume to the nearest cm³?

D8 Find the volume of a cone with height 8.9 cm and base radius 3.7 cm.

D9 (a) Write an expression for the area of a cone's base with radius r.

(b) Using your answer to (a), write an expression for the volume of the cone, given that its height is h.

D10 Find the total area of the four triangular faces of this square-based pyramid.

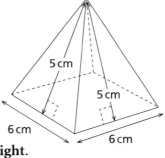

5 cm

5 cm

6 cm

6 cm

We call the length 5 cm in question D10 the **slant height**.

D11 A pyramid has a regular polygon for its base. Each triangular face has a base length of 2 cm and a slant height of 8 cm.

(a) What is the area of one of the triangular faces?

The base is a decagon, so there are 10 triangular faces.

(b) What is the total area of all the triangular faces?

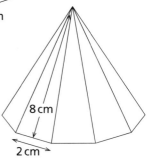

8 cm

2 cm

D12 A pyramid has a regular polygon for its base. Each triangular face has a base length of b cm and a slant height of l cm.

(a) Write an expression for the area of one of the triangular faces.

The base is an n-sided polygon, so there are n triangular faces.

(b) Write an expression for the total area of all the triangular faces.

In question D12 you should have found that the total area of the triangular faces is

$$\frac{n \times b \times l}{2} \quad \text{or} \quad \frac{nbl}{2}$$

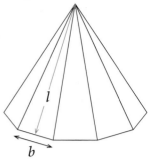

An expression for the perimeter of the polygonal base is

nb (the number of edges multiplied by the base length)

If you call the perimeter p, you can replace nb by p in the expression for the area of the triangular faces, giving

$$\text{Area} = \frac{pl}{2}$$

D13 Use this formula to work out the total area (to the nearest cm²) of the triangular faces of these regular polygonal pyramids.

(a) Base perimeter 20 cm   (b) Base perimeter 35 cm   (c) Base perimeter 18.8 cm
    Slant height 7 cm   Slant height 12 cm   Slant height 8.4 cm

D14 The total area of the triangular faces of a regular polygonal pyramid is 720 cm².
Its slant height is 18 cm.
Work out the perimeter of its base.

D15 A cone has a base radius of 25.0 cm.

(a) Work out the perimeter of the base (to the nearest 0.1 cm).

The slant height is 40.0 cm.

(b) Treating the cone as if it was a regular polygonal pyramid, work out the area of its curved surface, to the nearest cm².

D16 A cone has a base radius r.

(a) Write an expression (using π) for the perimeter of the base.

The slant height is l.

(b) Treating the cone as if it was a regular polygonal pyramid, write an expression for the area of its curved surface.

D17 Check the expression you wrote for D16(b) then use it as a formula to find the area of the curved surface for each of these cones.

(a) Base radius 5.0 cm   (b) Base radius 7.2 cm   (c) Base radius 15.9 cm
    Slant height 7.0 cm   Slant height 10.5 cm   Slant height 19.8 cm

D18 The curved surface of a cone has area 110 cm².
The radius of its base is 5.0 cm.
Find its slant height to the nearest 0.1 cm.

D19 The curved surface of a cone has area 370 cm².
Its slant height is 15.0 cm.
Find the radius of its base to the nearest 0.1 cm.

D20 This cone has a base radius of 5 cm and a **perpendicular** height of 12 cm.

(a) Use Pythagoras to find the slant height.

(b) Find the area of the curved surface of the cone.

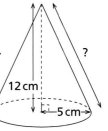

You have seen that the formula for the area of the **curved** surface of a cone is $\pi r l$, where r is the radius of the base and l is the slant height.

D21 (a) Write the formula for the total surface area of a cone, including its base.

(b) Use brackets to simplify this formula.

D22 Use the formula from D21(b) to work out the total surface area of these cones.

(a) Base radius 4.0 cm
    Slant height 11.0 cm

(b) Base radius 5.4 cm
    Slant height 8.9 cm

(c) Base radius 3.2 m
    Slant height 8.5 m

A **frustum** of a cone looks like this. Its base and its top are both circles and are parallel to one another.

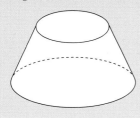

You can think of it as a large cone that has had a smaller cone cut off.

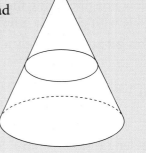

D23 This elephant stand is a frustum of a cone. Find its volume.

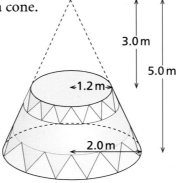

*D24

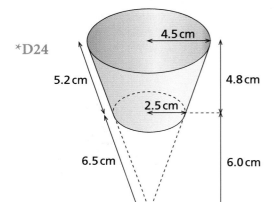

This plastic ice cream tub is a frustum of a cone.

(a) Calculate the volume of ice cream it can contain.

(b) Calculate the total area of the plastic for the tub and the lid. (Ignore the extra plastic needed at the edge of the lid.)

E *Sphere*

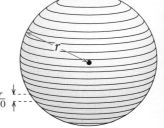

Here is a way to find the formula for the surface area of a sphere.

This sphere has radius r. It has been divided into 20 slices, each $\frac{r}{10}$ thick.

Concentrate on the surface of one of the slices.

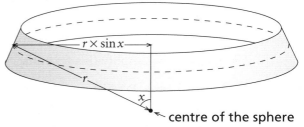

The radius of the dotted circle is $r \times \sin x$.
So its circumference is $2\pi r \times \sin x$.

The width, w, of the strip
is $\frac{r}{10} \div \sin x$.

centre of the sphere

Imagine cutting the curved strip into trapeziums.

Now imagine laying the trapeziums flat on the table with alternate ones turned upside down.

You get approximately a rectangle.

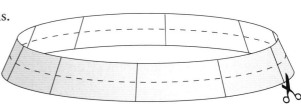

$\frac{r}{10} \div \sin x$

$2\pi r \times \sin x$

Its length is $2\pi r \times \sin x$ (the circumference of the dotted circle).
Its width is $\frac{r}{10} \div \sin x$ (the width of the curved strip).

So its area is $2\pi r \times \sin x \times \frac{r}{10} \div \sin x = \frac{\pi r^2}{5}$.

This approach works for all 20 slices, so the surface area of the sphere is approximately $20 \times \frac{\pi r^2}{5} = 4\pi r^2$.

The result is more accurate the thinner the slices used, and in fact the surface area of a sphere is exactly $4\pi r^2$.

Notice from the formula that the surface area of a sphere is four times the area of a circle with the same radius.
The construction of a tennis ball may help you see why.

Two pieces of fabric this shape wrap around each other to make the ball.	The total area of the pieces is roughly the total area of these four circles …	… which have the same radius as the ball.

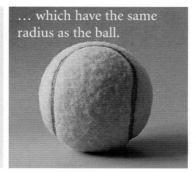

E1 Calculate the surface area of each of these spheres, to the nearest cm².

(a) Radius 10.0 cm   (b) Radius 7.8 cm   (c) Radius 15.6 cm

E2 The radius of the Earth is 6300 km.

(a) Calculate its surface area.

(b) Given that 70% of the Earth's surface is ocean,
what is the area of the Earth's oceans?

E3 A sphere has surface area 412 cm².
Calculate its radius to the nearest 0.1 cm.

You can think of a sphere as made up of many slender pyramids, each with
its vertex at the centre of the sphere and height r, the radius of the sphere.

So the volume of one pyramid is $\frac{1}{3}Ar$, where A the area of its base.

So the total volume of all the pyramids is

$\frac{1}{3}$ × the total area of the bases of all the pyramids × r

This equals $\frac{1}{3} \times 4\pi r^2 \times r$, because the total area of the bases is the surface area of the sphere.

Simplifying the expression, the volume of a sphere is $\frac{4}{3}\pi r^3$.

E4 Use the formula to find the volume of each of these spheres.

(a) Radius 5.0 cm   (b) Radius 8.8 cm   (c) Radius 14.7 cm

E5 A liquid fuel tank is a sphere with a radius of 80 cm.
How many litres of fuel can it contain?
(A litre is 1000 cm³.)

E6 A squash ball has a radius of 2.0 cm.
It is sold in a cubical box.
There is no extra space for it to move around in the box.
What percentage of the volume of the box is air?

E7 A tennis ball has radius 3.4 cm.
Some tennis balls are sold in threes in a cylindrical tube.
There is no space for them to move around in the tube.
What percentage of the volume of the tube is air?

*E8 An chemical engineer wants a spherical tank built to contain a million litres of liquid.
What radius will the tank need to have?

*E9 A Terry's chocolate orange is approximately a sphere of
radius 3.2 cm.
It separates into 20 pieces like this.

Calculate (a) the volume and (b) the surface area of one piece.

These commonly used formulas have been introduced in this unit.

For a **cylinder** with radius of cross-section r and length l,

$\quad\quad$ volume $= \pi r^2 l$ $\quad\quad\quad$ area of curved surface $= 2\pi r l$ $\quad\quad$ total surface area $= 2\pi r(l + r)$

For a **pyramid** with base area A and perpendicular height h,

$\quad\quad$ volume $= \frac{1}{3}Ah$

For a **cone** with base radius r, perpendicular height h and slant height l,

$\quad\quad$ volume $= \frac{1}{3}\pi r^2 h$ $\quad\quad\quad$ area of curved surface $= \pi r l$ $\quad\quad$ total surface area $= \pi r(l + r)$

For a **sphere** with radius r,

$\quad\quad$ volume $= \frac{4}{3}\pi r^3$ $\quad\quad\quad$ surface area $= 4\pi r^2$

Test yourself

Not drawn accurately

T1 The diagram shows a water butt.
The water butt is a cylinder with a base radius of 35 cm.
The height of the water butt is 130 cm.

Calculate the number of litres of water in the butt when it is full.
Give your answer to an appropriate degree of accuracy.

130 cm

35 cm $\quad$ AQA

T2 A sphere has a radius of 5.4 cm.
A cone has a height of 8 cm.
The volume of the sphere is equal to the volume of the cone.

Calculate the radius of the base of the cone.
Give your answer, in centimetres, correct to two significant figures. $\quad\quad$ Edexcel

T3 A metal sphere has a radius of 6 cm.

(a) Calculate the volume of the sphere. Leave your answer as a multiple of π.

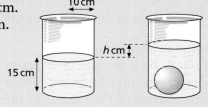

The diagram shows a cylinder of radius 10 cm.
It already contains water to a depth of 15 cm.
The metal sphere is placed in the water.

(b) Calculate the height, h cm,
that the water level rises. $\quad\quad$ OCR

T4 A party hat is made from card.
The hat is made in two parts: a cone on top of a ring.
The cone has a height of 20 cm and base radius of 7.5 cm.

The ring has an internal radius of 7.5 cm
and an external radius of 10 cm.

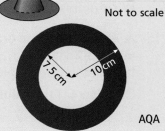

Not to scale

Calculate the area of the card used
in making the party hat.
Give your answer to an
appropriate degree of accuracy.

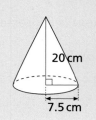

7.5 cm $\quad\quad$ AQA

Quadratic graphs

You should know how to

◆ plot graphs of functions like $y = 2x^2 + 3$

◆ rearrange simple formulas

◆ multiply out brackets like $(x - 2)(x + 3)$

This work will help you

◆ substitute into and interpret quadratic functions

◆ draw the graphs of quadratic functions and use them to solve simple problems

◆ use quadratic graphs and straight lines to solve related equations

A *Parabolic paths*

A1 A ball is made to bounce so that it clears a wall.
The diagram shows the path of the ball as it bounces over the wall.

The equation of the path of the ball for this bounce is $y = 1 - 4x^2$
(x and y are measured in metres).

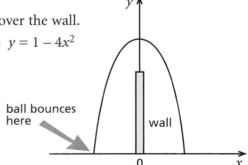

(a) (i) Find the value of y when $x = 0$.

   (ii) The wall is 80 centimetres high.
        By how much does the ball clear the wall?

(b) Find the two possible values of x when $y = 0$.
     What does this tell you?

(c) What is the height of the ball when
     it is halfway (measured horizontally)
     between where it bounces and the wall?

A2 A ball is kicked so that it clears a fence. The diagram shows the path of the ball.

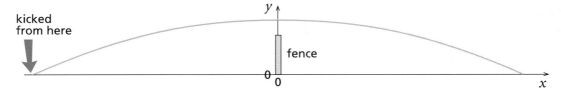

The equation of the path of the ball is $y = 4 - \dfrac{x^2}{144}$
(x and y are measured in metres).

(a) (i) Find the value of y when $x = 0$.

(ii) The fence is 2.5 metres high.
By how much does the ball clear the fence?

(b) How far from the fence is the ball kicked?

A3 A girl throws a stone from the top of a cliff, as in the diagram.

The equation of the path of the stone through the air
is $y = 50 - \dfrac{x^2}{2}$ (x and y are measured in metres).

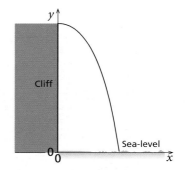

(a) What is the height of the cliff?

(b) How far from the bottom of the cliff
does the stone hit the sea?

(c) What is the value of y when x is 5?
What does this tell you?

(d) What is the value of y when x is 15?
Why is this information meaningless in this case?

*A4** A cannonball is fired from a cannon.
The equation of its path is $y = \dfrac{x}{2} - \dfrac{x^2}{80}$.

The cannonball just clears a wall which is
30 metres horizontally from the cannon,
and lands on the ground on the other side.

(a) How high is the wall?

(b) How far from the wall
does the cannonball land?

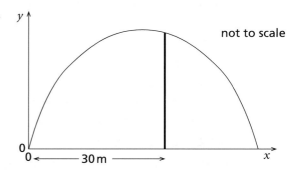

B *Plotting quadratic functions*

A **quadratic function** has an equation of the form $y = ax^2 + bx + c$.
For example, these are all quadratic functions: $y = 2x^2 - 3$, $y = {}^-5x^2 + 2x - 7$, $y = 8 - x^2$.

The graph of any quadratic function is called a **parabola**.

$y = x^2 + 1$

Solve $x^2 + 1 = 4$

Solve $x^2 + 1 = 8$

For each question, copy and complete the table and plot the points on suitable axes.

Join the points with a smooth curve then answer the questions about each graph.

B1 (a) Draw the graph of $y = x^2 - 2$ for $^-2 \leq x \leq 2$.

x	-2	-1	0	1	2
$y = x^2 - 2$	2			-1	

    (b) What is the equation of the line of symmetry of this graph?

    (c) Use the graph to solve $x^2 - 2 = 0$ to one decimal place.
       (That is, find the values of x for which $y = 0$.)

B2 (a) Draw the graph of $y = 2x^2 + 1$ for $^-2 \leq x \leq 2$.

x	-2	-1	0	1	2
x^2	4			1	
$2x^2$	8			2	
$y = 2x^2 + 1$	9			3	

    (b) What is the minimum value of y for this graph?
       How can you be sure your answer is right without plotting more points?

    (c) Use the graph to solve the equation $2x^2 + 1 = 4$ to one decimal place.

    (d) Explain why there is no solution to the equation $2x^2 + 1 = 0$.

B3 (a) Copy and complete the table below and use it to draw $y = x(5 - x)$ for $^-1 \leq x \leq 5$.

x	-1	0	1	2	3	4	5	6
$5 - x$			4	3	2			
$y = x(5 - x)$			4	6	6			

    (b) At what value of x is y a maximum?

    (c) (i) From your graph, what is the maximum value of y?

        (ii) Check your answer by substituting into the equation.

    (d) Use your graph to solve the equation $x(5 - x) = 5$.

C *Parabolas and solving quadratic equations*

To plot the graph of an equation like $y = (x-1)(x+3)$ you need first to draw up a table of values of x and y.

x	$(x-1)(x+3)$	y
$^-4$	$^-5 \times \ ^-1$	5
$^-3$	$^-4 \times \ 0$	0
$^-2$	$^-3 \times \ 1$	$^-3$
$^-1$	$^-2 \times \ 2$	$^-4$
0	$^-1 \times \ 3$	$^-3$
1	$0 \times \ 4$	0
2	$1 \times \ 5$	5

The expression $(x-1)(x+3)$ has the value 0 when either $(x-1)$ is 0 or $(x+3)$ is 0, that is when either x is 1 or x is $^-3$.

So if $(x-1)(x+3) = 0$, then either $x = 1$ or $x = \ ^-3$.

So when $x = 1$ and when $x = \ ^-3$, the value of y is 0.
The graph will cross the x-axis at $x = 1$ and $x = \ ^-3$.

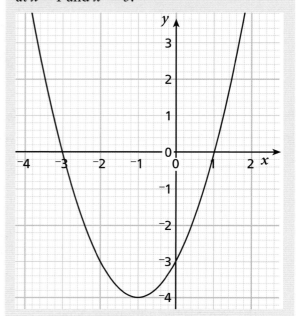

C1 Write down the values of x where the graphs of each of these functions cross the x-axis. (You do not need to draw the graphs.)

   (a) $y = (x-4)(x+5)$ (b) $y = (x+2)(x+4)$ (c) $y = (x-5)(x-6)$

C2 For the function $y = x(x-2)$…

   (a) What two values of x make $x(x-2) = 0$?

   (b) Write down the coordinates of the points where $y = x(x-2)$ crosses the x-axis.

C3 (a) How many values of x make $(x+2)^2 = 0$?

   (b) One of these sketches shows $y = (x+2)^2$. Which one?

   Explain your answer.

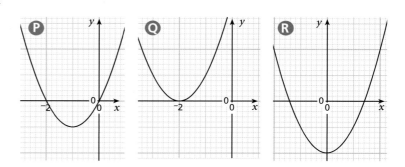

C4 For each of the following functions …
- factorise the right-hand side
- write down the values of x where the graph of the function crosses the x-axis

(a) $y = x^2 + 3x + 2$ (b) $y = x^2 - 5x$ (c) $y = x^2 - 5x + 6$

(d) $y = x^2 - 4x + 4$ (e) $y = x^2 - x - 6$ (f) $y = x^2 - 1$

C5 The diagram shows part of the graph of $y = x^2 - 2x + 2$.

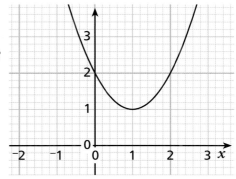

(a) How many times does the graph cross the x-axis?

(b) What does this tell you about the solutions of $x^2 - 2x + 2 = 0$?

(c) Use the graph to write down the solution to the equation $x^2 - 2x + 2 = 1$.

C6 The diagram below shows the graphs of $y = x^2 + x - 2$, $y = x^2 - 6x + 9$ and $y = x^2 - 3x + 3$.

(a) By substituting $x = 0$ (or any other value of x) into each equation, work out which graph corresponds to which equation.

(b) Write down the solutions to the equations $x^2 + x - 2 = 0$, $x^2 - 6x + 9 = 0$ and $x^2 - 3x + 3 = 0$.

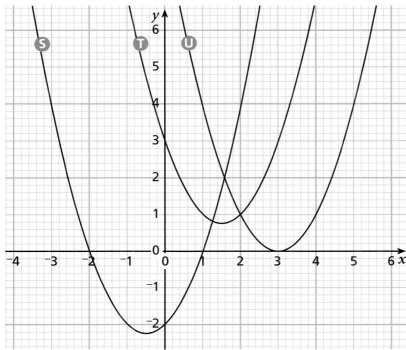

D *Quadratic models*

D1 The area of a circle is given roughly by the formula $A = 3r^2$.

    (a) Make a table for $0 \leq r \leq 3$ and graph this function.

    (b) Use your graph to write down the rough area of a circle of radius 2.3 m.

    (c) If a circle has an area of $10\,\text{m}^2$, about what is its radius?

    (d) Check your answers to parts (b) and (c)
        using the accurate version of the formula.

D2 A stone is dropped from a block of flats.
The height of the stone above the ground is given by $s = 125 - 5t^2$.
s is the height of the stone in metres,
t is the time in seconds from the release of the stone.

    (a) Draw up a table of suitable values of t and s.

    (b) Use the table to graph the function.

    (c) Use your graph to find the height of the stone after 2.5 seconds.
        Check your answer by calculation.

    (d) (i) Use your graph to solve the equation $125 - 5t^2 = 75$.

         (ii) What is the meaning of your answer?

D3 A farmer has 20 metres of fencing.
He wishes to use it to form a rectangular
enclosure in the corner of a field, as in the diagram.

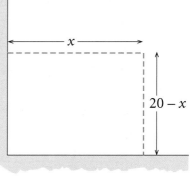

    (a) Write down an expression for the area, $A\,\text{m}^2$,
        enclosed by the fencing.

    (b) Plot the graph of A for values of x
        between 0 and 20.

    (c) For what values of x is the area $40\,\text{m}^2$?

    (d) What values of x give an enclosed area
        greater than $90\,\text{m}^2$?

    (e) What is the maximum area the farmer can enclose?
        What are the lengths of the fencing for this maximum area?

D4

Suppose that the farmer wishes to use 20 m of
fencing to form a rectangular enclosure against
one side of a field, as in the diagram.

Investigate the area that can be enclosed.

E *Further equations*

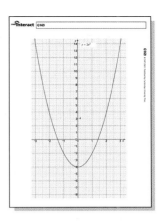

You need sheet G169.

On sheet G169 is the graph of $y = 2x^2 - 4$.

Draw and label suitable straight lines on the graph in order to solve

    (a) $2x^2 = 7$ (b) $2x^2 - 4 = x + 3$ (c) $2x^2 - 2x - 1 = 0$

You need another copy of sheet G169.

E1 Draw suitable lines on the sheet to solve these equations, giving answers to 1 d.p.

    (a) $2x^2 - 4 = 0$ (b) $2x^2 - 4 = x$ (c) $2x^2 - 4 = 2x$

E2 (a) Copy and complete this working to rearrange
       the equation $2x^2 - x - 2 = 0$ into the form $2x^2 - 4 = \ldots$

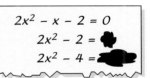

    (b) On sheet G169, draw a suitable line on the graph
       of $y = 2x^2 - 4$ to solve the equation $2x^2 - x - 2 = 0$.
       Write down the solutions to one decimal place.

E3 (a) Rearrange the equation $2x^2 + 2x - 5 = 0$ into the form $2x^2 - 4 = \ldots$

    (b) Draw a suitable line on sheet G169 to solve the
       equation $2x^2 + 2x - 5 = 0$, giving answers to 1 d.p.

E4 For each of these equations …

    • rearrange the equation into the form $2x^2 - 4 = \ldots$

    • write down the equation of the line you would draw to solve the equation

    You do not need to draw the lines or solve the equations.

    (a) $2x^2 = 3x + 1$ (b) $2x^2 + 2x = 3$ (c) $2x^2 + x + 1 = 0$

E5 If you are given the graph of $y = x^2 + x$, what line would
you need to draw to solve the equation

    (a) $x^2 + x = 2$ (b) $x^2 + x - 5 = 0$ (c) $x^2 + 2x - 5 = 0$ (d) $x^2 + 2x - 2 = 0$

E6 You need sheet G170.

    The graph of $y = x^2 - 2x - 4$ is drawn on the sheet.

    (a) On the same axes, draw the line with equation $y = 2x + 1$.

    (b) Hence solve the equation $x^2 - 2x - 4 = 2x + 1$.

    (c) Show that $x^2 - 2x - 4 = 2x + 1$ can be simplified to $x^2 - 4x - 5 = 0$.

    (d) You are going to use the graph of $y = x^2 - 2x - 4$ to solve the equation $x^2 - x - 8 = 0$.
       What is the equation of the line you would need to draw?

AQA(NEAB) 1998

E7 Suppose you have the graphs of $y = x^2 + 2x$ and $y = x - 1$.
Which of these equations has solutions where the two graphs intersect?

A $x^2 - x - 1 = 0$ B $x^2 + x + 1 = 0$ C $x^2 + 3x - 1 = 0$ D $x^2 + x = 1$

E8 The diagram on the opposite page shows $y = x^2 - 1$ and three straight lines.

(a) Write down the equation of each of the straight lines A, B and C.

(b) You can use one line and the curve $y = x^2 - 1$ to solve the equation $x^2 - x - 1 = 0$.

(i) Which line is it?

(ii) Use the graph to write down the solutions of the equation $x^2 - x - 1 = 0$ to 1 d.p.

(c) Use the appropriate line to solve each of these equations.

(i) $x^2 + x - 1 = 0$ (ii) $x^2 - 2x - 2 = 0$

Show all your working clearly.

E9 (a) On graph paper, draw and label axes with $^-3 \leq x \leq 3$ and $^-2 \leq y \leq 12$.
On your axes, draw accurately the graph of $y = x^2 - x$.

(b) Showing all your working, use your graph and suitable straight lines to solve

(i) $x^2 - x = 1$ (ii) $x^2 = x + 5$ (iii) $x^2 + x - 1 = 0$

E10 (a) Copy and complete the table of values for $y = x^2 - x - 1$.

x	$^-2$	$^-1$	0	1	2	3
y			$^-1$		1	

(b) On graph paper, using a scale where 2 cm stands for one unit,
draw the graph of $y = x^2 - x - 1$ for values of x from $^-2$ to 3.

(c) (i) Use your graph to solve the equation $x^2 - x - 1 = 1$.

(ii) Use your graph to solve the equation $x^2 - x - 1 = x$.

(iii) What happens if you try to solve the equation $x^2 - x - 1 = ^-2$?
What can you say about the equation $x^2 - x - 1 = ^-2$?

(d) The graph of $y = x^2 - x - 1$ and a graph of the form $y = mx + c$
can be used to solve the equation $x^2 - 4x + 2 = 0$.

Find the values of m and c needed.
(You do not have to draw the corresponding graph.)

*E11 (a) Copy and complete this working to rearrange
the equation $2x^2 + 4x - 3 = 0$ into the form $x^2 + x = \ldots$

(b) What line would you draw on the graph of $y = x^2 + x$
to solve the equation $2x^2 + 4x - 3 = 0$?

> $2x^2 + 4x - 3 = 0$
> $x^2 + 2x - \clubsuit = 0$
> $x^2 + x = \clubsuit - x$

*E12 Work out what line you would draw on the graph of $y = x^2 + x$
to solve the equation $2x^2 + x - 3 = 0$.

Graph for question E8

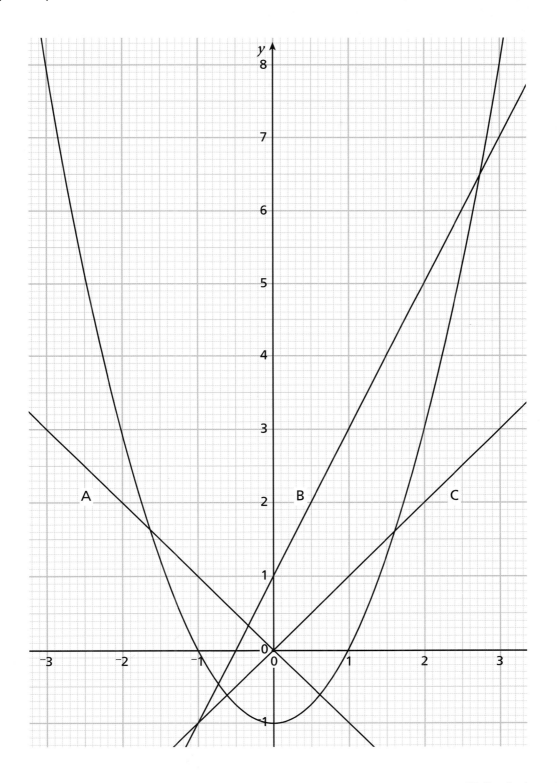

✖ **T1** (a) Complete the table which gives the values of $y = 2x^2 + 4x - 5$ for values of x ranging from $^-4$ to 3.

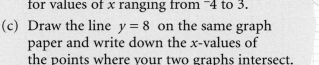

x	$^-4$	$^-3$	$^-2$	$^-1$	0	1	2	3
$y = 2x^2 + 4x - 5$	11	1		$^-7$	$^-5$	1		25

(b) On graph paper draw axes as shown. Draw the graph of $y = 2x^2 + 4x - 5$ for values of x ranging from $^-4$ to 3.

Go from $^-10$ to 30

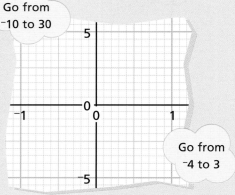

(c) Draw the line $y = 8$ on the same graph paper and write down the x-values of the points where your two graphs intersect.

(d) Write down the equation in x whose solutions are the x-values you found in (c).

Go from $^-4$ to 3

WJEC

T2 (a) Copy and complete the table of values for $y = 8x - x^2$.

x	0	1	2	3	4	5	6	7	8
y	0	7					12	7	0

(b) On graph paper, draw axes with $0 \le x \le 8$ and $0 \le y \le 16$. Draw the graph of $y = 8x - x^2$.

(c) Use your graph to solve the following equations.

    (i) $8x - x^2 = 8$ (ii) $x^2 - 8x + 3 = 0$ (iii) $6x - x^2 = 1$

T3 The graph paper for this question is A on sheet G171.

A rectangle has a perimeter of $12\,cm$ and an area of $7\,cm^2$.

Let the length of the rectangle be $x\,cm$.

$x\,cm$

(a) Show that $x^2 - 6x + 7 = 0$.

(b) (i) Complete the table of values for $y = x^2 - 6x + 7$.

x	0	1	2	3	4	5	6
y	7		$^-1$			2	

    (ii) Draw the graph of $y = x^2 - 6x + 7$.

(c) Use your graph to estimate the dimensions of the rectangle.   AQA(SEG) 1998

T4 This question is question B on sheet G171.

Review 6

1 (a) Find the equation of the line labelled *a*.

(b) Write down the three inequalities satisfied by points inside the region coloured green (including the boundaries).

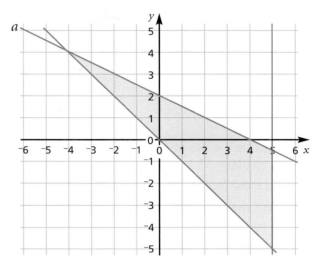

2 The table below shows the breakdown of a company's workforce by salary and by gender.

Salary (£1000s)	10–19.9	20–29.9	30–39.9	40+
Number of men	25	34	19	3
Number of women	40	28	17	5

Use this data to compare men's and women's salaries.

3 A vase is made from sheet metal in two pieces, as shown below.

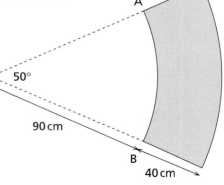

(a) Calculate the length of the arc AB, which is equal to the circumference of the base.

(b) Calculate the radius of the base.

(c) Calculate the radius of the top of the vase.

(d) Calculate, showing clearly all your working, the total area of sheet metal used to make the vase.

(e) The sheet metal weighs 6.2 kilograms per **square metre**. Calculate the weight of the vase, in kilograms.

4 The equations of graphs A and B are both of the form

$$y = kx^2$$

Find the value of k for each graph.

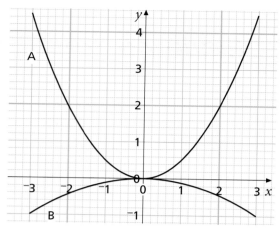

5 This is the net of a square-based pyramid. Calculate

(a) the length of each side of the base

(b) the total surface area of the pyramid

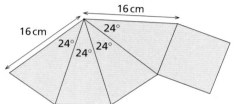

6 (a) Draw the graph of $y = 4 + x - \frac{1}{2}x^2$ for values of x from $^-3$ to 5.

(b) On the same axes, draw the line $y = x$.

(c) Use the graphs to estimate, to 1 d.p., the values of x for which $4 + x - \frac{1}{2}x^2 = x$.

(d) Simplify the equation $4 + x - \frac{1}{2}x^2 = x$ and solve it algebraically. Check that the result agrees with your answer to (c).

7 20 hectares of a farm yield 85 tonnes of a crop at harvest.

(a) What is the average yield in tonnes per hectare?

(b) If the average yield stays the same, how many hectares would need to be planted to yield 200 tonnes at harvest?

8 In this diagram a circle of radius 5 cm with centre O touches all three sides of an isosceles triangle ABC whose base BC = 14 cm. The circle touches the base at M.

(a) Calculate angle OBM and hence angle ABM.

(b) Calculate AM and hence the area of triangle ABC.

(c) What percentage of the area of the triangle ABC is the area of the circle? Give your answer to the nearest 1%.

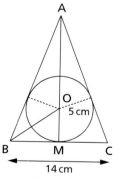

Challenge

Elaine has a journey to make.
If she travels at an average speed of 40 m.p.h. it will take 6 hours less than if she travels at an average speed of 25 m.p.h.
How long will the journey take if she travels at 50 m.p.h.?

You should be confident with

◆ the four operations on number fractions

◆ multiplying out brackets and factorising

This work will help you manipulate and simplify fractions with letters in them.

A Substituting and cancelling

Q:	$\dfrac{a^2}{b} \times \dfrac{b^2}{a}$	$\dfrac{ab^2}{ab}$	$a^2 \div ab$	$\dfrac{ab \times a^2b}{(ab)^2}$	$\dfrac{(ba)^2}{a^3b}$
A:	a	b	$\dfrac{b}{a}$	$\dfrac{a}{b}$	ab

A1 Find the value of the following when $p = 2$ and $q = 3$.

    (a) pq (b) $(p + q)^2$ (c) $4pq^2$ (d) $(pq)^2$ (e) $5p^2q$

A2 Find the value of the following when $p = \frac{1}{2}$ and $q = \frac{1}{3}$.

    (a) $\dfrac{1}{q}$ (b) $\dfrac{1}{p}$ (c) $\dfrac{1}{p} - \dfrac{1}{q}$ (d) $\dfrac{1}{p} - q$ (e) $\dfrac{p}{q}$

A3 Find the value of $\dfrac{a}{b} - \dfrac{b}{a}$ when $a = \frac{2}{5}$ and $b = {}^{-}2$.

A4 (a) If $x = {}^{-}4$, calculate the value of $3x^2$.

    (b) If $a = \frac{2}{3}$ and $b = \frac{2}{5}$, calculate the exact value of $a^2\left(1 - \dfrac{1}{b}\right)$.   AQA(NEAB) 1997

A5 If $p = 4$ and $q = 6$, calculate these (leaving your answers as fractions).

    (a) p^{-1} (b) q^{-2} (c) $\left(\dfrac{p}{q}\right)^{-1}$ (d) $\left(\dfrac{p}{q}\right)^{-2}$ (e) p^2q^{-2}

A6 Simplify these.

    (a) $t^3 \times t^5$ (b) $p^6 \div p^2$ (c) $\dfrac{a \times a^2}{a}$ AQA(NEAB) 1998

A7 Simplify these.

    (a) $\dfrac{p^2q^2}{pq}$ (b) $\dfrac{p^6q^4}{pq}$ (c) $p^6q^4 \div p^2q^5$ (d) $\dfrac{(p^3q^2)^2}{pq}$

A8 Simplify these.

    (a) $\dfrac{4pq^2}{pq}$ (b) $3p^3 \times 2pq^2$ (c) $(2ab^2)^{-3}$ (d) $(2p^2q^3)^2 \div p^4q^4$

A9 Simplify these as far as possible.

(a) $\dfrac{a^2b}{a} \times b^2$ 　　(b) $\dfrac{ab^4}{a^3} \times a^2$ 　　(c) $\dfrac{ab^2}{a^2} \times ab$ 　　(d) $\dfrac{a^4b^6}{a^2b^2} \times a^2b^4$

A10 Simplify as far as possible

(a) $\dfrac{b}{a} \times \dfrac{a^2}{b}$ 　　(b) $\dfrac{b^2}{a} \times \dfrac{a}{b^3}$ 　　(c) $\dfrac{ab}{a^2} \times \dfrac{a}{b}$ 　　(d) $\dfrac{2b}{a^2} \times \dfrac{a}{8b}$

B Roots

$6^2 = 36$ so $6 = \sqrt{36}$	$(ab)^2 = a^2b^2$ so $ab = \sqrt{a^2b^2}$	$(a^3b^2)^2 = a^6b^4$ so $a^3b^2 = \sqrt{a^6b^4}$	$\left(\dfrac{b^2}{a^3}\right)^2 = \dfrac{b^4}{a^6}$ so $\dfrac{b^2}{a^3} = \sqrt{\dfrac{b^4}{a^6}}$	$\sqrt{a^6b^4}$ $= \sqrt{a^6}\,\sqrt{b^4}$ $= a^3b^2$

B1 Find the value of each of these.

(a) $\sqrt{36 \times 25}$ 　　(b) $\sqrt{4900}$ 　　(c) $\sqrt{3^4 \times 2^6}$ 　　(d) $\sqrt{5^4 \times 4^2}$

B2 Find the value of each of these.

(a) $\sqrt{\dfrac{25}{64}}$ 　　(b) $\sqrt{\dfrac{49 \times 36}{25}}$ 　　(c) $\sqrt{\dfrac{2^6}{3^4}}$ 　　(d) $\sqrt{\dfrac{3^6}{2^{10}}}$

B3 Simplify.

(a) $\sqrt{(p^4q^6)}$ 　　(b) $\sqrt{(a^2b^4c^6)}$ 　　(c) $\sqrt{(m^{12}n^6)}$ 　　(d) $\sqrt{25a^8}$

B4 Simplify.

(a) $\sqrt{a^{-2}}$ 　　(b) $\sqrt{(p^{-4}q^6)}$ 　　(c) $\sqrt{(a^2b^{-6})}$ 　　(d) $\sqrt{64a^{-4}}$

B5 Simplify.

(a) $\sqrt{9a^6}$ 　　(b) $\sqrt{25p^4q^8}$ 　　(c) $\sqrt{16a^6b^{-2}}$ 　　(d) $\sqrt{\dfrac{p^4}{9q^2}}$

B6 Simplify.

(a) $\sqrt{(p^4q^6)}$ 　　(b) $\sqrt{\dfrac{p^8}{q^4}}$ 　　(c) $\sqrt{\dfrac{9p^{16}}{y^4}}$ 　　(d) $\sqrt{\dfrac{25p^2q^{10}}{r^4}}$

B7 The expression $a^2b^{-3} \times \sqrt{\dfrac{a^6}{b^2}}$ can be simplified to a^xb^y, where x and y are numbers.

Find the values of x and y.

OCR

C *Simplifying*

Q:

$$\frac{8ab^3 + 8b^4}{4b^3}$$

$$\frac{4a^3 + 2a^2b}{2a^2}$$

$$\frac{a^2 + 2ab}{a}$$

$$\frac{ab^2 + a^2b}{ab}$$

A:

$a + b$ $a + 2b$ $2a + b$ $2a + 2b$

C1 Multiply out each of these expressions.

    (a) $2(x - 12)$ (b) $2a(a + 3)$ (c) $3ab(a + b)$ (d) $\frac{1}{2}(4n - 6m)$

C2 Find the missing expressions.

    (a) $6(\rule{1cm}{0.3cm}) = 6h + 12$ (b) $2n(\rule{1cm}{0.3cm}) = 6n^2 + 4n$ (c) $4(\rule{1cm}{0.3cm}) = 8k + 2$

C3 Simplify these.

    (a) $\dfrac{12d - 18}{6}$ (b) $\dfrac{6d - 3}{6}$ (c) $\frac{1}{4}(12 + 2d)$ (d) $(8d - 32) \div 12$

C4 Simplify these.

    (a) $\dfrac{10d^2 - 5}{5}$ (b) $\dfrac{12d^2 - 3}{3}$ (c) $\frac{1}{6}(12d^3 - 30)$ (d) $\frac{1}{32}(32 - 64d^2)$

C5 Simplify each of these.

    (a) $\dfrac{8a + 12}{4}$ (b) $\dfrac{8a^2 + 12a}{a}$ (c) $(8a^2 + 12a) \div 4a$ (d) $(8a^2 + 12a^4) \div a^2$

C6 Find the missing expressions.

    (a) $\dfrac{(\rule{0.8cm}{0.3cm})}{4} = a + 1$ (b) $\dfrac{(\rule{0.8cm}{0.3cm})}{a} = a + 1$ (c) $\dfrac{(\rule{0.8cm}{0.3cm})}{2a} = a + 1$

C7 Find the missing expressions.

    (a)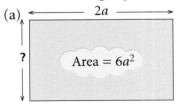

    (b)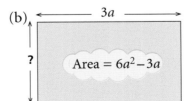

    (c)

(a) width $2a$, Area $= 6a^2$, height $?$; (b) width $3a$, Area $= 6a^2 - 3a$, height $?$; (c) width $a - b$, Area $= a^2 - b^2$, height $?$)

C8 Factorise these completely.

    (a) $6p^2 - 9pq$ (b) $2p^2 + 4p$ (c) $8p^2 - 2q^2$ (d) $12p^2q^2 - 6q^2$

C9 Simplify these.

    (a) $\dfrac{d^2 + d^4}{d^2}$ (b) $\dfrac{d^2 - d^4}{d^2}$ (c) $d^{-2}(12d^2 + d^6)$ (d) $(8d^2 + 4d^3) \div 2d^3$

C10 Find the missing expressions.

    (a) $\dfrac{(\rule{0.8cm}{0.3cm})}{d^2} = d^2 + d$ (b) $\dfrac{(\rule{0.8cm}{0.3cm})}{2d} = 2d + 1$ (c) $\dfrac{(\rule{0.8cm}{0.3cm})}{2d^2} = 8d^3 + 4d$

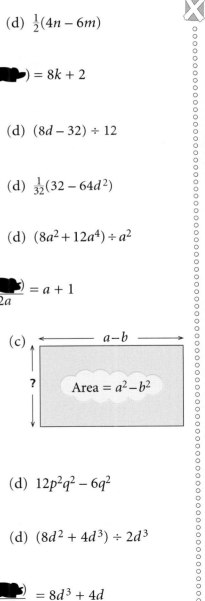

C11 Simplify these.

(a) $\dfrac{cd^2 + c^2 d}{cd}$

(b) $\dfrac{cd^2 + c^2 d}{c^2 d^2}$

(c) $\dfrac{4cd^2 + 8c^2 d}{12c^2 d^2}$

(d) $\dfrac{cd^2 + c^2 d}{c^4 d^4}$

C12 Find and simplify the missing expressions.

(a)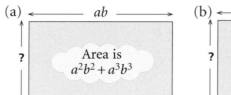

(b)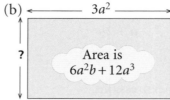

(c)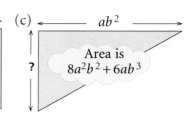

D *Combining fractions*

$$\dfrac{1}{6} + \dfrac{3}{4}$$

6 and 4 both go into 12 …

$$\dfrac{a}{6} + \dfrac{3b}{4}$$

$$= \dfrac{2}{12} + \dfrac{9}{12}$$

… so write each fraction with a denominator of 12.

$$= \dfrac{2a}{12} + \dfrac{9b}{12}$$

$$= \dfrac{11}{12}$$

$$= \dfrac{2a + 9b}{12}$$

? $\dfrac{1}{6a} + \dfrac{3}{4a}$

? $\dfrac{a+1}{6} + \dfrac{a-1}{4}$

? $\dfrac{6}{a} \times \dfrac{a}{4}$

D1 (a) Copy and complete this working to add together $\frac{1}{6}$ and $\frac{5}{8}$.

(b) Write as a single fraction

$$\dfrac{b}{6} + \dfrac{5b}{8}$$

Show all your working clearly.

$$\dfrac{1}{6} + \dfrac{5}{8}$$

$$= \dfrac{1 \times \clubsuit}{24} + \dfrac{5 \times \clubsuit}{24}$$

$$= \dfrac{\clubsuit + \clubsuit}{24} = \dfrac{\clubsuit}{24}$$

D2 Write each of these as a single fraction.

(a) $\dfrac{a}{2} + \dfrac{a}{4}$

(b) $\dfrac{a}{2} - \dfrac{a}{4}$

(c) $\dfrac{3a}{2} + \dfrac{a}{4}$

(d) $\dfrac{3a}{2} + \dfrac{5a}{4}$

D3 (a) Copy and complete this formula for the perimeter of this triangle.

$$Perimeter = \dfrac{a}{5} + \dfrac{a}{3} + \dots$$

(b) Write the expression for the perimeter as a single fraction.

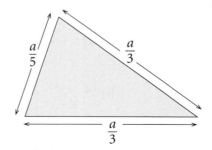

D4 Write each of these as a single fraction.

(a) $\dfrac{a+2}{2} + \dfrac{a}{4}$ 　　(b) $\dfrac{a+3}{3} + \dfrac{a}{6}$ 　　(c) $\dfrac{2a+1}{3} - \dfrac{a}{4}$ 　　(d) $\dfrac{3a+1}{2} - \dfrac{4a}{4}$

D5 Write down an expression for the perimeter of each of these shapes.
Write each expression as a single fraction.

(a)

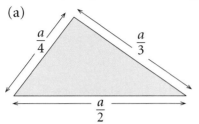

(b)

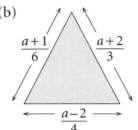

(c)

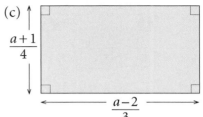

D6 By first writing 2 as $\dfrac{2}{1}$, find $2 + \dfrac{3a}{5}$ as a single fraction.

D7 Write each of these as a single fraction.

(a) $3 + \dfrac{a}{4}$ 　　(b) $1 + \dfrac{2a-1}{4}$ 　　(c) $2 + \dfrac{5a-6}{3}$ 　　(d) $5 - \dfrac{a-6}{2}$

D8 Copy and complete this working
to add together $\dfrac{1}{4a}$ and $\dfrac{5}{6a}$.

$$\dfrac{1}{4a} + \dfrac{5}{6a}$$

$$= \dfrac{1 \times \text{\ding{263}}}{12a} + \dfrac{5 \times \text{\ding{263}}}{12a}$$

$$= \dfrac{\text{\ding{263}} + \text{\ding{263}}}{12a} = \dfrac{\text{\ding{263}}}{12a}$$

D9 Write each of these as a single fraction.

(a) $\dfrac{1}{a} + \dfrac{1}{2a}$ 　　(b) $\dfrac{3}{2a} - \dfrac{1}{4a}$ 　　(c) $\dfrac{5}{4a} + \dfrac{1}{12a}$ 　　(d) $\dfrac{1}{3a} + \dfrac{1}{4a}$

D10 Write each of these as a single fraction.

(a) $\dfrac{a}{2} + \dfrac{2}{a}$ 　　(b) $\dfrac{3a}{2} - \dfrac{1}{4a}$ 　　(c) $\dfrac{5a}{4} + \dfrac{1}{8a}$ 　　(d) $\dfrac{2a}{3} + \dfrac{1}{6a}$

D11 Write each of these as a single fraction.

(a) $\dfrac{a+2}{2} + \dfrac{2a-1}{4}$ 　(b) $\dfrac{a+3}{3} + \dfrac{a+1}{6}$ 　(c) $\dfrac{2a+1}{2} - \dfrac{3a+1}{3}$ 　(d) $\dfrac{a^2+1}{a} - \dfrac{a}{2}$

(e) $\dfrac{a}{2} + \dfrac{a^2-1}{4a}$ 　(f) $\dfrac{2a-3}{4} + \dfrac{2a-1}{6}$ 　(g) $\dfrac{3a+1}{6} - \dfrac{2a-1}{9}$ 　(h) $\dfrac{2}{a} - \dfrac{a}{2}$

D12 There are four pairs of equivalent expressions here. Can you find them?

A $1 + \dfrac{2}{x}$ **B** $\dfrac{1+2x}{x}$ **C** $\dfrac{x-2}{x}$ **D** $1 - \dfrac{2}{x}$ **E** $\dfrac{2+x}{x}$ **F** $2 + \dfrac{1}{x}$ **G** $\dfrac{2x-1}{x}$ **H** $2 - \dfrac{1}{x}$

D13 Which of the expressions below are equivalent to $\dfrac{2a+1}{a} - \dfrac{a+4}{2}$?

A $\dfrac{2-a^2+8a}{2a}$ **B** $\dfrac{2-a^2}{2a}$ **C** $\dfrac{1}{a} - \dfrac{a}{2}$ **D** $\dfrac{3a-2}{2a}$

To multiply two fractions together, first cancel if you can
then multiply the numerators and multiply the denominators.

Simplify $\dfrac{a^2}{15} \times \dfrac{5}{4a}$ $\quad \dfrac{\cancel{a^2}}{\underset{3}{\cancel{15}}} \times \dfrac{\cancel{5}}{4\cancel{a}} = \dfrac{a}{12}$

D14 Simplify

(a) $\dfrac{a}{2} \times \dfrac{4}{a}$ (b) $\dfrac{3a}{2} \times \dfrac{a}{3}$ (c) $\dfrac{3a}{2} \times \dfrac{5a}{4}$ (d) $\dfrac{a}{5} \times \dfrac{15b}{4}$

(e) $\dfrac{3a}{2} \times \dfrac{1}{a}$ (f) $\dfrac{ab}{2} \times \dfrac{a}{b}$ (g) $\dfrac{ab^2}{2} \times \dfrac{5}{4ab}$ (h) $\dfrac{b}{2a} \times \dfrac{5a^2b}{4ab}$

To divide one fraction by a second, multiply the first
by the reciprocal of the second.

Simplify $\dfrac{3a^2}{16} \div \dfrac{9}{8a}$ $\quad \dfrac{3a^2}{16} \div \dfrac{9}{8a} = \dfrac{\cancel{3}a^2}{\underset{2}{\cancel{16}}} \times \dfrac{\cancel{8}a}{\underset{3}{\cancel{9}}} = \dfrac{a^3}{6}$

D15 Copy and complete this fraction division. $\dfrac{3a}{10} \div \dfrac{6a^2}{5} = \dfrac{3a}{10} \times \dfrac{5}{\blacksquare} = \ldots\ldots$

D16 (a) What is the reciprocal of a^2?

(b) Work out   (i) $\dfrac{a^4}{2} \div a^2$ (ii) $\dfrac{1}{2a^3} \div a^2$ (iii) $\dfrac{a^4}{b^2} \div a^2$

D17 Simplify these.

(a) $\dfrac{a}{2} \div \dfrac{4}{a}$ (b) $\dfrac{a}{2} \div \dfrac{a}{3}$ (c) $\dfrac{3a}{10} \div \dfrac{5}{a^2}$ (d) $\dfrac{ab}{3} \div \dfrac{5b}{4a}$

(e) $\dfrac{a^2}{2} \div \dfrac{1}{a}$ (f) $\dfrac{a}{2b^2} \div \dfrac{a}{b}$ (g) $\dfrac{b^2}{2a^2} \div \dfrac{b}{4a}$ (h) $\dfrac{bc}{a} \div \dfrac{abc}{4}$

Test yourself

T1 Find the exact value of $\frac{1}{a} + \frac{1}{b}$ when $a = \frac{1}{2}$ and $b = \frac{1}{3}$.   AQA 1999

T2 Find the value of $ab^2 + \frac{b}{a}$ when $a = \frac{4}{5}$ and $b = ^-3$.   AQA(NEAB) 1997

T3 Simplify these.

(a) $\frac{a^2 b}{ab}$ (b) $\frac{8a^6 b^4}{2ab}$ (c) $15a^6 b^2 \div 3a^2 b$ (d) $\frac{(2ab^2)^2}{6ab}$

T4 Simplify these.

(a) $\sqrt{a^2 b^4}$ (b) $\sqrt{a^{-4} b^{-2}}$ (c) $\sqrt{\frac{a^2}{b^6}}$ (d) $\sqrt{\frac{100a^2 b^2}{c^6}}$

T5 Simplify each of these.

(a) $\frac{1}{3}(6a + 9)$ (b) $\frac{2s^2 + 8s}{2s}$ (c) $\frac{2v^2 + 8v^3}{4v^2}$ (d) $\frac{p^2 - p^3 q^4}{p^2}$

T6 Find the missing expressions.

(a) $\frac{(\rule{1.2cm}{0.4cm})}{a^2} = a + 1$ (b) $\frac{(\rule{1.2cm}{0.4cm})}{a^2} = 3b^3 + a^2$ (c) $\frac{(\rule{1.2cm}{0.4cm})}{2ab} = 2ab^2 + 4ab$

T7 Find and simplify the missing expressions.

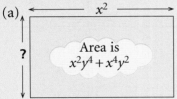 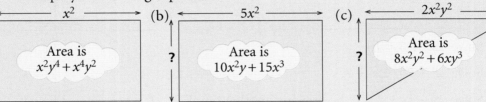 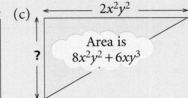

(a) x^2 — ? — Area is $x^2 y^4 + x^4 y^2$

(b) $5x^2$ — ? — Area is $10x^2 y + 15x^3$

(c) $2x^2 y^2$ — ? — Area is $8x^2 y^2 + 6xy^3$

T8 Write each of these as a single fraction.

(a) $\frac{x}{3} + \frac{x}{12}$ (b) $\frac{x}{3} - \frac{x}{12}$ (c) $\frac{3x}{4} - \frac{x}{8}$ (d) $\frac{3x}{5} + \frac{4x}{15}$

T9 Write each of these as a single fraction.

(a) $\frac{u+3}{6} - \frac{u}{9}$ (b) $\frac{u-5}{2} + \frac{u}{4}$ (c) $\frac{2u+9}{8} - \frac{u+1}{4}$ (d) $\frac{3u+1}{12} - \frac{2u-3}{8}$

(e) $\frac{u}{3} + \frac{2}{u}$ (f) $\frac{u}{2} - \frac{1}{3u}$ (g) $\frac{3u}{4} - \frac{1}{8u}$ (h) $\frac{5u}{9} - \frac{1}{6u}$

T10 Simplify these.

(a) $\frac{s}{3} \times \frac{6}{s}$ (b) $\frac{2s}{3} \times \frac{s}{4}$ (c) $\frac{3}{2s} \times \frac{s}{6}$ (d) $\frac{s}{3} \times \frac{15s}{7}$

(e) $\frac{s}{3} \div \frac{6}{s}$ (f) $\frac{2s}{3} \div 4s^2$ (g) $\frac{3}{2s} \div \frac{s}{6}$ (h) $\frac{s}{3} \div \frac{15s}{7}$

28 Further graphs

You will revise solving equations by trial and improvement methods.

This work will help you

◆ extend your understanding of cubic and quadratic graphs

◆ use graphs to solve equations involving x^3

◆ use graphs of functions such as $y = \dfrac{3}{x}$

◆ solve problems involving graphs that model real-life situations

A Quadratics and cubics

$y = 2x^2 + 3$		$y = x^3 + 2$					

x	$^-3$	$^-2$	$^-1$	0	1	2	3
$2x^2 + 3$	21						
$x^3 + 2$	$^-25$						

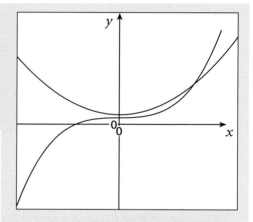

- Solve $2x^2 + 3 = 10$.

- Solve $x^3 + 2 = 2x^2 + 3$.

- Solve $x^3 - 2x + 2 = 0$ by drawing another graph. Use trial and improvement to find the solution to two decimal places.

A1 (a) Copy and complete this table of values for $y = x^3 - 2x^2$.

(b) Draw the graph of $y = x^3 - 2x^2$ on suitable axes.

(c) Use your graph to write down the solutions to 1 d.p. of

x	$^-2$	$^-1$	0	1	2	3
x^3	$^-8$				8	
$-2x^2$	$^-8$				$^-8$	
$y = x^3 - 2x^2$	$^-16$				0	

    (i) $x^3 - 2x^2 = 4$ (ii) $x^3 - 2x^2 = {}^-1$ (iii) $x^3 - 2x^2 = {}^-3$

(d) (i) What line could you draw on your graph to solve $x^3 - 2x^2 = x$?

    (ii) Draw the line and find the solutions to $x^3 - 2x^2 = x$ to 1 d.p.

(e) (i) Use your graph to estimate a solution to $x^3 - 2x^2 = 1$.

    (ii) By using a method of trial and improvement, improve your solution to give an answer to two decimal places.

    If possible, check your solution by zooming in on a graphical calculator.

A2 (a) Draw up a table of values for $^-3 \le x \le 3$, and draw the graphs
of $y = 2x^2 - 4$ and $y = 10 - x^3$ on the same axes.

(b) Use your graphs to write down the solutions to 1 d.p. of
   (i) $2x^2 - 4 = 5$ (ii) $10 - x^3 = 15$ (iii) $2x^2 - 4 = 10 - x^3$

(c) Rearrange the equation $2x^2 - 2x - 7 = 0$ into the form $2x^2 - 4 = \ldots$
By drawing a straight line on your graph, solve $2x^2 - 2x - 7 = 0$ to 1 d.p.

(d) Rearrange the equation $9 + 2x - x^3 = 0$ into the form $10 - x^3 = \ldots$
Draw another straight line on your graph to solve $9 + 2x - x^3 = 0$.

(e) By using a method of trial and improvement, improve your solution
to part (d) to give an answer to two decimal places.

If possible, check your answer by using the 'zoom' facility on a
graph-plotting program or graphical calculator.

A3 This question is on sheet G172.   OCR

A4 The curve shown is that of $y = x^3 - 4x^2 + 4$.

On the grid are drawn three straight lines.

(a) Which straight line would you use to
solve each of the equations below?
Show all your working clearly.

J
$x^3 - 4x^2 - x + 4 = 0$

K
$x^3 - 4x^2 + x + 4 = 0$

L
$x^3 - 4x^2 + x = 0$

(b) Use the correct straight line to solve
each of the equations, giving your
solutions to one decimal place.

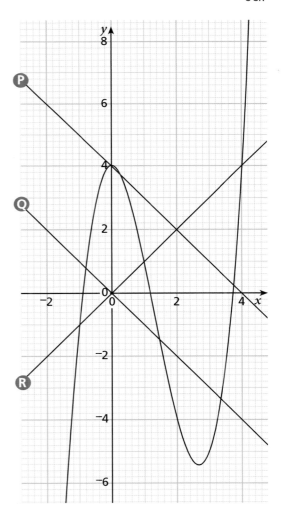

A5 You do not need to draw any graphs for this question.

Suppose you have the graph of $y = x^3 + 3x^2 + x$.
Work out the equation of the straight line you would need
to draw to solve each of the following equations.

(a) $x^3 + 3x^2 + x = 4$ (b) $x^3 + 3x^2 + x - 6 = 0$

(c) $x^3 + 3x^2 + x = 3x$ (d) $x^3 + 3x^2 + 2x = 0$

(e) $x^3 + 3x^2 + 2x - 3 = 0$ (f) $x^3 + 3x^2 + 6x - 5 = 0$

A6 (a) Draw up a table of values for the functions $y = 2x^2 - 3x - 2$ and $y = \frac{1}{2}x^3 - 12$.
(Add lines to the table, such as a line for x^2, for $2x^2$, for ^-3x and so on, if it helps.)

x	$^-2$	$^-1$	0	1	2	3	4
$2x^2 - 3x - 2$	12	3	$^-2$				
$\frac{1}{2}x^3 - 12$	$^-16$	$^-12\frac{1}{2}$	$^-12$				

Use your table to draw the graphs of $y = 2x^2 - 3x - 2$ and
$y = \frac{1}{2}x^3 - 12$ on the same axes.

(b) (i) Estimate the value of x that makes $2x^2 - 3x - 2$ a minimum.

    (ii) What is the minimum value of $2x^2 - 3x - 2$?

(c) Use your graphs to write down the solutions to 1 d.p. of $2x^2 - 3x - 2 = 2$.

(d) Use your graphs to write down the solution to 1 d.p. of $\frac{1}{2}x^3 - 12 = 2$.

(e) How many solutions are there to $2x^2 - 3x - 2 = {}^-5$?

*$\!^*$**A7** To solve the equation $x^3 - 4x - 2 = 0$ using the graph of $y = \frac{1}{2}x^3 - 12$
we need to rearrange $x^3 - 4x - 2 = 0$ into the form $\frac{1}{2}x^3 - 12 = \ldots$

(a) Copy and complete this working to
rearrange the equation.

(b) Draw a suitable straight line on
the graph of $y = \frac{1}{2}x^3 - 12$ that you drew
in question A6 to solve $x^3 - 4x - 2 = 0$.
Give your answers to 1 d.p.

$x^3 - 4x - 2 = 0$
$\frac{1}{2}x^3 - \blacklozenge x - \blacklozenge = 0$
$\frac{1}{2}x^3 - \blacklozenge = \blacklozenge$
$\frac{1}{2}x^3 - 12 = \blacklozenge x - \blacklozenge$

(c) Use trial and improvement to find the positive solution
of $x^3 - 4x - 2 = 0$ to two decimal places.

If possible, check your answer by using the 'zoom' facility on a
graph-plotting program or graphical calculator.

B *Reciprocal and other functions*

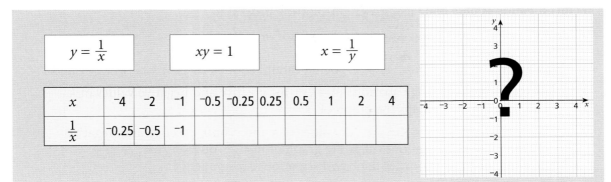

$$y = \frac{1}{x}$$

$$xy = 1$$

$$x = \frac{1}{y}$$

x	⁻4	⁻2	⁻1	⁻0.5	⁻0.25	0.25	0.5	1	2	4
$\frac{1}{x}$	⁻0.25	⁻0.5	⁻1							

B1 (a) How many lines of symmetry are there for the graph of $y = \frac{1}{x}$?

(b) Write down the equation of each line of symmetry.

(c) What is the order of rotation symmetry of the graph?

(d) Draw the line $y = x$ on your graph.
Use the line to solve the equation $x = \frac{1}{x}$.

(e) Draw the line $y = {}^-x$ on your graph.
Use the line to explain why there is no solution to the equation $\frac{1}{x} = {}^-x$.

B2 (a) On a new set of axes, draw the graphs of $y = \frac{1}{x}$ and $y = \frac{-1}{x}$.
(If you need to, draw up a table of values for $y = \frac{-1}{x}$.)

(b) What transformation maps the graph of $y = \frac{1}{x}$ on to the graph of $y = \frac{-1}{x}$?
Is there more than one answer?

B3 (a) Copy and complete this table of values for the equation $y = \frac{2}{x}$.

x	⁻4	⁻2	⁻1	⁻0.5	⁻0.25	0.25	0.5	1	2	4
$\frac{2}{x}$	⁻0.5	⁻1	⁻2	⁻4						

(b) Choose suitable scales and draw the graph of $y = \frac{2}{x}$.

C Investigating graphs

This investigation is best done using a graph-plotting program.

- Investigate graphs having equations $y = x^n$ for $n = 1, 2, 3, \ldots$
  What is the graph of $y = x^0$?
  Investigate $y = x^n$ for $n = {}^-1, {}^-2, \ldots$ being careful about values of x between $^-1$ and 1.

- Investigate graphs having equations $y = kx^2$ for positive and negative values of k.
  What transformation maps the graph of $y = x^2$ on to $y = kx^2$?

- Investigate graphs of $y = kx^2 + c$ for values of k and c, both positive and negative.
  What pair of transformations maps the graph of $y = x^2$ on to $y = kx^2 + c$?
  Does it matter in which order the transformations are applied?

- Investigate graphs of $y = kx^3 + c$ for values of k and c, both positive and negative.
  What pair of transformations maps the graph of $y = x^3$ on to $y = kx^3 + c$?

- Investigate graphs of functions of the form $y = \dfrac{k}{x} + c$.

When identifying graphs, think about the following.

- Is the graph …?

Quadratic	Cubic	Reciprocal
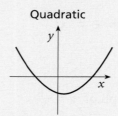	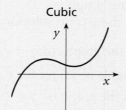	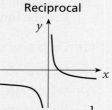
This involves x^2.	This involves x^3.	This involves $\dfrac{1}{x}$.

- What happens to y when x is very large and positive… negative…?
- Does the graph go through $(0, 0)$?

C1 (a) Which sketch graph fits which equation?
    (b) Draw a sketch of the missing graph.

$$y = x^2 - x \qquad y = x^2 - 2 \qquad y = 2 - x^2 \qquad y = x^2 + 2$$

Ⓐ 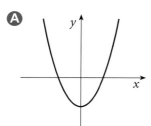 Ⓑ 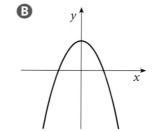 Ⓒ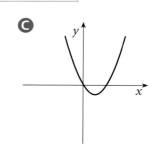

C2 Which sketch graph fits which equation?

$y = x^3 - 1$ $y = 1 - x^3$ $y = x^3 + 1$

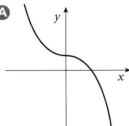   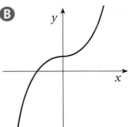   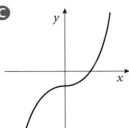

C3 Which sketch graph fits which equation?

$y = \dfrac{1}{x} - 1$ $y = 1 - \dfrac{1}{x}$ $y = 1 + \dfrac{1}{x}$

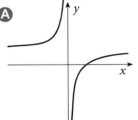   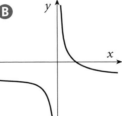   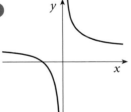

C4 Which sketch graph fits which equation?

$y = x^3 - x^2$ $y = 1 - x^2$ $y = x - x^2$

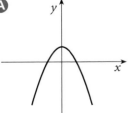   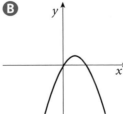   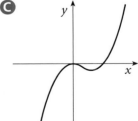

C5 **Ⓐ** 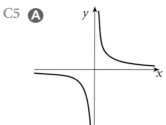 **Ⓑ** 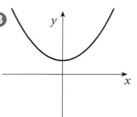 **Ⓒ**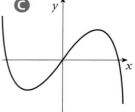

Ⓓ 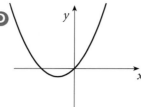 **Ⓔ** 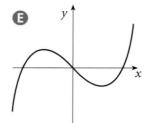 **Ⓕ**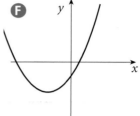

Each of the equations below represents one of the graphs A to F.
For each equation, write down the letter of the correct graph.

$y = x^2 + 3x$ $\qquad$ $y = x - x^3$ $\qquad$ $y = x^3 - 2x$

$y = x^2 + 2x - 4$ $\qquad$ $y = \dfrac{4}{x}$ $\qquad$ $y = x^2 + 3$ $\qquad$ Edexcel

C6 Write down a possible equation for each of these graphs.
The numbers you use in your equations should be integers.

(a) 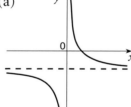 (b)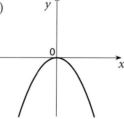

(c) 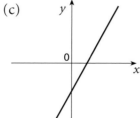 (d)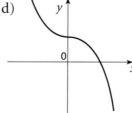

OCR

D Graphs to solve problems

D1 A children's toy is designed to project a small teddy bear
vertically into the air when a button is pressed.
For safety reasons, it is important that the bear should
not reach too great a height.

The vertical height, h metres, of the bear above the firing point
t seconds after firing is given by the equation $h = 6t - 5t^2$.

(a) Work out the values of h for $t = 0, 0.25, 0.5, \ldots$ up to $t = 1.5$.

(b) Plot the graph of h against t for your values.

(c) What is the maximum height of the teddy bear above the firing point?

(d) How long after firing does it take the bear to reach this height?

(e) The bear is fired from a table 1 metre above the floor.
How long does it take to hit the floor?

D2 A rectangular pen is to be built against a retaining wall.
The pen needs to enclose an area of $64\,\text{m}^2$.
The width of the pen is x metres.

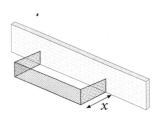

(a) Show that the length of fencing needed, L m, is
given by the formula $L = 2x + \dfrac{64}{x}$.

(b) Draw the graph of L against x for suitable values of x.

(c) Hence find (to 1 d.p.) the minimum length of fencing
and the dimensions of the pen.

(d) Use a method of trial and improvement to find the
value of x that gives the minimum length of fencing to 2 d.p.

D3 A canvas wind-shelter has two square ends, each of side x metres,
and a rectangular back of length l metres.
The area of canvas is $9\,\text{m}^2$.

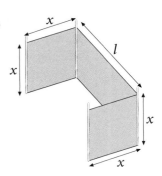

(a) Show that $l = \dfrac{9}{x} - 2x$, and hence that
the enclosed volume, $V\,\text{m}^3$, is $9x - 2x^3$.

(b) Plot the graph of V against x for sensible values of x.

(c) (i) Use your graph to find the value of x that gives
the largest possible volume, to 1 d.p.

(ii) From your graph, what is the largest volume?

(d) (i) Use trial and improvement to find this value of x to 2 d.p.

(ii) Hence find the corresponding value of l
and check your answer to part (c)(ii).

D4 A piece of paper measures 21 cm by 30 cm.
From each corner, a square of side x cm is cut,
as shown in the diagram.
The paper is then folded up to make a box without a lid.

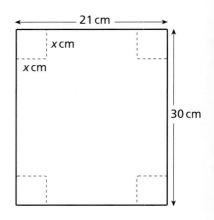

(a) Show that the volume V of the box, in cm³,
is given by the formula $V = x(21 - 2x)(30 - 2x)$.

(b) Draw the graph of V against x for $0 \leq x \leq 6$.

(c) Use your graph to find the value of x that
gives the maximum volume.
What is the maximum volume?

(d) Use trial and improvement to find
this value of x to 2 d.p.

(e) Find the value of V when $x = 20$ by substituting in the formula.
Is this value of V higher or lower than the value you found in part (c)?
Why does this value of x actually not give a larger volume than that in part (c)?

D5 A design company is asked to design a plastic box.
The box is to have square ends, and its surface area
is to be exactly 1 square metre.

The volume is required to be as large as possible.

(a) Write down an expression for the total surface area
of the box, in terms of s and l.
(s and l are measured in metres.)

(b) Put your expression in part (a) equal to 1,
and make l the subject.

(c) Write down a formula for the volume of the box, V m³, in terms of s and l,
and show that $V = \dfrac{s(1 - 2s^2)}{4}$.

(d) Draw a graph of V against s for values of s between 0 and 0.8.
Find the value of s that gives the maximum volume.

(e) Work out the dimensions of the box, and thus its volume.
Check that this volume agrees with the value on your graph.

Test yourself

The axes for questions T1 and T2 are on sheet G173.

T1 The graph of $y = x^3 - 2x^2 - 4x$ has been drawn on grid A on sheet G173.

(a) Use the graph to find estimates of the solutions to the equation

    (i) $x^3 - 2x^2 - 4x = 0$ (ii) $x^3 - 2x^2 - 4x = 1$

(b) By drawing a clearly labelled straight line on the grid, find estimates of the solutions to the equation $x^3 - 2x^2 - 6x = 1$.   Edexcel

T2 (a) Copy and complete this table of values for $y = x^3 + 3x^2 - 24x$.

x	0	1	2	3	4	5
y						

(b) Draw the graph of $y = x^3 + 3x^2 - 24x$ on grid B on sheet G173.

(c) Use your graph to find a solution of the equation $x^3 + 3x^2 - 24x = 20$.

(d) The equation $x^3 + 3x^2 - 28x = 20$ can be solved by adding a straight line to the graph.

    (i) Find the equation of this line.

    (ii) Draw this line on the graph and use it to solve the equation.   OCR(MEG)

T3 (a) Below are three graphs.
Match each graph with one of the following equations.

    Equation A: $y = 3x - p$ Equation B: $y = x^2 + p$

    Equation C: $3x + 4y = p$ Equation D: $y = px^3$

In each case p is a positive number.

(i) 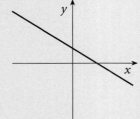   (ii) 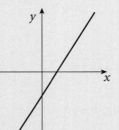   (iii)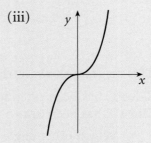

(b) Sketch a graph of the equation you have not yet chosen.   AQA(NEAB) 1998

29 Angles and circles

This work will help you understand
- what is meant by proving a statement
- the relationships between angles connected with circles, and how they are proved

A Angles and triangles

It is a well-known fact that the angles of a triangle add up to 180°.

But how can we be sure that this is true for **all** triangles?
We could draw some triangles and measure their angles accurately.
But we could never draw **every** possible triangle.
And in any case, we can never be absolutely accurate in measuring.

In mathematics, we try to **prove** that statements are true.
Proving a statement means showing that it follows from other statements
which are already known to be true.

For example, suppose these three statements are known to be true:

> It is impossible to get from London to Birmingham in less than an hour.
> Stanley's offices in Birmingham were broken into between 5:15 p.m. and 5:45 p.m.
> Pete was in London at 5 p.m.

- Does it follow that Pete did not break into Stanley's offices?

Proving that the angles of a triangle add up to 180°

We shall assume for the moment that these three statements are known to be true.
They are labelled **a** (for 'assumption').

a Angles on a straight line add up to 180°.	**a** Corresponding angles made with parallel lines are equal.	**a** Alternate angles made with parallel lines are equal.

This diagram shows a triangle ABC.

Side AC has been extended to D.
Line CE has been drawn parallel to AB.

- Which other angle is equal to *x*, and why?
- Which other angle is equal to *y*, and why?
- Explain why the three angles of triangle ABC add up to 180°.

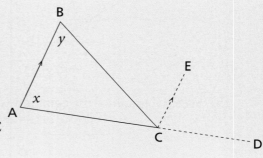

If one side of a triangle is extended, the angle formed is called an **exterior angle** of the triangle.

In the diagram at the foot of the previous page, the exterior angle BCD is proved to be equal to $x + y$.

This fact is usually stated as follows (it is labelled **p** to show it has been proved):

> **p** **An exterior angle of a triangle is equal to the sum of the other two interior angles.**

Isosceles triangles

We shall assume for the moment that this statement is also known to be true:

> **a** In an isosceles triangle, the angles opposite the equal sides are equal.

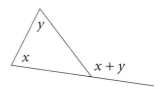

A1 Find the value of each letter. Explain your method in each case.

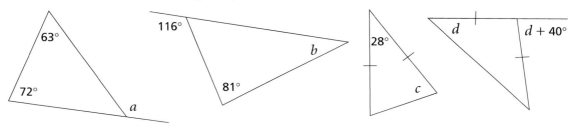

A2 Use the fact that the angles of a triangle add up to 180° to prove that the angles of a quadrilateral add up to 360°.

B *Angles in circles*

Draw a circle, radius about 5 cm.

Mark two points A, B on it and join them to the centre, O.

Mark a point P on the circle. Join A and B to P.

- Measure angles AOB and APB.

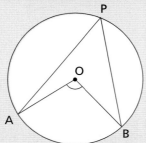

Choose different positions for A and B and repeat.

Include cases where angle AOB is reflex.

- Collect everyone's results together. What do you notice?

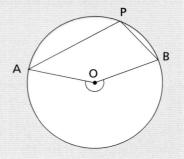

In this diagram, P has been joined to O and line PO has been extended to C.

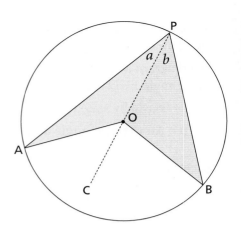

- What kind of triangle is each of the triangles POA and POB?
- What can you say about angle OAP?
- What can you say about angle AOC?
- What can you say about angle BOC?
- Explain why angle AOB is twice angle APB.

In this proof it is assumed that

    a All radii of a circle are equal.

This is, of course, how a circle is defined.

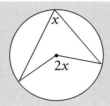

You have now proved this fact:

p Angle at centre
= twice angle at circumference

B1 Find the angles marked with letters. Explain how you worked out each one.

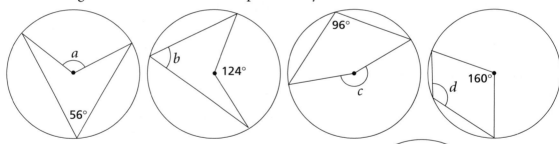

B2 In this diagram AOB is a diameter of the circle.

(a) What is the size of angle AOB?

(b) Explain why angle APB is a right angle.

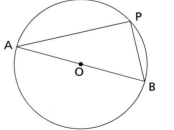

In B2 you proved this:

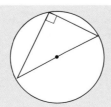

p The angle in a semicircle is a right angle.

B3 In the diagram here, imagine that A and B are fixed, but that P moves on the circle between A and B.

P_1 and P_2 are two possible positions of P.
Explain why angles AP_1B and AP_2B are equal.

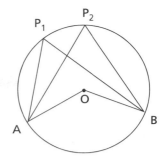

In B3 you proved this:

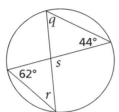

p Angles in same segment are equal.
(The shaded part is called a **segment** of the circle.)

B4 Find the angles marked with letters, giving reasons for each step of working.

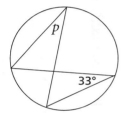 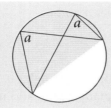 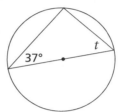 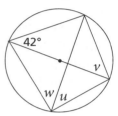

C *Cyclic quadrilaterals*

Quadrilateral ABCD, whose vertices are all on a circle, is called a **cyclic quadrilateral** ('cyclic quad' for short).

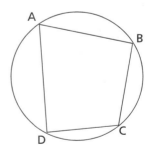

C1 In the diagram on the left, vertices B and D of the cyclic quad have been joined to the centre of the circle, making angles x and y at the centre.

(a) Work out angles x, y and z giving the reasons.

(b) Suppose angle BAD were 50° instead of 40°. Work out x, y and z in this case.

(c) Now repeat for when angle BAD is 60°.

(d) Do you notice any connection between angles BAD and BCD?

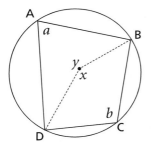

C2 In the diagram on the left, vertices B and D of the cyclic quad have been joined to the centre of the circle, making angles x and y at the centre.

(a) State the relationship between angles x and a.

(b) State the relationship between angles y and b.

(c) State the relationship between angles x and y.

(d) From what you have stated, deduce the relationship between angles a and b.

In C2 you proved this:

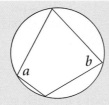

p Opposite angles a, b of a cyclic quadrilateral add up to 180°.

C3 Here side DC has been extended.
The angle marked e is an exterior angle of the cyclic quad.

From the result you got in C2, deduce the relationship between angles e and a.

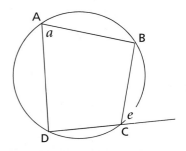

In C3 you proved this:

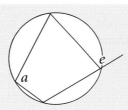

p Exterior angle e of a cyclic quadrilateral = opposite interior angle a

C4 Calculate angles a, b, c, ..., giving reasons for each step of working.

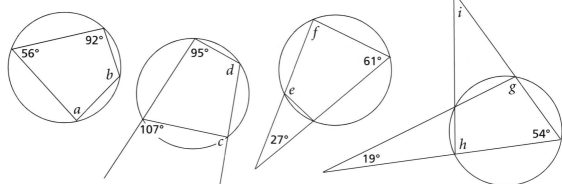

C5 Calculate angle BAC, giving reasons for each step of working.

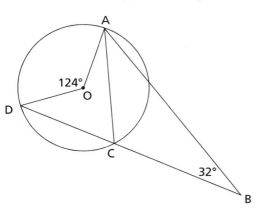

C6 Calculate angle POQ, giving reasons for each step of working.

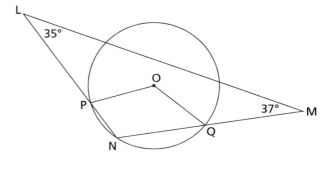

C7 Calculate angles *a*, *b* and *c*, giving reasons for each step of working.

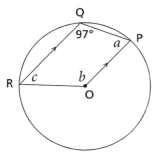

C8 Calculate angle ADB, giving reasons for each step of working.

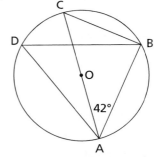

C9 Calculate angles LNM and LOM, giving reasons for each step of working.

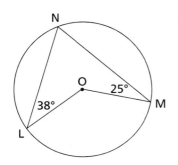

C10 (a) Find angle BEF, giving reasons for your answer.

(b) Explain why lines CD and EF are parallel, no matter what the size of angle BCG.

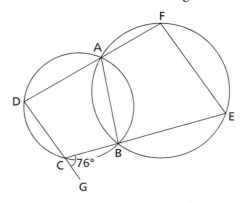

D Tangents

A **tangent** is a line which touches a circle at one point only.

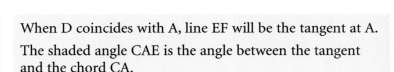

We will assume that this statement is known to be true:

a A tangent is perpendicular to the radius at the point of contact.

Think of a wheel touching horizontal ground.
The centre of the wheel is vertically above the point of contact,
so the radius is at right angles to the horizontal tangent.

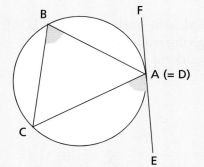

If the entire diagram is rotated it will make no difference
to the fact that the angle is a right angle.

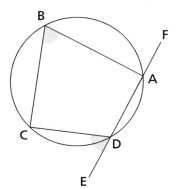

This diagram shows a cyclic quad ABCD.
Side AD has been extended in both directions to make line EF.

Exterior angle CDE is equal to opposite interior angle CBA.

Imagine that point D moves
along the circle towards A.

The shaded angles stay equal.

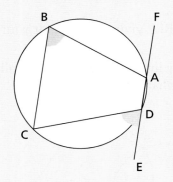

When D coincides with A, line EF will be the tangent at A.

The shaded angle CAE is the angle between the tangent
and the chord CA.

The shaded angle CBA is called the angle in the 'alternate
segment' (the segment on the other side
of the chord from the shaded angle CAE).

The result above is usually stated as:

The angle between tangent and chord is equal to the angle in the alternate segment.

In questions D1 to D10, you must **give the reason** for each step of working.

D1 Find angle ADE.

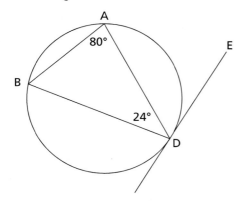

D2 Find angle QRS.

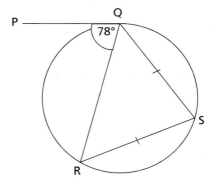

D3 Find angle DEB.

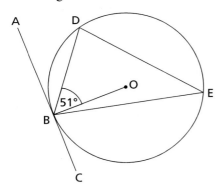

D4 Find angle TQU.

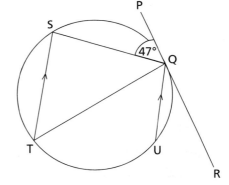

D5 (a) Find angle BDF.
   (b) Find angle DFE.

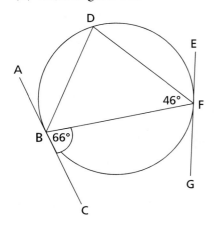

D6 Find angle QSR.

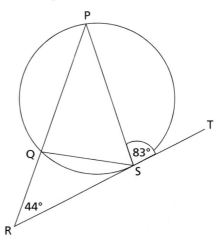

D7 Find angle STU.

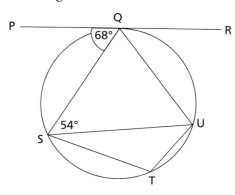

D8 (a) Find angle ACB.
(b) Find angle ABC.

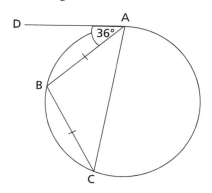

D9 Find angle BDC.

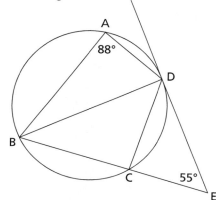

D10 Find angle SPR.

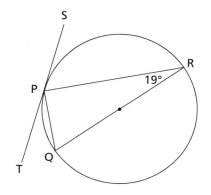

D11 **Proving the 'alternate segment' statement**

In this diagram, TA is the tangent to the circle at A,
AB is a chord, and angle ACB is in the 'alternate segment'
to angle TAB.

Extra lines have been drawn from A and B to the centre O.

Let x be the size of angle TAB.

We need to prove that angle ACB $= x$.

(a) Explain why the angles marked y are equal.

(b) Explain why $x + y = 90°$.

(c) Find an expression for angle AOB in terms of y, giving the reason.

(d) Find an expression for angle ACB in terms of y, giving the reason.

(e) Hence explain why angle ACB $= x$.

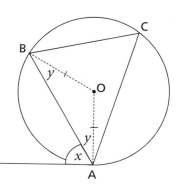

E Mixed questions

E1 Find the value of each letter.
(The diagrams are printed on sheet G174.)

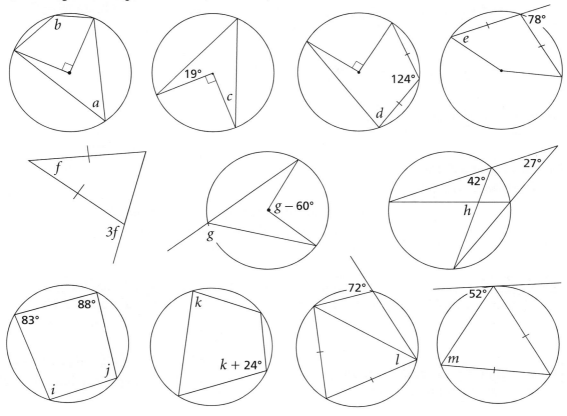

*E2 X and Y are the centres of two circles of
equal radius.
PQR is a straight line.

Calculate angles PXY and RYX, showing
your method.

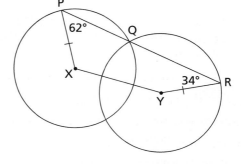

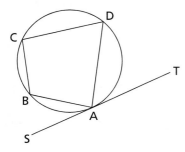

*E3 A, B, C, D are four points on a circle.
The line SAT is the tangent to the circle at A.

Explain why

angle SAB + angle TAD = angle BCD

F *A deductive network*

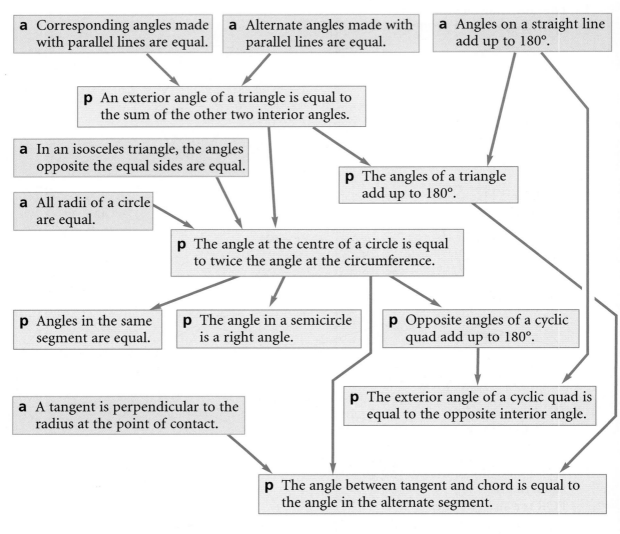

a Corresponding angles made with parallel lines are equal.

a Alternate angles made with parallel lines are equal.

a Angles on a straight line add up to 180°.

p An exterior angle of a triangle is equal to the sum of the other two interior angles.

a In an isosceles triangle, the angles opposite the equal sides are equal.

p The angles of a triangle add up to 180°.

a All radii of a circle are equal.

p The angle at the centre of a circle is equal to twice the angle at the circumference.

p Angles in the same segment are equal.

p The angle in a semicircle is a right angle.

p Opposite angles of a cyclic quad add up to 180°.

a A tangent is perpendicular to the radius at the point of contact.

p The exterior angle of a cyclic quad is equal to the opposite interior angle.

p The angle between tangent and chord is equal to the angle in the alternate segment.

Test yourself

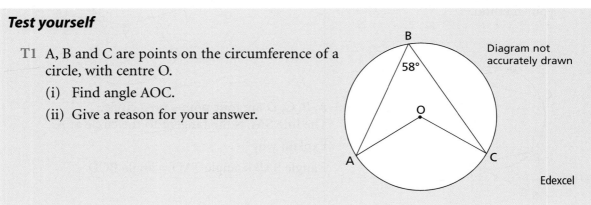

T1 A, B and C are points on the circumference of a circle, with centre O.

(i)   Find angle AOC.

(ii) Give a reason for your answer.

B

58°

O

A

C

Diagram not
accurately drawn

Edexcel

T2 A, B, C and D are points on the circumference of a circle. BD is a diameter of the circle and PA is a tangent to the circle at A.

Angle ADB = 25° and angle CDB = 18°.

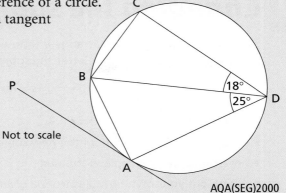

Not to scale

(a) Write down the value of angle

    (i) BCD (ii) PAB

(b) Calculate angle ABC.

AQA(SEG)2000

T3 A, B, C and D are four points on the circumference of a circle.

TA is the tangent to the circle at A.

Angle DAT = 30°.
Angle ADC = 132°.

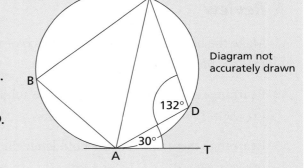

Diagram not accurately drawn

(a) (i) Calculate the size of angle ABC.

    (ii) Explain your method.

(b) (i) Calculate the size of angle CBD.

    (ii) Explain your method.

(c) Explain why AC cannot be a diameter of the circle.

Edexcel

T4 A, B, C and D are four points on the circumference of a circle centre O. AC is a straight line passing through the centre of the circle.

The tangent PT meets the circle at D.

Given that angle AOD = 62°, find **each** of the following angles. Give reasons for your answers.

(a) Angle ABD   (b) Angle ADC

(c) Angle CAD   (d) Angle CDP

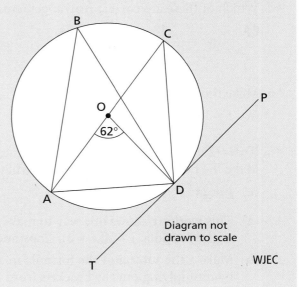

Diagram not drawn to scale

WJEC

30 Changing the subject 2

You should know how to rearrange a formula where the new subject appears once, for example making the bold letter the subject of each of these types of formula.

$$a = b + \boldsymbol{c}d, \quad h = \frac{ij\boldsymbol{k}}{80}, \quad p = 2 - \frac{\boldsymbol{q}}{d}, \quad s = 6\boldsymbol{p}^2, \quad p = a(\boldsymbol{k} + h)$$

This work will help you

- ◆ rearrange a formula where the new subject appears more than once
- ◆ rearrange more complex formulas
- ◆ work out your own formulas from a given situation

A Review

A1 Make the bold letter the subject of each of these formulas.

    (a) $a = 6\boldsymbol{b} - 3$ (b) $w = 5\boldsymbol{d} + 10$ (c) $2h = 5\boldsymbol{j} - 3$ (d) $3e = 21 + 4\boldsymbol{f}$

A2 Rearrange each of these to make x the subject.

    (a) $2x + 3y = 12$ (b) $y = 10 - 2x$ (c) $4y - 3x = 24$ (d) $2y = 8 - 3x$

A3 Rearrange these to make the bold letter the subject.

    (a) $a = w\boldsymbol{b} - c$ (b) $s = c\boldsymbol{d} + f$ (c) $ah = ef - b\boldsymbol{j}$ (d) $e = p + ab\boldsymbol{f}$

A4 Which of these are correct rearrangements of $ax + by = c$?

Ⓐ $a = \dfrac{by + c}{x}$ Ⓑ $by = c - ax$ Ⓒ $b = \dfrac{c - ax}{y}$ Ⓓ $b = \dfrac{c + ax}{y}$ Ⓔ $y = \dfrac{c - ax}{b}$ Ⓕ $ax = c + by$

A5 Make the bold letter the subject.

    (a) $t = 12 + \dfrac{\boldsymbol{s}}{3}$ (b) $h = \dfrac{\boldsymbol{j} - 2}{5}$ (c) $x = \dfrac{3\boldsymbol{y}}{2} - 1$ (d) $a = b - \dfrac{\boldsymbol{c}}{d}$

A6 If $x^2 = 25$, then $x = 5$ or $^-5$; we write $x = \pm 5$. Similarly, if $a^2 = k$, then $a = \pm\sqrt{k}$.
Make the bold letter the subject. You will need to use $\pm$ in each of your answers.

    (a) $y = 2\boldsymbol{x}^2$ (b) $m = 2\boldsymbol{n}^2 l$ (c) $ad = \boldsymbol{b}^2 c^2$ (d) $\boldsymbol{x}^2 + y^2 = c^2$

A7 (a) This working shows one way to make x the subject of the formula $y = 4(x + 2)$. Copy and complete it.

    (b) Make x the subject of the formula in a different way by multiplying out the brackets from $y = 4(x + 2)$ first.

$$y = 4(x + 2)$$
$$\frac{y}{\clubsuit} = x + 2$$
$$x = \frac{y}{\clubsuit} - \clubsuit$$

A8 Make x the subject of each of these.

    (a) $y = 2(1 - x)$ (b) $a(x + y) = b$ (c) $y = c(d + x)$ (d) $y - y_1 = 2(x - x_1)$

B Brackets and fractions

Make *a* the subject of the formula

$$k = \frac{a}{b} - 2$$

$$k + 2 = \frac{a}{b}$$

$$b(k + 2) = a$$

$$\text{or } a = b(k + 2)$$

Make *v* the subject of the formula

$$s = \frac{u}{v} + 10$$

$$s - 10 = \frac{u}{v}$$

$$v(s - 10) = u$$

$$v = \frac{u}{s - 10}$$

B1 The formula $h = \frac{a}{k} + j$ gives h in terms of a, k and j.

Which of the following are correct rearrangements of the formula?

A $\quad a = hk - j$ **B** $\quad a = k(h - j)$ **C** $\quad a = jk - kh$ **D** $\quad a = \frac{k}{h - j}$ **E** $\quad a = hk - jk$

B2 Rearrange each of these formulas to make the bold letter the subject.

(a) $l = \frac{2a}{\boldsymbol{b}}$ (b) $m = \frac{ab}{\boldsymbol{c}}$ (c) $r = \frac{\pi}{x\boldsymbol{y}z}$ (d) $h = 1 + \frac{d}{\boldsymbol{e}}$

(e) $s = 1 - \frac{1}{\boldsymbol{t}}$ (f) $a = b - \frac{d}{\boldsymbol{c}}$ (g) $z = ab + \frac{2\pi}{r\boldsymbol{\theta}}$ (h) $e = \frac{d}{4} + \frac{f}{\boldsymbol{h}}$

B3 Which of the following are correct rearrangements of $s = w - \frac{g}{r}$?

A $\quad w = s - \frac{g}{r}$ **B** $\quad g = r(s - w)$ **C** $\quad r = \frac{g}{s - w}$ **D** $\quad r = \frac{g}{w - s}$ **E** $\quad w = \frac{g}{r} + s$ **F** $\quad g = r(w - s)$

B4 Copy and complete this working to make x the subject of $y = \frac{a}{(x-c)}$.

$$y = \frac{a}{(x - c)}$$

$$y(x - c) = \clubsuit$$

$$x - c = \frac{a}{\clubsuit}$$

$$x = \frac{a}{\clubsuit} \, \clubsuit$$

B5 Make the bold letter the subject.

(a) $a = \frac{b}{(c + \boldsymbol{d})}$ (b) $5x = \frac{2z}{6 - \boldsymbol{y}}$ (c) $x + 3 = \frac{2z}{5 + \boldsymbol{y}}$ (d) $w = \frac{t + 2}{3 + \boldsymbol{u}}$

C *More than once*

Sometimes the variable you want to make into the subject occurs
in more than one place in the original formula.
You need to get the variable you want in one place only, often by factorising.

Make c the subject of $a = 4bc - cd^2$.

$$a = 4bc - cd^2$$
$$a = c(4b - d^2)$$
$$\frac{a}{4b - d^2} = c \quad \text{or} \quad c = \frac{a}{4b - d^2}$$

Make r the subject of $a = \frac{br + t}{r - s}$.

$$a(r - s) = br + t$$
$$ar - as = br + t$$
$$ar - br = t + as$$
$$r(a - b) = t + as$$
$$r = \frac{t + as}{a - b}$$

C1 Make n the subject in each of these.

(a) $f = an + dn$ (b) $3T = nT_0 + nk$ (c) $3n + s = kn + p$ (d) $rn = 3(n - e)$

C2 Make f the subject of each of these.

(a) $3f = a(f + 2)$ (b) $3(f + u) = 5(f - u)$
(c) $k_1(2f + 1) = k_2(2 - f)$ (d) $2(f - 3) = k(1 + f)$

C3 Rearrange each of these formulas to make k the subject.

(a) $s = \frac{3k}{k - 1}$ (b) $t = \frac{k + 4}{k}$ (c) $k_0 = \frac{3 - k}{k - 1}$ (d) $ab = \frac{dk}{k - e}$

C4 Make the letter in square brackets the subject.

(a) $mv - mu = Ft$ $[m]$ (b) $T = P + \frac{PRT}{100}$ $[P]$

(c) $xt^2 - \frac{k(t + 1)}{x} = 0$ $[x]$ (d) $m^2 = \frac{1}{p^2}(n^2 - p^2)$ $[p]$

C5 Make r the subject of $3(r - 4) = s(7 - 2r)$.   OCR

C6 $t = \frac{8(p + q)}{pq}$ $p = 2.71, \ q = {}^-3.97$.

(a) Calculate the value of t.
   Give your answer to a suitable degree of accuracy.

(b) Make q the subject of the formula $t = \frac{8(p + q)}{pq}$.   Edexcel

C7 Make v the subject of this formula. $3v + 5 = av + b$ Edexcel

C8 $m = \frac{cab}{a - b}$ Express b in terms of a, c and m.   AQA(NEAB) 1997

D *Forming and manipulating formulas*

Find a formula for the perimeter, P, of a rectangle given its area, A,
and the length, l, of one side.

First jot down what you know. Draw a sketch if it helps.	Can you work out anything else from what you are given? Mark it on the sketch.	Can you now work out what you need? Simplify your answer if possible.
We know the area and one of the sides. 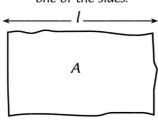	*Since you get the area of a rectangle by multiplying two of the sides, the unknown side must be equal to $\frac{A}{l}$.* 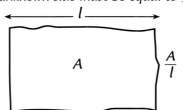	*We can see that the perimeter is* $$\frac{A}{l} + \frac{A}{l} + l + l$$ $$= 2\left(\frac{A}{l} + l\right)$$

D1 The diagram shows a trapezium.

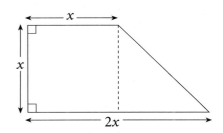

(a) Show that the area, A, of the trapezium
   is given by $A = \frac{3}{2}x^2$.

(b) Rearrange this formula to make x the subject.

(c) Find the value of x for a trapezium of area $20\,\text{cm}^2$.

D2

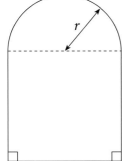

This doorway is formed by a semicircle
on top of a square.

(a) Show that a formula for the perimeter, P,
   of the doorway can be written $P = \pi r + 6r$.

(b) Make r the subject of this formula.

(c) What is the height of a doorway
   with a perimeter of 10 metres?

D3 A rectangle has perimeter $16\,\text{cm}$ and area $12\,\text{cm}^2$.
The length of one side is p cm.

(a) Show that the length of the other side is $8 - p$.

(b) Hence show that $p^2 - 8p + 12 = 0$.

(c) Factorise to solve this equation and find the dimensions of the rectangle.

D4 The box shown is a cuboid, with edges
of lengths a, $2a$ and b.

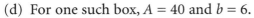

(a) Show that the surface area, A, of the box
is given by the formula $A = 4a^2 + 6ab$.

(b) Rearrange this formula to make b the subject.

(c) Work out the value of b for a box with
surface area 100 and $a = 2$.

(d) For one such box, $A = 40$ and $b = 6$.

(i) Show that $a^2 + 9a - 10 = 0$.

(ii) Solve the equation and hence write down the dimensions of this box.

D5 Maria has two brothers, Juan and Angel. Juan is 3 years older than Maria.
Angel is 2 years younger than Maria. The product of the brothers' ages is 126.

(a) If Maria's age is x years, show that $x^2 + x - 132 = 0$.

(b) Solve this equation and find the ages of the three children.

D6 A cylinder has radius x and height $2x$.   A sphere has radius r.

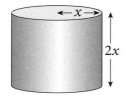

The **total** surface area of the cylinder is equal to the surface area of the sphere.

(a) Show that $r^2 = \frac{3}{2}x^2$.

(b) When $r = 2$, find the value of x. Give your answer to 1 d.p.

D7 Jan measures the diagonal of a square field as 80 m.
The side of the field has length x m.

(a) Explain why $2x^2 = 6400$.

(b) Calculate the value of x.

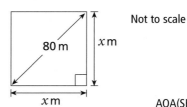

Not to scale

AQA(SEG) 1998

***D8** (a) Given that $s = \left(\frac{u+v}{2}\right)t$ and that $v = u + at$ show that $s = ut + \frac{1}{2}at^2$.

(b) Rearrange $s = ut + \frac{1}{2}at^2$ to make a the subject.

(c) Find the value of a if $u = 10$, $t = 4$ and $s = 100$.

E *Mixed questions*

E1 The following formula gives distance s
in terms of acceleration a and speeds u and v.

$$s = \frac{v^2 - u^2}{2a}$$

(a) Find s when $a = {}^-8$, $v = 1$ and $u = 15.5$.

(b) Rearrange the formula to make u the subject.   AQA(SEG) 1998

E2 You are given the formula $P = \dfrac{V^2}{R}$.

(a) Work out the value of P when $V = 3.85$ and $R = \frac{8}{5}$.

(b) Rearrange the formula to give V in terms of P and R.   AQA 1999

E3 (a) Rearrange the formula $v = u + 0.1t$ to make t the subject.

(b) Substitute this expression for t in the formula $s = ut + 0.05t^2$
to obtain a formula connecting u, v and s.

Show that this formula may be rearranged as $v^2 = u^2 + 0.2s$.   OCR

E4 The formula $T = 2\pi\sqrt{\dfrac{l}{g}}$ gives the time, T seconds, for the period of a pendulum.

l is the length of the pendulum in metres, g is a constant.

(a) Rearrange the formula to make l the subject.

(b) Find the length of a pendulum with a period of 1 second at Washington, USA,
where $g = 9.80$. Give your answer to a suitable degree of accuracy.

E5 The Earth is a sphere of radius 6.35×10^6 m.

(a) Calculate the volume of the Earth. Give your answer in standard form.
State the units of your answer.

(b) Rearrange the formula $V = \dfrac{4\pi r^3}{3}$ to give r in terms of V.   AQA(SEG) 2000

E6 (a) Factorise $x^2 + 3x - 18$.

(b) Make v the subject of the formula $E = \frac{1}{2}mv^2 + mgh$.   OCR(MEG)

E7 Make x the subject of the formula $y = \dfrac{x^2 + 4}{5}$.   Edexcel

E8 The formula for the surface area of this cone is
$$A = \pi r^2 + \pi rl$$

(a) Factorise fully $\pi r^2 + \pi rl$.

(b) Rewrite the formula with l as the subject.   OCR

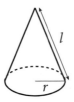

Test yourself

T1 Which of the following are correct rearrangements of $a = bc - \dfrac{d}{e}$?

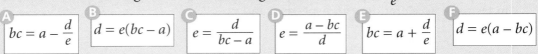

A $\quad bc = a - \dfrac{d}{e}$
B $\quad d = e(bc - a)$
C $\quad e = \dfrac{d}{bc - a}$
D $\quad e = \dfrac{a - bc}{d}$
E $\quad bc = a + \dfrac{d}{e}$
F $\quad d = e(a - bc)$

T2 Rearrange the formula $A = 2\pi rh + \pi r^2$ to give h in terms of A and r.

T3 Rearrange the formula $w = \dfrac{a + b}{2}$ to express a in terms of w and b.

T4 Make k the subject of $P = k^2 h$.

T5 $v^2 = \dfrac{GM}{R}$ $\qquad G = 6.6 \times 10^{-11},\ M = 6 \times 10^{24},\ R = 6\,800\,000.$

    (a) Calculate the value of v.
       Give your answer in standard form, correct to two significant figures.

    (b) Rearrange the formula $v^2 = \dfrac{GM}{R}$ to make M the subject.   *Edexcel*

T6 Make the bold letter the subject in each of these.

    (a) $a = \dfrac{f}{2\boldsymbol{e}}$ (b) $k = \dfrac{5t}{\boldsymbol{p}}$ (c) $s = \dfrac{2\pi}{\boldsymbol{h}l}$ (d) $g = 5 - \dfrac{4a}{\boldsymbol{b}}$

    (e) $x = \dfrac{y}{(2 + \boldsymbol{z})}$ (f) $ah = \dfrac{10f}{5 - \boldsymbol{g}}$ (g) $w - 1 = \dfrac{5a}{1 + \boldsymbol{b}}$ (h) $l = \dfrac{u + 4}{4 + \boldsymbol{v}}$

T7 Rearrange this formula to make c the subject. $\quad a = \dfrac{c}{2 - c}$ *AQA(NEAB) 1998*

T8 Make a the subject of the formula $\quad 6(a + 2b) = 4a + 7.$ *OCR*

T9 Make the letter in square brackets the subject.

    (a) $5s + 2 = a(s + 2)$ $[s]$ (b) $k = \dfrac{h}{h - 2}$ $[h]$ (c) $6(n + m) = 3(n - m)$ $[n]$

    (d) $T = \dfrac{a - c}{c - d}$ $[c]$ (e) $L = L_0 + \dfrac{\alpha L_0 t}{100}$ $[L_0]$ (f) $x^2 y^2 = x^2 - y^2$ $[x]$

T10 Given that $ay - b = cy - d$, express y in terms of a, b, c and d.   *OCR*

T11 The diagram shows a waste-bin in the shape of a
hemisphere of radius r cm on top of a cylinder of height $2r$ cm.

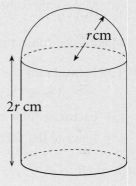

    (a) Show that the total volume, V cm^3 of the bin
       is given by $V = \frac{8}{3}\pi r^3$.

    (b) Rearrange this formula to make r the subject.

    (c) The company wishes to make waste-bins with
       a total volume of 35 **litres**.
       Calculate the value of r for this size bin.

       Give your answer to a reasonable degree of accuracy.

Review 7

1 Write each of these as a single algebraic fraction.

(a) $\frac{a}{2} + \frac{a}{4}$ (b) $\frac{2}{a} + \frac{4}{a}$ (c) $\frac{2}{a} + \frac{a}{4}$ (d) $\frac{1}{2a} + \frac{1}{4a}$ (e) $\frac{2}{a} \div \frac{a}{4}$

2 (a) Show that the equation
$$x^3 + x - 1 = 0$$
can be written in the form
$$x^2 + 1 = \frac{1}{x}.$$

(b) From the diagram, estimate
a solution of the equation
$$x^3 + x - 1 = 0.$$

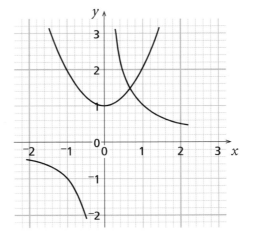

3 Calculate angles a, b and c, giving reasons for each step of working.

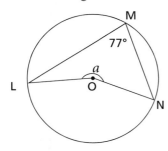 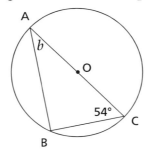 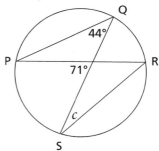

4 Make v the subject of the formula $w = \dfrac{v - 2}{v}$.

5 These box plots show the distributions of test scores in two subjects.

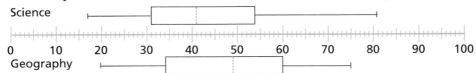

(a) What is the range of the science scores?

(b) What is the interquartile range of the geography scores?

(c) Write a couple of sentences comparing the two distributions.

6 An **open** cardboard box of height h cm has a square base of side x cm.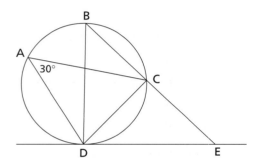

   (a) Write down an expression in terms of x and h for the total area of cardboard.

   (b) The area of cardboard is $50 \, \text{cm}^2$.

      Show that $h = \dfrac{50 - x^2}{4x}$.

   (c) Show that if $V \, \text{cm}^3$ is the volume of the box,

      then $V = \dfrac{x(50 - x^2)}{4}$.

   (d) Draw a graph of V against x for values of x from 0 to 7.

   (e) Use your graph to estimate the value of x for which V is a maximum and hence the maximum volume of the box.

7 Write each of these as a single algebraic fraction.

   (a) $\dfrac{x+5}{3} + \dfrac{x-2}{2}$ (b) $\dfrac{x}{2} - \dfrac{x-1}{4}$ (c) $a + \dfrac{2a-3}{5}$

   (d) $\dfrac{a}{x} \div \dfrac{x}{b}$ (e) $\dfrac{p}{2q} - \dfrac{q}{6}$ (f) $\dfrac{a}{b} - \dfrac{b}{a}$

8 A, B, C and D are four points on the circumference of a circle.
BD is a diameter of the circle
and DE is a tangent to the circle at D.
Angle DAC = 30°.

Calculate angle BED,
giving the reason for each step of your working.

9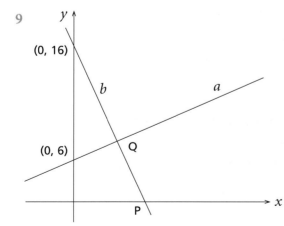

   (a) The gradient of line a is $\frac{1}{2}$.
      What is the equation of line a?

   (b) Line b is perpendicular to line a.
      What is the gradient of line b?

   (c) What is the equation of line b?

   (d) Find the coordinates of the point P.

   (e) Find the coordinates of the point Q.

10 The cost of cinema tickets for 2 adults and 3 children is £20.
    The cost of cinema tickets for 1 adult and 4 children is £18.75.

    By forming and solving a pair of simultaneous equations, find the cost of

    (a) an adult ticket   (b) a child's ticket

11 Calculate angles p, q, r and s, giving
    reasons for each step of working.

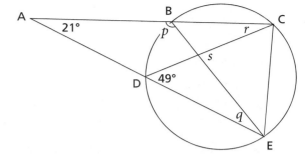

12 Rearrange each of these formulas to make t the subject.

    (a) $b = \dfrac{a}{a+t}$ (b) $b = \dfrac{a+t}{t}$ (c) $b = \dfrac{at}{a+t}$ (d) $b = \dfrac{a+t}{a-t}$

13 This diagram shows part of a regular polygon with 15 sides.
    O is the centre of the polygon.
    The perimeter of the polygon is 660 cm.

    Calculate

    (a) the length of each side of the polygon

    (b) the size of angle a

    (c) the area of the polygon, in m^2 (to three significant figures)

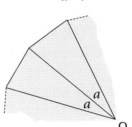

14 Ben and Asad each have some money.
    Ben gives Asad £7. Asad now has 3 times as much as Ben.
    Ben gives Asad another £3. Asad now has 5 times as much as Ben.

    How much did Ben and Asad have to start with?
    Show how you get your answer.

15 Class 10X consists of 18 girls and 12 boys.
    Class 10Y consists of 15 girls and 9 boys.

    A teacher picks a student at random from each of the two classes.
    Calculate the probability that the teacher picks

    (a) two girls   (b) one student of each sex

16 The three sides of a right-angled triangle, in cm, are
    x, $x + 7$ and $x + 9$.

    Write down an equation in x, simplify it and solve it.

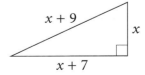

17 A, B, C and D are four points on the
   circumference of a circle.
   DE is the tangent to the circle at D.

   Given that angle CDE = 35°
   and angle BDC = 70°,
   calculate each of the following angles.

   Give reasons for your answers.

   (a) Angle BCD

   (b) Angle BAD

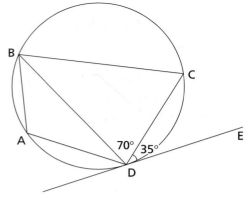

*18 A 'pseudo random' sequence is a sequence which looks like a random sequence
   but is in fact produced by a process whose results are predictable.

   One method for generating a pseudo-random sequence is as follows.

   • Start with a two-digit number, say 38.
     Square it: 1444. Use the middle pair of digits as the next number: 44.
     Square 44: 1936. Middle pair: 93, and so on.

   • This produces the sequence: 38, 44, 93, 64, …

   • If squaring produces fewer than four digits, initial zeros are included.
     So 256 becomes 0256, 49 becomes 0049, and so on.

   (a) Start with 62 and use this process to generate a sequence.
       What happens?

   (b) Explain why it is that whatever two-digit number you start with, eventually the
       sequence will recur.

Challenge

In a regular polygon all the diagonals that start from one vertex
are drawn.

   • Prove that the angles between the diagonals are all equal.

     (Hint: draw the circle through all the vertices of the
     regular polygon.)

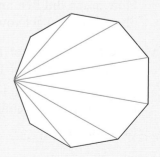

31 Accuracy

This work will help you

◆ find the upper and lower bounds of the interval within which a number must lie when you are given an approximate value

◆ find the upper and lower bounds of the result of a calculation involving approximate values

A Rounding and intervals

For class or group discussion

• What is the height of each person, to the nearest 10 centimetres?

  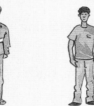   

178 cm   181 cm   177 cm   180 cm   183 cm   176 cm

• What other heights give the same result when rounded to the nearest 10 cm?

• What is the length of each pencil, to the nearest centimetre?

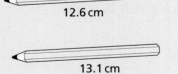

12.6 cm

12.9 cm

13.4 cm

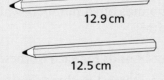

13.1 cm   12.5 cm   12.8 cm

• What other lengths give the result 13 cm when rounded to the nearest centimetre?

Steve weighed a parcel and rounded the weight to the nearest 0.1 kg.
He gave the weight as 1.3 kg.

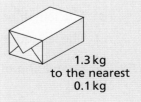

1.3 kg
to the nearest
0.1 kg

• Which of these could its actual weight have been?

1.24 kg   1.248 kg   1.25 kg   1.268 kg   1.31 kg

1.345 02 kg   1.35 kg   1.3725 kg   1.39 kg

You are told that the height of a building is 47.2 m, to the nearest 0.1 m.
• What could the actual height be? Write down five possibilities.

B Upper and lower bounds

This interval contains all the numbers
for which 6.8 is the nearest tenth.

6.75 is called the **lower bound** of the interval, and 6.85 the **upper bound**.
(Strictly speaking, 6.85 itself is not in the interval, because 6.85 would be
rounded to 6.9, not 6.8. The interval can be described as $6.75 \leq x < 6.85$.)

B1 Think of all the numbers for which 7.1 is the nearest tenth.
What are the lower and upper bounds of this interval?

B2 What are the lower and upper bounds of the interval containing
all the numbers for which 6.1 is the nearest tenth?

B3 What are the lower and upper bounds of all numbers for which
7.0 is the nearest tenth?

B4 The length of a bolt is given as 5.7 cm, to two significant figures.
What are the upper and lower bounds of its actual length?

B5 The weight of a truck is 3400 kg, to two significant figures.
What are the upper and lower bounds of the weight of the truck?

B6 What are the lower and upper bounds of all numbers which give 3.28 when
rounded to the nearest hundredth?

B7 Write down the lower and upper bounds for each of these.
 (a) The height of a post is 3.46 m, correct to three significant figures.
 (b) The weight of a sack is 3.8 kg, correct to two significant figures.
 (c) The capacity of a tank is 1530 litres, correct to the nearest 10 litres.
 (d) The length of a path, to the nearest tenth of a km, is 4.0 km.
 (e) The length of a pipe is 4.58 m, correct to the nearest cm.

B8 The population of a town is given as 74 200, correct to the nearest hundred.
What is
 (a) the maximum possible population   (b) the minimum possible population

B9 The resistance of an electrical component is given as 3.50 ohm, correct to 2 d.p.
What are the upper and lower bounds of the actual resistance?

C Calculating with approximations

For class or group discussion

What can you say about ...

1 ... the total length of three rods like these?

| 3.6 m, to the nearest 0.1 m | 3.6 m, to the nearest 0.1 m | 3.6 m, to the nearest 0.1 m |

2 ... the total length of these rods?

| 3.6 m, to the nearest 0.1 m | 2.3 m, to the nearest 0.1 m |

3 ... the perimeter of this rectangle?   2.8 m, to the nearest 0.1 m

1.3 m, to the
nearest 0.1 m

4 ... the area of the rectangle?

5 ... the difference in length between these two rods?

| 7.4 m, to the nearest 0.1 m |
| 4.2 m, to the nearest 0.1 m |

6 ... the average speed of this plane?

Distance travelled: 4284 km, to the nearest km
Time taken: 8 hours, to the nearest hour

C1 Find the upper and lower bounds of the total weight of four bottles,
each weighing 325 g to the nearest gram.

C2 Find the upper and lower bounds of the total length of five rods,
each of length 3.9 m to the nearest 0.1 m.

C3 A torch contains three batteries.
The torch itself, without batteries, weighs 86 g, correct to the nearest gram.
Each battery weighs 54 g, correct to the nearest gram.
Find the upper and lower bounds of the total weight of the torch and batteries.

Example

Find the upper and lower bounds of the perimeter and the area of a rectangle
whose length is 3.3 m and width 2.6 m, both to the nearest 0.1 m.

3.3 m

2.6 m

> The lower and upper bounds for the length are 3.25 m and 3.35 m.
> The lower and upper bounds for the width are 2.55 m and 2.65 m.
>
> Lower bound for perimeter = $2 \times 3.25 + 2 \times 2.55 = 6.5 + 5.1 =$ **11.6 cm**
> Upper bound for perimeter = $2 \times 3.35 + 2 \times 2.65 = 6.7 + 5.3 =$ **12.0 cm**
>
> Lower bound for area = $3.25 \times 2.55 =$ **8.2875 cm^2**
> Upper bound for area = $3.35 \times 2.65 =$ **8.8775 cm^2**

C4 The sides of a triangle are of length 12 cm, 17 cm and 9 cm, each correct to
the nearest centimetre.
Find the upper and lower bounds of the perimeter of the triangle.

C5 The dimensions of a cuboid, to the nearest 0.1 cm, are 7.6 cm, 9.5 cm and 12.0 cm.
Find the upper and lower bounds of the volume of the cuboid.

C6 In this right-angled triangle, side $a = 13$ cm to the nearest cm
and angle $\theta = 43°$ to the nearest degree.

Calculate upper and lower bounds for the height h.
Give each answer to four significant figures.

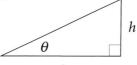

h

θ

a

D *Subtraction*

Example

Saul weighed 85 kg before training and 81 kg afterwards.
Both weights are to the nearest kilogram.
Find the lower and upper bounds for his weight loss.

> Before training: lower bound 84.5 kg, upper bound 85.5 kg
>
> After training: lower bound 80.5 kg, upper bound 81.5 kg
>
> The sketch below shows these intervals.

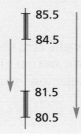

85.5

84.5

81.5

80.5

The loss would be smallest if he
went from the lowest starting weight
to the highest finishing weight.

So the lower bound of his weight
loss is $84.5 - 81.5 =$ **3 kg**.

The loss would be biggest if he went
from the highest starting weight
to the lowest finishing weight.

So the upper bound of his weight
loss is $85.5 - 80.5 =$ **5 kg**.

D1 A beaker of water weighed 426 g, to the nearest gram.
After the water had been poured out, the beaker weighed 173 g, to the nearest gram.
Find the lower and upper bounds of the weight of the water.

D2 A piece weighing 260 g, to the nearest 10 g, is cut from a cheese weighing 840 g, to the nearest 10 g.
Find the lower and upper bounds of the weight of the cheese that remains.

D3 A cable is 50 m long, correct to the nearest metre.
Barrie cuts off a piece which is 15 m long, correct to the nearest 10 cm.
Calculate the maximum length of the remaining cable. OCR

D4 (a) Plastic bricks are made in the form of cubes.
The length of each brick is 11.3 cm, measured correct to the nearest mm.
Write down the least and greatest values of the length of a brick.

(b) The length of a shelf between two walls is 91 cm,
measured correct to the nearest cm.
Explain, showing all your calculations, why it is
not always possible to place eight bricks on the shelf.

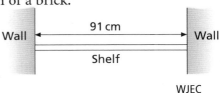

WJEC

E *Division*

Example

A runner covers 75 m (to the nearest metre) in 7 seconds (to the nearest second).
Find the lower and upper bounds for the runner's average speed.

The lower and upper bounds for the distance are 74.5 m and 75.5 m.
The lower and upper bounds for the time are 6.5 seconds and 7.5 seconds.

The speed would be lowest if the runner covered the shortest distance in the longest time. So the lower bound of the average speed is 74.5/7.5 = **9.933... m/s.**

The speed would be highest if the runner covered the longest distance in the shortest time. So the upper bound of the average speed is 75.5/6.5 = **11.615... m/s.**

E1 A pump delivers 3260 litres of water (to the nearest 10 litres) in
17 minutes (to the nearest minute).

Find the upper and lower bounds of the rate, in litres per minute, at which the pump is working. (Give your answers to four significant figures.)

E2 Kirstie measures the mass and volume of a piece of metal.
The mass is 141 g, to the nearest gram; the volume is 18 cm³, to the nearest cm³.

Find the upper and lower bounds of the density of the metal in g/cm³.
(Give your answers to four significant figures.)

E3 The magnification of a lens is given by the formula $m = \frac{v}{u}$.
In an experiment, u is measured as 8.5 cm and v is measured as 14.0 cm, both correct to the nearest 0.1 cm.

Find the least possible value of m.
You **must** show full details of your calculation.   OCR

E4 The length of a rectangle is a centimetres.
Correct to two decimal places, $a = 6.37$.

(a) For this value of a, write down (i) the upper bound (ii) the lower bound

Correct to one significant figure, the area of the same rectangle is 20 cm^2.

(b) Calculate the upper bound for the width of the rectangle.
Write down all the figures on your calculator display.

A diagonal of the rectangle makes an angle of $\theta°$ with one of the sides of length a cm.

(c) Calculate the upper bound for the value of tan $\theta°$.
Write down all the figures on your calculator display.   Edexcel

E5 A hoist uses a cable with a breaking load of 1400 kg, measured to two significant figures.

It is used to lift boxes with a weight of 40 kg, measured to two significant figures.

What is the greatest number of boxes that can be lifted at one time to be sure that the cable does not break?

AQA(NEAB)1999

*E6 The formula $\frac{1}{u} + \frac{1}{v} = \frac{1}{f}$ is used in physics.

u and v are the distances from a lens of an object and its image.
f is a quantity called the 'focal length' of the lens.

(a) In an experiment with a lens, u and v are measured as 64 cm and 25 cm.
Calculate, correct to the nearest 0.1 cm, the value of f obtained when these measurements are substituted into the formula.

(b) In fact the values of u and v given in (a) are measured correct to the nearest centimetre.
Calculate, correct to 2 d.p., upper and lower bounds for the value of f.

(c) In another experiment a different lens is used.
The focal length of this lens is 25 cm, correct to the nearest cm.
The value of u is measured as 40 cm, correct to the nearest cm.
Calculate, correct to 2 d.p., the upper bound for the value of v.

Test yourself

⊠ **T1** The table shows the number of hours of sunshine and the rainfall,
in centimetres, in each of six places one day in December.

	Number of hours of sunshine	Rainfall in cm
Anglesey	5.7	0.06
Birmingham	4.4	0.35
Folkestone	3.6	0.42
Guernsey	7.2	0.08
Jersey	6.5	0.17
Torquay	6.9	0.04

The number of hours of sunshine is given correct to one decimal place.

The rainfall is given to the nearest 0.01 cm.

(a) Write down

    (i) the lower bound of the number of hours of sunshine in Anglesey

    (ii) the upper bound of the rainfall in Torquay

(b) Calculate the lower bound of the **sum** of the number of hours of sunshine in Birmingham and in Folkestone.

(c) Calculate the greatest possible difference between the rainfall in Guernsey and the rainfall in Jersey.   Edexcel

T2 Given that $x = 3.4$ and $y = 4.8$, both to one decimal place, find the upper and lower bounds of the following.

(a) $x + y$ (b) $2x + y$ (c) $2x - y$ (d) xy (e) $\frac{x}{y}$

T3 The diagram shows a map of the M1 motorway between junctions 20 and 22.

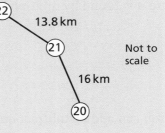

The distance between junction 20 and junction 21 is 16 km, to the nearest kilometre.
The distance between junction 21 and junction 22 is 13.8 km, to the nearest 0.1 km.

(a) What is the lower bound of the total distance between junction 20 and junction 22?

A car takes 17 minutes, to the nearest minute, to travel between junction 20 and junction 22.

(b) Calculate the lower bound of the average speed in km/minute of the car between junctions 20 and 22.   AQA(SEG) 2000

32 Dimensions

This work will help you

◆ work out the dimension of an expression

◆ use dimensions to check that an expression is sensible

The area of a triangle is given by $\frac{1}{2}$ **base** × **height**.

In this expression,

the base is a ***length***,

the height is a ***length***,

the $\frac{1}{2}$ is a ***pure number***.

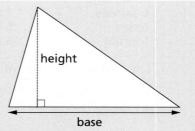

We say the **dimension** of the area is **[*length*]²** because it is found by multiplying *length* × *length* × *a number*.

(This is also shown by the units, for example cm for length and cm² for area.)

A length has dimension $[length]^1$.

A pure number (including, for example, π) has no dimension.

The volume of a cylinder of radius r and height h is $\pi r^2 h$.

The dimension of the expression $\pi r^2 h$ is **[*length*]³** because it is found by multiplying *number* × *(length)²* × *length*.

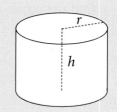

- You can add or subtract quantities with the same dimension.

    $4\,\text{cm} + 2\,\text{cm} = 6\,\text{cm}$

    $6\,\text{cm}^2 + 3\,\text{cm}^2 = 9\,\text{cm}^2$

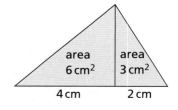

But it does not make sense to add or subtract quantities with different dimensions. For example, you can't add an area of 6 cm² to a length of 4 cm.

    $6\,\text{cm}^2 + 4\,\text{cm} = \text{nonsense!}$

By checking dimensions we can see whether an expression could represent a length, an area or a volume (or none of these).

Examples

Suppose a, b, c represent lengths. What, if anything, could these represent?

(a) $(a + b)^2$

> The dimension of $(a + b)$ is [*length*],
> so the dimension of $(a + b)^2$ is [*length*]^2.
> So the expression could represent an **area**.

(b) $a(b - 2c)$

> The dimension of $2c$ is [*length*],
> so the dimension of $(b - 2c)$ is also [*length*].
> The dimension of $a(b - 2c)$ is [*length*] × [*length*] = [*length*]^2.
> So the expression could represent an **area**.

(c) $\dfrac{a^3b}{c}$

> The dimension of a^3b is [*length*]^3 × [*length*] = [*length*]^4.
> The dimension of $\dfrac{a^3b}{c}$ is $\dfrac{[\textit{length}]^4}{[\textit{length}]}$ = [*length*]^3.
> So the expression could represent a **volume**.

(d) $b^2 + c$

> The dimension of b^2 is [*length*]^2. The dimension of c is [*length*].
> It is impossible to add a length to an area.
> So the expression is **dimensionally inconsistent**.
> It does not represent anything.

In these questions, the letters a, b, c, d, h, r, … represent lengths.

1 Write an expression for each of the following.
   Check that your expression is dimensionally correct.

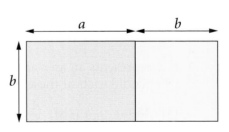

(a) The area of the green rectangle

(b) The perimeter of the green rectangle

(c) The perimeter of the whole coloured rectangle

(d) The total area of the coloured rectangle

(e) The difference between the green and yellow areas

2 What is the dimension of each of these expressions?

(a) abc (b) $a + b$ (c) ab^2 (d) $\dfrac{ab}{c}$ (e) $2a - b$

3 One of the expressions below could represent a length, one an area, one a volume and one a pure number. Which is which?

$$a^2b \qquad \frac{ab}{c^2} \qquad 2ab \qquad a + 3b$$

4 What is the dimension of each of these expressions?

(a) $a(b + c)$ (b) $\dfrac{a - b}{2}$ (c) $\dfrac{a^2 - b^2}{4}$ (d) $(2a - b)^3$

5 Say whether each of these expressions could represent a length, an area, a volume or a pure number, or whether it is dimensionally inconsistent.

(a) $ab + cd$ (b) $\dfrac{a^2 + b^2}{c}$ (c) $\dfrac{abc}{d^3}$ (d) $ab^2 - 4c^3$ (e) $a^2 + 3b$

6 A book on engineering includes the formulas below, which relate to cylinders, cones and spheres.

For each formula, say whether it could represent a length, an area, a volume, or a pure number.

(a) $4\pi r^2$ (b) $\dfrac{\pi(b^2 - a^2)}{4}$ (c) $2\pi r(r + h)$ (d) $\dfrac{4\pi r^3}{3}$

(e) $\pi h(a + b)$ (f) $\dfrac{ah}{b - a}$ (g) $\frac{1}{3}\pi h(a^2 + ab + b^2)$

The length of the hypotenuse of this right-angled triangle is
$$\sqrt{a^2 + b^2}$$
We can check that this expression has the correct dimension:

a^2 and b^2 are both $[length]^2$

So $a^2 + b^2$ is also $[length]^2$

So $\sqrt{a^2 + b^2}$ is $[length]$.

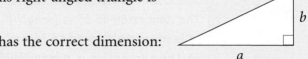

7 What is the dimension of each of these expressions?
(a) $\sqrt{ab}$ (b) $a\sqrt{bc}$ (c) $\sqrt{b^2 - 2c^2}$ (d) $\dfrac{a^3}{\sqrt{bc}}$ (e) $\dfrac{2\sqrt{ab}}{c}$

*8 If A represents an area and b and c represent lengths, what could each of these expressions represent?

(a) Ab (b) $\dfrac{A}{b}$ (c) $\dfrac{Ab}{c}$ (d) $\sqrt{\dfrac{A}{\pi}}$ (e) $\dfrac{\sqrt{2A + b^2}}{c}$

Test yourself

T1 The diagram shows a prism.

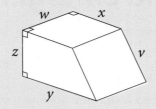

Not to scale

The following formulas represent certain quantities connected with the prism.

$$wx + wy \qquad \tfrac{1}{2}z(x + y)w \qquad \frac{z(x + y)}{2} \qquad 2(v + 2w + x + y + z)$$

(a) Which of these formulas represents length?

(b) Which of these formulas represents volume?

AQA(SEG) 2000

T2 In the following expressions r, a and b represent lengths.
For each expression state whether it represents

a **length**   an **area**   a **volume**   or **none** of these

(a) πab (b) $\pi r^2 a + 2\pi r$ (c) $\dfrac{\pi r a^3}{b}$

AQA 2000

T3 Here are some expressions.

$$\frac{ab}{h} \quad 2\pi b^2 \quad (a + b)ch \quad 2\pi a^3 \quad \pi ab \quad 2(a^2 + b^2) \quad \pi a^2 b$$

The letters a, b, c and h represent lengths.
π and 2 are numbers that have no dimensions.

Three of the expressions could represent areas. Which are they?

Edexcel

T4 The letters f, g and h all represent lengths.
For each of the following expressions, state whether it could represent a length,
an area, a volume or none of these.

(a) $f^2(h + g)$ (b) $\sqrt{h^2 gf}$ (c) $\pi(3f + 2g)$

OCR

T5

A	B	C	D
$p^3 + 3q^3$	$p^2 + 2q$	$2p + 3q$	$3p^2 + 2pq$

The boxes A, B, C and D show four expressions.
The letters p and q represent lengths.
2 and 3 are numbers which have no dimension.

(a) Write one of the letters A, B C or D for the expression which represents

    (i) an area   (ii) a length

The box X shows an expression.
The letters p and q represent lengths. n is a number.
The expression represents a volume.

X
$p^n(p + q)$

(b) Find the value of n.

Edexcel

33 *Similarity and enlargement*

You will revise finding and using the scale factor of an enlargement.

This work will help you

◆ solve problems involving similar triangles

◆ see how the scale factor, area factor and volume factor of an enlargement are related

A *Similar shapes*

When a shape is enlarged, the result is said to be **similar** to the original shape.
All lengths in the shape are multiplied by the same number, called the **scale factor**.
If the scale factor is less than 1, the result is smaller than the original,
but we often still use the word 'enlargement'.

A1   Shape Q is similar to shape P.
The scale factor of the
enlargement is 2.5.
Find *a*, *b*, *c* and *d*.

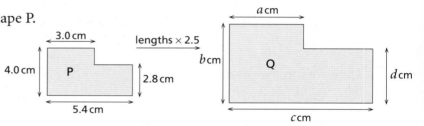

A2   A shop offers to enlarge photos 10 cm by 15 cm into posters 50 cm by 75 cm.
What is the scale factor of the enlargement?

A3   Shape B is similar to shape A.

(a) What is the scale factor
of the enlargement?

(b) Find *a* and *b*.

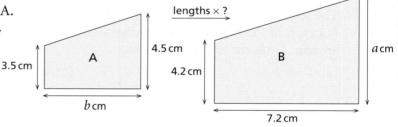

A4   Shape D is similar to shape C.

(a) What is the scale factor?

(b) Find *a*, *b* and *c*.

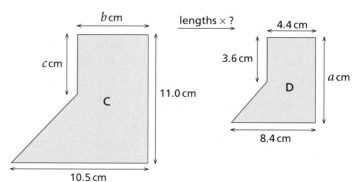

A5 Shape X is enlarged using a scale factor 3 to produce shape Y.
Y is then reduced by scale factor 0.8 to produce shape Z.
With what single scale factor could shape Z be produced directly from shape X?

A6 An architecture student draws a building to a scale of 1:50.
How big **in millimetres** would she draw a door that is 1 m by 2.2 m in real life?

A7 On a scale drawing a window 1.6 m tall is drawn 64 mm tall.
  (a) What is the scale of the drawing?
  (b) The window is drawn 42 mm wide. How wide is that, in metres, in real life?

A8 A map uses a scale of 1:25 000.
  (a) The distance between two farmhouses on the map is 5 cm.
      How far apart are the actual farmhouses, in km?
  (b) How long, in cm, will the route of a 15 km walk be on the map?

A9 From a junction the direct route to a town is given as 12 km. A man chooses an alternative route that is 9 cm on his road map. The scale of the road map is 1:200 000. How much further is his alternative route than the signposted one?

*A10 To set the scaling on a photocopier you use a percentage.
For example, 125% means a scale factor of 1.25.
  (a) What percentage setting should you use if you want to increase the width of a drawing from 15.4 cm to 20.0 cm? Give your answer to the nearest 1%.
  (b) Paul wants to reduce a picture on a photocopier so that all the lengths are 36% of the original. The smallest the photocopier will do is 50%. What should he do?

B Similar triangles

Are any of these triangles similar to each other?
If so, which pairs are similar?

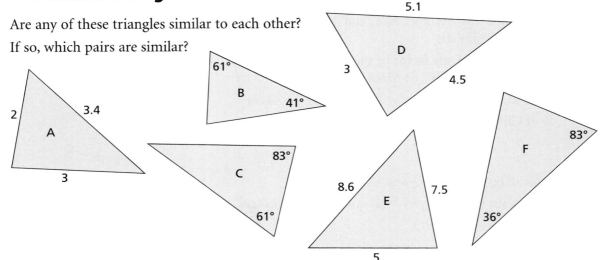

When triangles are similar ...

- Corresponding angles are equal.

- The ratio of each side in one triangle
  to its corresponding side in the other
  is equal to the scale factor of the enlargement.

- The ratio of a pair of sides in one triangle
  is equal to the ratio of the corresponding pair
  of sides in the other triangle.

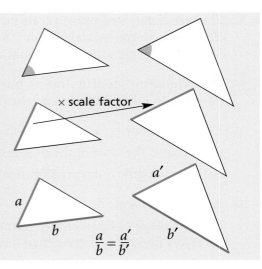

$$\frac{a}{b} = \frac{a'}{b'}$$

B1 In each of these pairs of similar triangles, find the values of
the lengths labelled with letters.

(a)

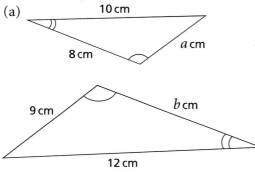

(b)

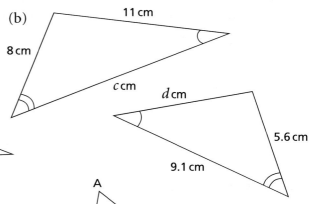

B2 In this diagram BC is parallel to DE.

(a) Explain why triangles ABC and ADE
must be similar.

(b) What is the scale factor of the enlargement
from triangle ABC to ADE?

(c) Find the length BC.

(d) If CE = 6 cm, how long is AE?

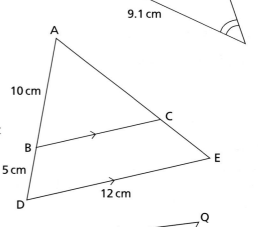

B3 In this diagram lines PQ and ST are parallel.

(a) Explain why triangles PQR and STR are similar.

(b) Calculate length PQ.

(c) PT = 12.5 cm. How long is PR?

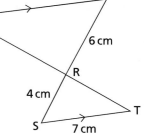

B4 BC is parallel to DE.
AB is twice as long as BD.
AD = 36 cm and AC = 27 cm

(a) Work out the length of AB.

(b) Work out the length of AE.

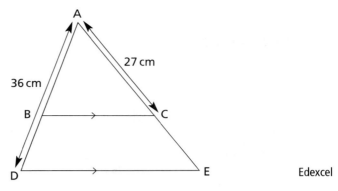

Edexcel

B5 JKLM is a trapezium with JK parallel to ML.

(a) Explain why triangles JKM and KLM are similar.

(b) Find the length of JK.

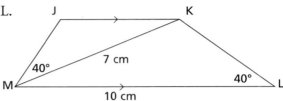

B6 In this circle chords PR and QS intersect at T.

(a) Prove that triangles PQT and SRT are similar.

(b) Find length TS.

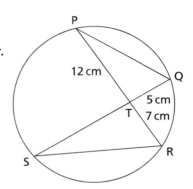

Tessellate to enlarge

Draw this L-shape on square dotty paper.

Four of these L-shapes can be put together without cutting to make an enlargement with scale factor 2.

How many of the original L-shapes would be needed to make an enlargement with scale factor 3? Can it be done without cutting?

What about bigger enlargements?

Experiment in the same way with these shapes and other simple shapes of your own.

C Enlargement and area

C1 (a) Calculate the area of this shape.

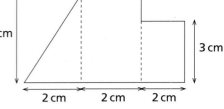

(b) The shape is to be enlarged with scale factor 2.
Sketch the enlarged shape and mark the new
dimensions on it.

(c) Calculate the area of the enlarged shape.
By what number has the area been multiplied as
a result of the enlargement? This number is called the **area factor**.

(d) Repeat steps (b) and (c) for an enlargement of the original shape with scale factor 3.

(e) Repeat (b) and (c) for an enlargement of the original with scale factor 1.5 .

(f) What do you notice about the area factors?

When rectangle Q is scaled down (reduced) with
scale factor $\frac{1}{2}$, you get rectangle P.

The area of Q is $8\,cm^2$ and the area of P is $2\,cm^2$.

So the area factor is $\frac{1}{4}$.

C2 (a) Copy and complete this table.

Original shape	Scaled shape	Scale factor	Original area	Scaled area	Area factor
Q	P	$\frac{1}{2}$	$8\,cm^2$	$2\,cm^2$	$\frac{1}{4}$
S	P				
R	P				
R	Q				
S	Q				
S	R				

(b) How is an area factor
related to the
scale factor?

C3 This sketch shows a rectangle being enlarged with scale factor k.

(a) What is the area (in square units) of the original rectangle?

(b) Copy the lower rectangle.
Label its width with an expression that has k in it.
Label its height with an expression that has k in it.

(c) Multiply these expressions together to get an expression for the area of the enlarged rectangle.

(d) By what area factor has the area been multiplied as a result of the enlargement?

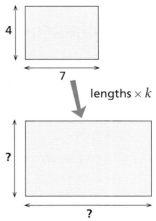

C4 This sketch shows a circle being enlarged with scale factor k.
The area of the original circle can be written $\pi \times 3^2 = 9\pi$.

(a) Write down the radius of the enlarged circle as an expression with k in it.

(b) Write an expression for the area of the enlarged circle. (Keep the symbol π in the expression: do not replace it by an approximate value of π.)

(c) By what area factor has the area been multiplied as a result of the enlargement?

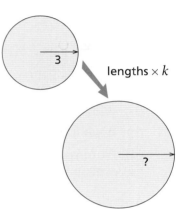

C5 Here a shape is being enlarged with scale factor k.

(a) What is the area in square units of the original shape?

(b) Copy the lower shape. Label the missing dimensions with expressions that have k in them.

(c) Carefully work out expressions involving k for the areas of the rectangles and triangle in the lower shape. Add them together to get an expression for the total area.

(d) By what area factor has the area been multiplied as a result of the enlargement?

(e) Does it matter whether k is greater than or less than 1?

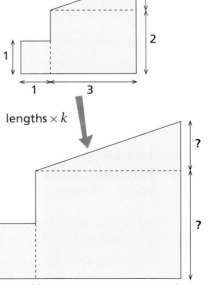

What you have seen for some shapes is true for any shape.

If a shape is enlarged by scale factor k, its area is multiplied by k^2.

We say: k^2 is the **area factor** of the enlargement.

This happens because every unit length becomes k units long,

and every square unit becomes k^2 square units in area.

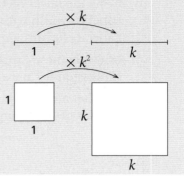

C6 On squared paper, draw axes from 0 to 10 in both directions.

   (a) Draw the pentagon with vertices at $(5, 1)$, $(4, 2)$, $(5, 3)$, $(7, 3)$, $(7, 2)$.
Draw an enlargement of the pentagon, scale factor 3, centre $(6, 0)$.

   (b) What is the area in square units of the original pentagon?

   (c) What is the area of the enlargement?

   (d) From your answers to (b) and (c), calculate the area factor and explain whether this is the expected result.

C7 A single picture on ordinary negative colour film is a rectangle 36 mm by 24 mm.
A standard colour print is about 150 mm by 100 mm.

   (a) What is the scale factor for the enlargement when a print is made?

   (b) What is the area factor?

C8 What is the area factor of an enlargement whose scale factor is

   (a) 5   (b) 9   (c) 100   (d) 4.5   (e) $\sqrt{3}$

C9 A design for a mosaic has an area of 176 cm^2.
The actual mosaic is an enlargement of the design with scale factor 2.5.
What is the area of the completed mosaic?

C10 An artist does a preparatory drawing for a mural to a scale of $1:10$.
A red circle on the drawing has an area of 35 cm^2.
What is the area of that circle on the full-size mural?

C11 (a) If you want to enlarge a photo so its area is multiplied by 4, what scale factor do you use for the enlargement?

    (b) What scale factor do you use if you want the area to be doubled?

C12 What is the scale factor of an enlargement whose area factor is

    (a) 16   (b) 100   (c) 42.25   (d) 40 000   (e) 5

C13 The area of a window is 400 times the area of the
rectangle representing it on an architect's drawing.
What is the scale of the architect's drawing?

C14 On a designer's drawing for a watch the watch face has an area of $105\,\text{cm}^2$.
The actual watch face has an area of $4.2\,\text{cm}^2$.

(a) What is the scale factor of the enlargement from the actual watch to the drawing?

(b) What is the scale factor of the scaling down from the drawing to the actual watch?

C15 What is the area factor of a scaling down whose scale factor is

(a) $\frac{1}{4}$ (b) 0.6   (c) $\frac{3}{4}$ (d) 0.2   (e) 0.1

C16 The sails of the Astra sailing dinghy have a total area of $12.8\,\text{m}^2$.

(a) The manufacturers produce another dinghy called the Comet,
which is the Astra design scaled down with a scale factor of 0.85.
What is the area of the Comet's sails?

(b) The Meteor is also a scaled-down version of the Astra.
Its sails have an area of $7.2\,\text{m}^2$.
What scale factor is used to scale the Astra down to the Meteor?

C17 (a) If you want to scale down a picture so its area is $\frac{1}{25}$ of
what it was before, what scale factor should you use?

(b) What scale factor should you use if you want a picture's area to be
$\frac{1}{10}$ of what it was before?

C18 (a) Two circles have areas in the ratio $9:4$. What is the ratio of their diameters?

(b) Two similar triangles have areas in the ratio $6.25:1$.
What is the ratio of their heights?

C19 (a) Which of these three 'shields' do you think has half its area coloured blue?
Estimate the fraction shaded in the other two shields.

A 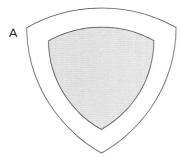   B 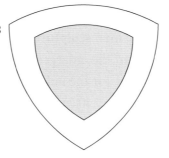   C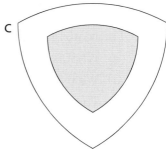

(b) Each shaded part is a reduction of the outline of the shield.
Take measurements and work out the area factor of each reduction.
Compare these answers with your estimates.

D Enlargement and volume

D1 (a) Calculate the volume of this cuboid.

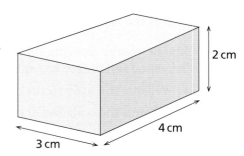

(b) The cuboid is to be enlarged with scale factor 2. Sketch the enlarged cuboid and mark the new dimensions on it.

(c) Calculate the volume of the enlarged cuboid. By what volume factor has the volume been multiplied as a result of the enlargement?

Repeat steps (b) and (c) for

(d) an enlargement of the original cuboid with scale factor 3

(e) an enlargement of the original with scale factor 1.5

(f) a scaling down of the original with scale factor 0.5

(g) a scaling of the original with scale factor k

If a shape is enlarged by scale factor k, its volume is multiplied by k^3.

We say: k^3 is the **volume factor** of the enlargement.

This happens because every unit length becomes k units long,

and every cubic unit becomes k^3 cubic units of volume.

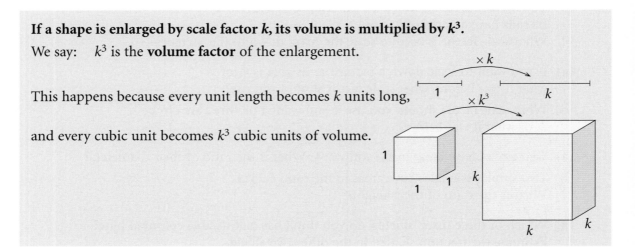

D2 A sculptor makes a small-scale clay model of a sculpture she is planning. The model is 40 cm tall. The final sculpture will be 100 cm tall.

(a) What is the scale factor of the enlargement?

(b) What is the volume factor for the enlargement?

(c) The volume of the model is 240 cm³.
What will be the volume of the final sculpture?

D3 Find the volume factor for an enlargement whose scale factor is

(a) 4   (b) 5   (c) 27   (d) 10   (e) 50

D4 A glass bottle factory makes two types of bottle.
The larger type is an enlargement of the smaller type with scale factor 3.
The smaller type has a volume of $10\,cm^3$.
What is the volume of the larger type?

D5 Another factory makes two types of bottle.
The larger type is an enlargement of the smaller type with scale factor 1.8.

   (a) Calculate the volume factor of the enlargement.

   (b) The smaller type of bottle has a volume of $60\,cm^3$.
      Calculate the volume of the larger type to the nearest cm^3.

D6 This carton contains $250\,g$ of soft cheese.
The manufacturer decides to enlarge the carton.
All the lengths will be increased by 10%.

   (a) What is the volume factor for the enlargement?

   (b) What weight of cheese will the new carton hold?

D7 A model road tanker is made to a scale of $1:10$.

   (a) The model tanker holds 75 litres of liquid.
      How much does the real tanker hold?

   (b) The inside of the model tanker has a surface area of $1.1\,m^2$.
      What is the inside surface area on the real tanker?

D8 An enlarged version of a statue weighs 216 times as much as the original.
Both the original and the enlargement are made of the same metal.
Find the scale factor of the enlargement.

D9 Find the scale factor of an enlargement whose volume factor is

   (a) 1000   (b) 343   (c) 8000   (d) 1 000 000   (e) 125 000

D10 A king had an altar of solid gold, in the shape of a cube.
He wanted a new altar with a volume exactly twice that of the existing altar.

The problem was to decide what the scale factor of the enlargement should be.
1.5 would be too much because the volume factor would be 3.375.

Use trial and improvement to find the scale factor needed.

D11 This carton contains ice-cream.
The manufacturers want a larger, similar-shaped, carton
that they can label '20% more ice-cream'.

   (a) What is the volume factor for this enlargement?

   (b) Use trial and improvement to find the scale factor needed.

D12 The young fish in this picture is similar in shape to the adult fish but scaled down by a factor of about 0.4.

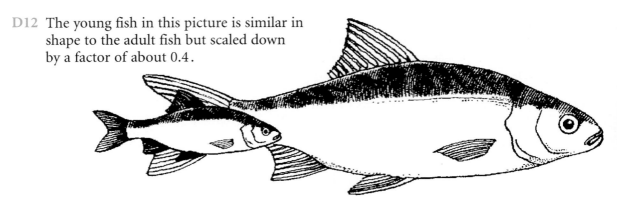

(a) Calculate the volume factor for this reduction.

(b) If the adult fish weighs 800 g, how much would you expect the young fish to weigh?

D13 Find the volume factor for an enlargement whose scale factor is

(a) $\frac{1}{2}$ (b) $\frac{1}{3}$ (c) $\frac{3}{4}$ (d) 0.8   (e) 0.1

D14 These are two similar-shaped mugs. The larger mug holds 324 cm³ of liquid. What does the smaller one hold?

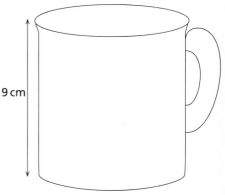

6 cm   9 cm

D15 Which of these pictures do you think shows the glass half full? Estimate the fraction of the glass filled in the other two pictures.

A    B    C

D15 (continued)

Each 'cone' of liquid can be treated as a similar-shaped reduction of the conical glass itself.

Use these measurements to work out the volume factor for each reduction.
Compare these answers with your estimates.

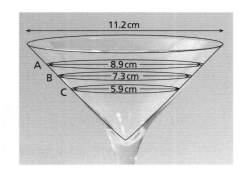

D16 Assume that these four eggs are similar in shape.

(a) The volume of the jackdaw's egg is 6.00 cm³.
Calculate the volumes of the other eggs.

(b) The surface area of the jackdaw's egg is 21.4 cm².
Calculate the surface areas of the other eggs.

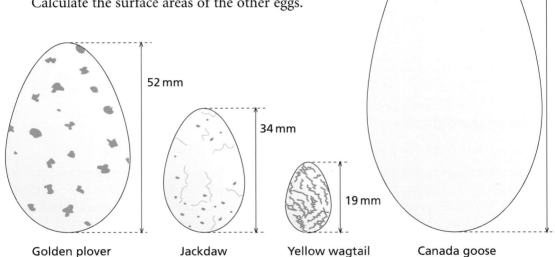

| Golden plover | Jackdaw | Yellow wagtail | Canada goose |

D17 A dolls' house is a reduction (scaled-down copy) of a real house.
Copy and complete this table of information about the two houses.

	Dolls' house	Real house
Height of front door	6 cm	3 m
Number of windows	4	
Total area of windows	96 cm²	
Height of chimney		1 m
Capacity of water tank	12 ml	
Number of roof tiles		1600
Area of one roof tile	0.45 cm²	
Volume of roof space	3024 cm³	
Percentage of floor carpeted		80%
Total area of carpet		185 m²

D18 Two spheres have volumes in the ratio $27:343$.
What is the ratio of their diameters?

D19 Two similar cones have surface areas in the ratio $25:9$.

(a) What is the ratio of their volumes?

(b) If the volume of the larger cone is $500\,cm^3$, what is the volume of the smaller cone?

D20 This picture appeared in a magazine in 1979.
The increasing size of the barrels is supposed to
represent the increasing price of a barrel of crude oil.

From your own measurements and calculations
say whether you think this is a fair representation
of the changing price.

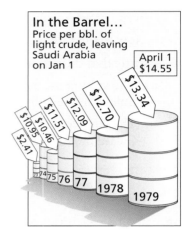

In the Barrel...
Price per bbl. of
light crude, leaving
Saudi Arabia
on Jan 1

April 1
$14.55

$13.34

$12.70

$12.09

$11.51

$10.46

$10.95

$2.41

74 75 76 77 1978 1979

On being the right size

These extracts come from an article by the biologist J B S Haldane (1892–1964).

Consider a giant man sixty feet high – about the height of Giant Pope and Giant
Pagan in the illustrated *Pilgrim's Progress* of my childhood. These monsters were not
only ten times as high as Christian, but ten times as wide and ten times as thick, so
their total weight was a thousand times his, or about eighty to ninety tons.
Unfortunately the cross-sections of their bones were only a hundred times those of
Christian, so that every square inch of giant bone had to support ten times the weight
borne by a square inch of human bone. As the human thigh-bone breaks under about
ten times the human weight, Pope and Pagan would have broken their thighs every
time they took a step. This was doubtless why they were sitting down in the picture I
remember. But it lessens one's respect for Christian and Jack the Giant Killer.

Gravity, a mere nuisance to Christian, was a terror to Pope and Pagan. To the mouse
and any smaller animal it presents practically no dangers. You can drop a mouse
down a three thousand foot mine shaft, and, on arriving at the bottom, it gets a slight
shock and walks away. A rat would probably be killed, though it can fall safely from
the eleventh floor of a building; a man is killed, a horse splashes. For the resistance
presented to movement by the air is proportional to the surface of the moving object.
Divide an animal's length, breadth and height each by ten; its weight is reduced to a
thousandth, but its surface only to a hundredth. So the resistance to falling in the case
of the small animal is relatively ten times greater than the driving force.

E *Mixed units*

This diagram shows that there are
100×100 square centimetres in a square metre.
That is 10 000 cm^2 in a m^2.

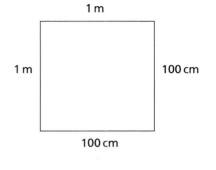

E1 (a) How many m^2 are there in a km^2?

(b) How many cm^2 are there in a km^2?

(c) How many mm^2 are there in a cm^2?

E2 (a) How many mm^3 are there in a cm^3?

(b) How many cm^3 are there in a m^3?

E3 (a) How many cm^3 are there in one litre?

(b) How many litres are there in a m^3?

It is sometimes necessary to convert the units of measurement.

Examples

A plan of a house is drawn to a scale of $1:50$.
The living room has an area of 15 m^2.
What area in cm^2 will this be
represented by on the plan?

A metre in real life is represented by 2 cm on the plan.
So a square metre can be represented by a
square 2 cm by 2 cm. This has an area of 4 cm^2.

So 15 m^2 is represented by 15×4 cm^2,
which is 60 cm^2.

A map is drawn to a scale of $1:50\,000$.
A lake on the map has an area of 2.6 cm^2.
What is the area of the real lake?

Scale factor = 50 000
Area factor = $50\,000^2$ = 2 500 000 000
Area of real lake = $2.6 \times 2\,500\,000\,000$ cm^2

1 km^2 = 10 000 000 000 cm^2

So area of real lake = $\dfrac{2.6 \times 2\,500\,000\,000}{10\,000\,000\,000}$ km^2

$= 0.65$ km^2

E4 A plan of a small garden is drawn to a scale of $1:25$.

(a) A pond covers 24 cm^2 on the plan.
What is the area in m^2 of the real pond?

(b) The real lawn has an area of 17 m^2.
What area in cm^2 will it cover on the plan?

E5 A street plan is drawn to a scale of $1:10\,000$.
A housing estate occupies $60\,cm^2$ on the plan.
What is the area of the real housing estate

(a) in m^2 (b) in km^2

E6 The area of Ullswater lake in Cumbria is $9\,km^2$.
What area in cm^2 would it occupy on a map with

(a) a scale of $1:50\,000$ (b) a scale of $1:25\,000$

E7 Orange fizz comes in three similar bottles containing 1 litre, 2 litres and 2.5 litres.
The base diameter of the 2 litre bottle is $12\,cm$.

(a) What is the base diameter of the 1 litre bottle, to the nearest $0.1\,cm$?

(b) What is the base diameter of the 2.5 litre bottle, to the nearest $0.1\,cm$?

E8 The Eiffel Tower is $300\,m$ high.
A model of the Eiffel Tower is $15\,cm$ high and its base area is $25.5\,cm^2$.
What is the base area of the actual tower?

E9 An ornamental pond has width $8\,m$ and contains $50\,m^3$ of water.
How much water will be contained by a model of the pond with width $20\,cm$?

E10 This is a model of a water trough.
When full, the surface area of the water in it is $400\,cm^2$.
The surface area of the water in the real trough is $1.2\,m^2$ when it is full.
The real trough can hold $0.1\,m^3$ of water.

How much water can the model hold?
Give your answer to a reasonable degree of accuracy and make your units clear.

E11 **The gold statue puzzle**

A group of eight explorers found a solid gold statue $40\,cm$ high with volume $0.012\,m^3$.
They couldn't decide who should keep the statue, so it was melted down
and exact replicas were made from the gold, one for each of them.
All the gold was used.
How tall were the replicas?

*E12 Ravi has a map of India that has no scale.
He knows that India has an area of about 3 million km^2.
He estimates that India covers $300\,cm^2$ on this map.

Calculate an estimate of the map scale.

*E13 A model railway is made to a scale of $1:72$.
One of its model goods wagons can hold $25\,cm^3$ of goods.
What volume, in m^3 to 1 d.p., does this represent on a real goods wagon?

Test yourself

T1 A sheet of drawing paper is mathematically similar to a sheet of A5 paper.
A sheet of A5 paper is a rectangle 210 mm long and 148 mm wide.
The sheet of drawing paper is 450 mm long.

Calculate the width of the sheet of drawing paper.
Give your answer correct to 3 significant figures.

Edexcel

T2 AB is parallel to CD.
The lines AD and BC intersect at point O.
AB = 11 cm,   AO = 8 cm,   OD = 6 cm.

Calculate the length of CD.

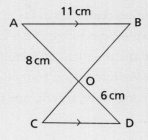

Edexcel

T3 All the dimensions of a design for a suitcase are increased by 20%.

(a) By what percentage has the surface area increased?

(b) By what percentage has the volume increased?

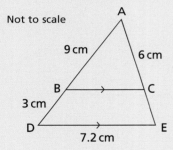

T4 In the diagram, BC is parallel to DE, and the triangles ABC and ADE are similar.
AB = 9 cm, AC = 6 cm, BD = 3 cm and DE = 7.2 cm.

Showing all your working, find the length of

(a) BC   (b) AE

WJEC

T5 x ml of paint are needed to paint the surface area of a cone of radius 3 cm.
nx ml of paint are needed to paint the surface area of a similar cone of radius 4.5 cm.

What is the value of n?

AQA 2002

T6 Two cones are similar in shape and have surface areas of 60 cm^2 and 135 cm^2.
The height of the smaller cone is 5 cm.
Find the height of the larger cone.

WJEC

T7 The diagram shows two **similar** cylinders.
The radius of the smaller cylinder is half
the radius of the larger cylinder.
The volume of the smaller cylinder is 200 cm^3.

Find the volume of the larger cylinder.

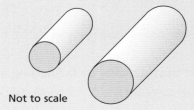

Not to scale

WJEC

34 Brackets and quadratic equations 2

You should know how to
- multiply out brackets such as $(x + 1)(x − 3)$ and factorise simple quadratic expressions
- solve quadratic equations such as $x^2 + 4x + 3 = 0$

This work will help you
- multiply out brackets such as $(4x + 1)(2x − 3)$
- factorise expressions such as $5x^2 − 9x − 2$
- solve quadratic equations by factorising, using perfect squares and using the formula
- solve problems by forming and solving a quadratic equation
- solve simultaneous equations, one linear and one quadratic

A *Simple quadratic equations*

Example

The shapes below have the same area.
Find the dimensions of the rectangle.

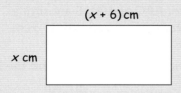

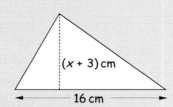

The area of the rectangle is $x(x + 6)$ cm^2

The area of the triangle is $\dfrac{16(x + 3)}{2} = 8(x + 3)$ cm^2

The areas are equal so $\qquad x(x + 6) = 8(x + 3)$

$x^2 + 6x = 8x + 24$ *Rearrange this to get 0 on one side*

$x^2 − 2x − 24 = 0$ *Factorise*

$(x − 6)(x + 4) = 0$

So $x = 6$ or $x = {}^-4$

But a length cannot be negative so $x = 6$ and the rectangle measures 6 cm by 12 cm.

A1 Solve the following by factorising.
- (a) $x^2 + 3x − 18 = 0$
- (b) $x^2 + 7x + 12 = 0$
- (c) $x^2 + 11x = 0$
- (d) $x^2 − 9x − 10 = 0$
- (e) $x^2 + 6x + 9 = 0$
- (f) $x^2 − 13x + 40 = 0$
- (g) $x^2 + x − 12 = 0$
- (h) $x^2 + 5x + 6 = 0$
- (i) $x^2 − 4x + 4 = 0$

A2 Solve these equations by rearranging and factorising.

    (a) $x^2 + 3x = 4$ (b) $x^2 + 16 = 8x$ (c) $x^2 = x$

    (d) $x^2 = 5x - 6$ (e) $x^2 = 11 + 10x$ (f) $3x^2 = 9x$

A3 (a) Show that the area of this rectangle in cm^2
      is equivalent to $x^2 + 8x + 15$.

    (b) If the area of the rectangle is $35\,\text{cm}^2\ldots$

      (i) Form an equation in x and solve it.

      (ii) Write down the length and width of the rectangle.

A4 The area of this rectangle is $42\,\text{cm}^2$.

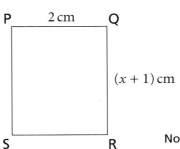

    (a) Show that $x^2 + 3x - 70 = 0$.

    (b) Solve the equation to find the dimensions of the rectangle.

A5 These two rectangles have the same area.

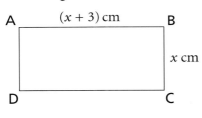

Not to scale

    (a) Form an equation in x and show that it can be simplified to

        $x^2 + x - 2 = 0$

    (b) Solve the equation $x^2 + x - 2 = 0$ to find the length of BC.   AQA(SEG) 2000

A6 The length of this rectangle is
7 cm longer than the width.

    (a) Write an expression for

      (i) the length of the rectangle

      (ii) the area of the rectangle

    (b) The area of the rectangle is $44\,\text{cm}^2$.

      (i) Form an equation in x and solve it.

      (ii) What is the perimeter of this rectangle?

B Further brackets

×	$4n$	5
$2n$	$8n^2$	$10n$
-3	$-12n$	-15

$(2n - 3)(4n + 5)$ ➡ ➡ $(2n - 3)(4n + 5)$
$= 8n^2 + 10n - 12n - 15$
$= 8n^2 - 2n - 15$

B1 Multiply out and simplify these.

(a) $(2x + 4)(x + 2)$ (b) $(3y + 1)(2y + 7)$ (c) $(3k + 5)^2$

(d) $(3z + 1)(4z - 5)$ (e) $(2p - 3)(p + 1)$ (f) $(7q - 4)(5q - 2)$

(g) $(2r - 5)(4r - 1)$ (h) $(2v - 3)^2$ (i) $(1 + 5m)(m - 10)$

(j) $(2n - 3)(2n + 3)$ (k) $(1 - 3a)(1 - 2a)$ (l) $(1 - 4w)^2$

(m) $(3b + 1)(3b - 1)$ (n) $(5 - 2c)(3c + 4)$ (o) $(3d - 5)(3d + 5)$

B2 The diagram shows a rectangular garden.
The area of the rectangular pond is $21\,\text{m}^2$.

Show that the area of the lawn in m^2 is
$$5x^2 + 38x$$

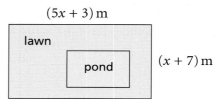

$(5x + 3)\,\text{m}$

lawn

pond

$(x + 7)\,\text{m}$

B3 Copy and complete these.

(a) $(2x + 1)(\rule{2cm}{0.4pt}) = 2x^2 + 7x + 3$

(b) $(\rule{2cm}{0.4pt})(2x + 5) = 6x^2 + 21x + 15$

(c) $(\rule{2cm}{0.4pt})(3x - 1) = 6x^2 + x - 1$

(d) $(x - 5)(\rule{2cm}{0.4pt}) = 4x^2 - 17x - 15$

(e) $(3x - 2)(\rule{2cm}{0.4pt}) = 15x^2 - 22x + 8$

×	$3x$	$-2y$
$2x$	$6x^2$	$-4xy$
y	$3xy$	$-2y^2$

$(2x + y)(3x - 2y)$ ➡ ➡ $(2x + y)(3x - 2y)$
$= 6x^2 - 4xy + 3xy - 2y^2$
$= 6x^2 - xy - 2y^2$

B4 Expand these and simplify them where possible.

(a) $(x + y)(2x + 3y)$ (b) $(x + 2)(2y + 3)$ (c) $(3x + y)(x - 2y)$

(d) $(2x - 3y)(x + 5y)$ (e) $(x - y)(2x - y)$ (f) $(3y - 2)(x - 5)$

(g) $(5x + 2y)(2x + 3)$ (h) $(2y + x)^2$ (i) $(2y - 3x)^2$

C *Factorising*

Can you factorise these?

A $3x^2 + 7x + 2$ **B** $3x^2 + 5x + 2$ **C** $3x^2 + 6x + 2$

D $2x^2 + 7x + 6$ **E** $6x^2 + 11x + 5$ **F** $6x^2 + 21x + 9$ **D** $4x^2 + 4x + 1$

C1 Find pairs of expressions from the bubble that multiply to give

(a) $3x^2 + 16x + 5$ (b) $3x^2 + 8x + 5$

(c) $2x^2 + 15x + 25$ (d) $6x^2 + 5x + 1$

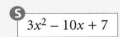

$x + 1$ $x + 5$ $3x + 5$ $2x + 1$ $2x + 5$ $3x + 1$

C2 Factorise these.

(a) $2x^2 + 15x + 7$ (b) $3x^2 + 10x + 3$ (c) $2x^2 + 9x + 4$

(d) $3x^2 + 8x + 4$ (e) $3x^2 + 7x + 4$ (f) $5x^2 + 17x + 6$

C3 Factorise these.

(a) $4k^2 + 8k + 3$ (b) $4m^2 + 13m + 3$ (c) $10n^2 + 17n + 7$

(d) $16p^2 + 8p + 1$ (e) $8x^2 + 24x + 18$ (f) $6y^2 + 23y + 10$

Can you factorise these?

P $2x^2 + 9x - 5$ **Q** $2x^2 - 3x - 5$ **R** $2x^2 + 6x - 5$

S $3x^2 - 10x + 7$ **T** $4x^2 - 21x + 5$ **U** $4x^2 + 2x - 12$ **V** $9x^2 - 6x + 1$

C4 Find pairs of expressions from the bubble that multiply to give

(a) $2x^2 + x - 3$ (b) $2x^2 - x - 1$

(c) $2x^2 - 3x + 1$ (d) $4x^2 - 8x + 3$

$x + 1$ $x - 1$ $2x - 3$ $2x - 1$ $2x + 3$ $2x + 1$

C5 Factorise these.

(a) $2x^2 + 3x - 5$ (b) $3x^2 + 8x - 3$ (c) $3x^2 - 2x - 5$

(d) $5x^2 - 9x - 2$ (e) $2x^2 - 7x + 3$ (f) $3x^2 - 22x + 7$

C6 Factorise these.

(a) $6k^2 + 7k - 5$ (b) $4m^2 - 4m + 1$ (c) $3n^2 - 5n - 8$

(d) $2p^2 - 13p + 6$ (e) $4x^2 - 16x + 15$ (f) $6y^2 - 5y - 6$

C7 (a) Expand the following.

(i) $(2x + 1)(2x - 1)$ (ii) $(3x + 2)(3x - 2)$ (iii) $(4x + 3)(4x - 3)$

(b) What do you notice about your results?

(c) Factorise   (i) $4x^2 - 9$ (ii) $9x^2 - 1$ (iii) $25x^2 - 16$

Any expression of the form $x^2 - y^2$ is known as the **difference of two squares**.

$$x^2 - y^2 = (x + y)(x - y)$$
$$16p^2 - 25q^2 = (4p + 5q)(4p - 5q)$$

C8 Factorise these.

(a) $a^2 - 9b^2$ (b) $4y^2 - x^2$ (c) $9m^2 - 4n^2$

C9 The nth term of a sequence is $9n^2 + 6n + 1$.

(a) Work out the first three terms of this sequence.

(b) Show that **every** term in the sequence must be a square number.

C10 Factorise these.

(a) $3x^2 + 7xy + 2y^2$ (b) $2a^2 - ab - b^2$ (c) $6p^2 + pq - 2q^2$

C11 (a) Expand $(a + b)(5 + b)$.

(b) Factorise $p^2 + pq + 3p + 3q$.

C12 Factorise these.

(a) $p^2 + p + qp + q$ (b) $2p^2 - 2pq - 3p + 3q$ (c) $3p^2 + 10q + 5p + 6pq$

Factorising a quadratic expression

- Look for common factors first.

$$2x^2 + 8x = 2x(x + 4)$$
$$2x^2 + 4x - 6 = 2(x^2 + 2x - 3)$$
$$= 2(x + 3)(x - 1)$$

- $ax^2 + bx + c = (?x + ?)(?x + ?)$

$$2x^2 + 7x + 3 = (2x + 1)(x + 3)$$

+ and + means **both** brackets have +

- $ax^2 - bx + c = (?x - ?)(?x - ?)$

$$2x^2 - 7x + 3 = (2x - 1)(x - 3)$$

− and + means **both** brackets have −

- $ax^2 \pm bx - c = (?x + ?)(?x - ?)$

$$2x^2 + 3x - 5 = (2x + 5)(x - 1)$$
$$2x^2 - 3x - 5 = (2x - 5)(x + 1)$$

− means brackets have + and −

- For expressions of the form $ax^2 - c$ look for the 'difference of two squares'.

$$x^2 - 49 = (x + 7)(x - 7)$$
$$25x^2 - 36 = (5x + 6)(5x - 6)$$

D Solving by factorising

Examples

Solve $2x^2 + 7x - 4 = 0$.

$$2x^2 + 7x - 4 = 0$$
$$(2x - 1)(x + 4) = 0$$
$$\text{Either } (2x - 1) = 0 \text{ or } (x + 4) = 0$$
$$\text{So } x = \frac{1}{2} \text{ or } x = {}^-4$$

Solve $3k^2 + 6 = 16k + 1$.

$$3k^2 + 6 = 16k + 1$$
$$3k^2 - 16k + 5 = 0$$
$$(3k - 1)(k - 5) = 0$$
$$\text{Either } (3k - 1) = 0 \text{ or } (k - 5) = 0$$
$$\text{So } k = \frac{1}{3} \text{ or } k = 5$$

D1 Solve the following equations by factorising.

(a) $2x^2 + 9x - 5 = 0$ (b) $2x^2 + 11x + 5 = 0$ (c) $3x^2 + 20x - 7 = 0$

(d) $2x^2 - 6x = 0$ (e) $2x^2 - 23x + 11 = 0$ (f) $5x^2 - 2x - 7 = 0$

D2 Solve the following equations by factorising.

(a) $4x^2 + 5x + 1 = 0$ (b) $4x^2 - 8x + 3 = 0$ (c) $2x^2 - 13x + 15 = 0$

(d) $3x^2 - x - 2 = 0$ (e) $16x^2 - 8x + 1 = 0$ (f) $4x^2 + 14x = 0$

(g) $4x^2 + 14x - 18 = 0$ (h) $4x^2 - 16x + 15 = 0$ (i) $4x^2 - 35x - 9 = 0$

D3 Solve the following equations by rearranging first.

(a) $2x^2 + 11x = 6$ (b) $3y^2 + 13y + 10 = 6$ (c) $5m^2 + 30m = 21 - 2m$

(d) $30k + 1 = 1 - 5k^2$ (e) $10p^2 = 28p + 6$ (f) $6 - 10a - 6a^2 = 6a$

D4 (a) Show that the area of this rectangle in cm^2 is equivalent to $2x^2 + 5x + 2$.

(b) If the area of the rectangle is $14\,\text{cm}^2$...

    (i) Form an equation in x and solve it.

    (ii) Write down the length and width of the rectangle.

$(2x + 1)\,\text{cm}$

$(x + 2)\,\text{cm}$

D5 The area of this rectangle is $24\,\text{cm}^2$. Find the length and width of the rectangle.

$(3x - 4)\,\text{cm}$

$(x - 1)\,\text{cm}$

D6 (a) (i) Factorise $4x^2 - 37x + 9$.

    (ii) Hence, or otherwise, solve the equation

$$4x^2 - 37x + 9 = 0$$

(b) By considering your answers to part (a), find all the solutions to the equation

$$4y^4 - 37y^2 + 9 = 0$$

AQA 1999

E *Perfect squares*

- 9 is a perfect square because we can write 9 as 3×3 or 3^2.
- $x^2 + 18x + 81$ is a perfect square because we can write it as $(x + 9)^2$.
- $4n^2 - 20n + 25$ is also a perfect square because we can write is as $(2n - 5)^2$.

E1 Expand and simplify these perfect squares.

(a) $(x + 5)^2$
(b) $(x - 4)^2$
(c) $(x - 3)^2$
(d) $(x + 12)^2$

E2 Copy and complete the following identities.

(a) $(x - \blacksquare)^2 = x^2 - 10x + 25$
(b) $(x + \blacksquare)^2 = x^2 + 12x + 36$

(c) $(x + \blacksquare)^2 = x^2 + 8x + \blacksquare$
(d) $(x - \blacksquare)^2 = x^2 - 4x + \blacksquare$

(e) $(x + \blacksquare)^2 = x^2 + 16x + \blacksquare$
(f) $(x - \blacksquare)^2 = x^2 - 14x + \blacksquare$

E3 Which of the following expressions are perfect squares?

A $x^2 + 2x + 1$

B $x^2 - 12x + 12$

C $x^2 - 12x + 36$

D $x^2 - 18x + 81$

E $x^2 + 6x + 9$

F $x^2 + 6x + 36$

E4 What number do you have to add to each expression to make it a perfect square?

(a) $x^2 + 4x$
(b) $x^2 + 6x$
(c) $x^2 + 8x$
(d) $x^2 + 2x$

(e) $x^2 - 4x$
(f) $x^2 - 6x$
(g) $x^2 - 2x$
(h) $x^2 - 10x$

E5 Expand
(a) $(x + a)^2$
(b) $(x - c)^2$

F *Using perfect squares to solve quadratic equations*

Can you solve these?

A $(x + 4)^2 = 25$

B $x^2 + 6x + 9 = 0$

C 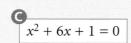 $x^2 + 6x + 1 = 0$

D $x^2 + 8x + 5 = 0$

E $x^2 + 10x - 2 = 0$

What about these?

P $x^2 - 8x + 16 = 0$

Q $x^2 - 8x - 1 = 0$

R $x^2 - 12x + 5 = 0$

F1 Solve the following equations using perfect squares.
Give each answer correct to three decimal places.

(a) $x^2 + 6x - 2 = 0$
(b) $x^2 + 10x - 5 = 0$
(c) $x^2 + 12x + 8 = 0$

(d) $x^2 + 4x + 1 = 0$
(e) $x^2 + 14x + 7 = 0$
(f) $x^2 + 8x - 7 = 0$

F2 Try to solve $x^2 + 2x + 3 = 0$. What happens?

Example

Solve $x^2 - 6x + 4 = 0$.

$$x^2 - 6x + 4 = 0$$
$$x^2 - 6x = {}^-4$$
$$x^2 - 6x + 9 = 5$$
$$(x - 3)^2 = 5$$
$$x - 3 = \pm\sqrt{5}$$

So $x = 3 + \sqrt{5}$ or $x = 3 - \sqrt{5}$

Rounded to three decimal places,

$$x = 5.236 \text{ or } x = 0.764$$

$x^2 - 6x + 4$ doesn't factorise. Subtract 4 from each side to get the constant on one side.

Half of $^-6$ is $^-3$ and $(^-3)^2 = 9$ so add 9 to each side to make a perfect square on the left.

Factorise the perfect square.

Remember the positive **and** negative square root.

This method is sometimes called '**completing the square**'.

F3 Solve the following equations using perfect squares.
Give each answer correct to three decimal places.

(a) $x^2 - 4x - 2 = 0$ (b) $x^2 - 12x - 5 = 0$ (c) $x^2 - 6x + 7 = 0$

(d) $x^2 - 10x + 1 = 0$ (e) $x^2 - 2x - 1 = 0$ (f) $x^2 - 8x + 9 = 0$

F4 Try to solve $x^2 - 4x + 8 = 0$. What happens?

F5 (a) Expand and simplify $(x + \frac{3}{2})^2$.

(b) Hence solve the equation $x^2 + 3x - 1 = 0$.
Give each answer correct to 2 d.p.

F6 (a) Expand and simplify $(x - \frac{1}{2})^2$.

(b) Hence solve the equation $x^2 - x - 3 = 0$.
Give each answer correct to 2 d.p.

F7 Solve the following equations, giving each answer correct to 3 d.p.

(a) $x^2 + 3x - 2 = 0$ (b) $x^2 - 5x - 3 = 0$

F8 Copy and complete this working to
solve the equation $2x^2 + 8x + 2 = 0$.

$$2x^2 + 8x + 2 = 0$$
$$x^2 + 4x + 1 = 0$$

If the coefficient of x^2 is not 1 you can divide both sides to make it 1.

F9 Solve these equations.
(First, divide both sides to make 1 the coefficient of x^2.)

(a) $2x^2 + 12x + 4 = 0$ (b) $3x^2 - 24x - 3 = 0$ (c) $5x^2 - 15x + 5 = 0$

*F10 Solve the following equations using perfect squares.
Give each answer correct to three decimal places.

(a) $2x^2 - 4x - 3 = 0$ (b) $2x^2 + 5x - 1 = 0$ (c) $3x^2 + x - 3 = 0$

G *Using a formula to solve quadratic equations*

If a quadratic equation is in the form
$$ax^2 + bx + c = 0$$
it can be shown that $x = \dfrac{-b \pm \sqrt{b^2 - 4ac}}{2a}$

Example

Solve $3x^2 - 8x + 2 = 0$.

$$3x^2 - 8x + 2 = 0$$

So $a = 3$, $b = {}^-8$ and $c = 2$

$$\text{giving} \quad x = \frac{{}^-(^-8) \pm \sqrt{(^-8)^2 - 4 \times 3 \times 2}}{2 \times 3}$$

$$= \frac{8 \pm \sqrt{40}}{6} = 2.387 \text{ or } 0.279 \text{ (to 3 d.p.)}$$

G1 Use the formula to solve each equation.

(a) $x^2 + 5x + 3 = 0$ (b) $x^2 - 3x - 1 = 0$ (c) $x^2 - 11x + 9 = 0$

(d) $3x^2 + 6x - 2 = 0$ (e) $2x^2 + 10x = 5$ (f) $5x^2 - 8x + 2 = 0$

(g) $3x^2 + 2x = 7$ (h) $2x^2 = 4x - 1$ (i) $2x + 4 = x^2$

G2 (a) Solve the following equations by factorising.

(i) $x^2 + 11x + 30 = 0$ (ii) $3x^2 - 4x + 1 = 0$

(b) Now solve the equations in part (a) by using the formula.
Which method do you prefer?

G3 Solve the following equations.
Choose your own method each time (factorising, using perfect squares or the formula).

(a) $x^2 + 6x - 7 = 0$ (b) $3x^2 + 7x + 1 = 0$ (c) $3x^2 + 10x + 3 = 0$

(d) $6x^2 - 11x = 30$ (e) $5x^2 - 8x = 21$ (f) $2x^2 + 2 = 5x$

G4 Here is a sketch of the graph of $y = x^2 - 5x - 6$.

(a) Explain why the x-coordinates of
the points A and B are found by
solving $x^2 - 5x - 6 = 0$.

(b) Find the coordinates of points A and B.

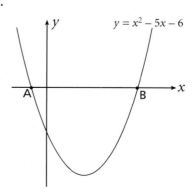

G5 Find the coordinates of the points where the graph of $y = 5x^2 + x - 4$ crosses the x-axis.

G6 Where does the graph of $y = 3x^2 + 26x - 9$ cross the x-axis?

G7 (a) Solve $4x^2 - 12x + 9 = 0$.

(b) Use your solution to choose the
correct sketch of $y = 4x^2 - 12x + 9$.

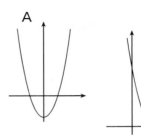
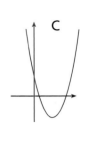

G8 (a) What happens when you use the formula to solve $x^2 - 2x + 5 = 0$?

(b) What does this tell you about the graph of $y = x^2 - 2x + 5$?

G9 The diagram shows a rectangle with
length $x + 4$ and width $x - 1$.
All measurements are given in centimetres.

The area of the rectangle is A square centimetres.

(a) Show that $A = x^2 + 3x - 4$.

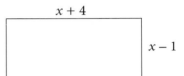

When $A = 10$, x satisfies the equation $x^2 + 3x - 14 = 0$.

(b) Find the length of the rectangle.
Give your answer correct to two decimal places.

Edexcel

G10 ABC is a right-angled triangle.

$AB = (3x + 4)\,$cm
$BC = (x + 2)\,$cm
$AC = (2x + 6)\,$cm

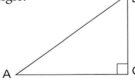

(a) (i) Use Pythagoras' theorem to write down an equation in x.

(ii) Show that this equation simplifies to $x^2 - x - 6 = 0$.

(b) (i) Solve this equation.

(ii) Write down the length of AB.

AQA(NEAB) 1997

G11 The perimeter of this rectangle is 26 cm.
Its area is 35 cm^2.

y cm

(a) Use the perimeter to find an expression,
in terms of y, for the length of this rectangle.

(b) Form an equation in y and show it can
be simplified to $y^2 - 13y + 35 = 0$.

(c) Solve the equation to find the length and width, correct to 2 d.p.

G12 The diagram shows a square-based cuboid.

The square ends have side length x cm.
The length of the cuboid is y cm.

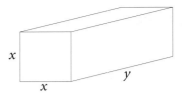

(a) The surface area of the cuboid is given by the expression $2x^2 + 4xy$.
Factorise this expression completely.

The surface area of the cuboid is $80\,\text{cm}^2$.

(b) (i) Show that $y = \dfrac{40 - x^2}{2x}$.

(ii) Calculate x when $y = 8$.
Hence calculate the volume of the cuboid.
Give your answers to one decimal place.   AQA(SEG) 2000

G13 A farmer wants to build a fence to make a rectangular pen.
He is using his barn as one side
of the pen, as shown.

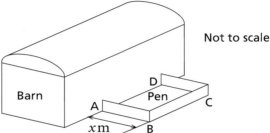

He has 60 metres of fencing and he wants the area of the pen to be $300\,\text{m}^2$.
The length of AB is x metres.

(a) Show that if the area is $300\,\text{m}^2$, x must satisfy the equation
$$x^2 - 30x + 150 = 0$$

(b) Solve the equation in (a) to find the **two** values of x that give the pen an area
of $300\,\text{m}^2$. Give your answers to an appropriate degree of accuracy.   AQA(SEG) 1998

G14 The base of a triangle is 5 cm longer than its height.
The area of the triangle is $30\,\text{cm}^2$.

Find the length of the base of the triangle correct to three significant figures.

G15 In a right-angled triangle the hypotenuse is 3 cm longer than one side and 5 cm longer
than the other.

Find the length of the hypotenuse correct to three significant figures.

H Simultaneous equations

Example

Find the points of intersection of the curve $y = x^2 - 3x + 3$ and $2x - y = 1$.

Rearranging the second equation gives $y = 2x - 1$.
The y-values are equal at the points of intersection so

$$2x - 1 = x^2 - 3x + 3$$
$$x^2 - 5x + 4 = 0$$
$$(x - 4)(x - 1) = 0$$

So $x = 4$ or $x = 1$

When $x = 4$ then $y = 2 \times 4 - 1 = 7$ and
when $x = 1$ then $y = 2 \times 1 - 1 = 1$.

So the points of intersection are $(1, 1)$ and $(4, 7)$.

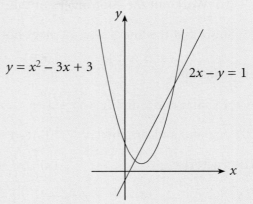

H1   Solve these pairs of simultaneous equations.

(a) $y = 5x - 6$
     $y = x^2$

(b) $y = x^2 + 7$
     $y = 3x + 17$

(c) $y = x^2 - 5$
     $y - 2x = 3$

(d) $y = 5x + 7$
     $y = 10 - 2x^2$

H2   Find the points of intersection for each pair of graphs.

(a)
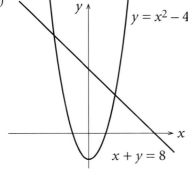

(b)
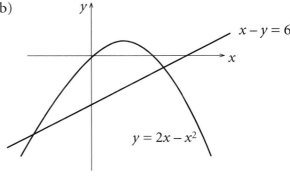

(c)
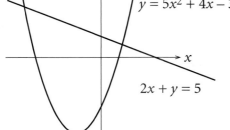

(d)
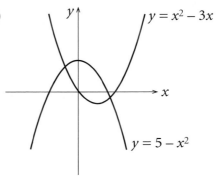

H3 (a) Solve $4x - 8 = x^2 - 4$.

(b) What does your solution show about the line $y = 4x - 8$ and the curve $y = x^2 - 4$?

H4 (a) Show that the line $y = x - 1$ touches the curve $y = 4x^2 + 5x$ at one point only.

(b) Work out the coordinates of this point.

H5 Show that the line $y = x - 3$ does not meet the curve $y = x^2 + 2x - 1$.

Test yourself

T1 Expand and simplify $(2p + 5)(p - 2)$.   OCR

T2 Expand and simplify   (a) $(3n - 5)^2$ (b) $(3x + 5y)(x - 6y)$

T3 Factorise   (a) $2n^2 - 11n + 12$ (b) $5x^2 + 17x - 12$

T4 Solve   (a) $2y^2 + y - 3 = 0$ (b) $2x^2 - 5x = 0$ (c) $9t^2 - 64 = 0$

T5 (a) Factorise $6x^2 - x - 12$.

(b) Solve the equation $x^2 - 7x - 6 = 0$ giving your
answers correct to two decimal places.   AQA(NEAB) 1998

T6 (a) What must be added to $x^2 - 12x$ so that it can be written as a perfect square in
the form $(x - a)^2$?

(b) Hence solve the equation $x^2 - 12x = 3$ giving your answers correct to 2 d.p.

T7 The diagram represents a flag which
measures 80 cm by 60 cm.
The flag has two axes of symmetry.

The flag has two colours, grey and white.
The bands of white are of equal width, x cm.

The white area forms half the area of the flag.

(a) Show that x satisfies the
equation $x^2 - 140x + 2400 = 0$.

(b) Solve the equation $x^2 - 140x + 2400 = 0$ to find the width of the white bands.
       AQA(NEAB) 1998

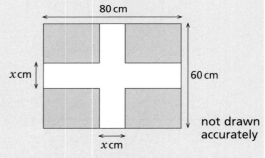

T8 The surface area of a cuboid with length x cm,
width $(x - 1)$ cm and height 3 cm is 63 cm².

(a) Show that x satisfies the equation $2x^2 + 10x - 69 = 0$.

(b) (i) Solve the equation $2x^2 + 10x - 69 = 0$, giving solutions to two decimal places.

(ii) Hence write down the dimensions of the cuboid.   WJEC

T9 Solve the simultaneous equations $y + x = 7$ and $y = 2x^2 + 12x - 28$.
Give your solutions to 1 d.p.

Review 8

1 The amount of money raised in an appeal by a charity is given as £157 000, correct to the nearest thousand pounds.

   What is

   (a) the maximum possible amount raised

   (b) the minimum possible amount raised

2 The letters a, b, h stand for lengths.
   Which of the expressions on the right could represent

   (a) an area   (b) a volume

   (c) a length   (d) a pure number

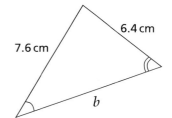

$$3h(a^2 + ab + b^2)$$

$$h\sqrt{a^2 + b^2}$$

$$\frac{ah}{b-a} \qquad \frac{a+b}{2h}$$

$$\frac{h(a+b)}{2} \qquad \frac{a^2+b^2}{h^2}$$

3 This diagram shows a pair of similar triangles.
   Calculate the lengths marked a and b.

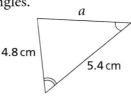

4 Factorise the following.

   (a) $x^2 + 5x + 6$ (b) $6y^2 + y - 12$ (c) $2m^2 + 6mn + 4n^2$

5 Greg wants to find the height of the tree in his garden.
   He measures the angle of elevation of the top of the tree
   as 64°, to the nearest degree, at a distance of 500 cm,
   to the nearest 10 cm, from the base of the tree.

   Calculate the maximum possible height of the tree.

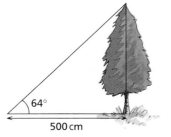

6 A map has a scale of $1:50\,000$.

   (a) Majid measures the distance of Catsfield from Ninfield as
       4.3 cm at a bearing of 054°.
       What is the actual distance and bearing of Catsfield from Ninfield?

   (b) He estimates the area of Arlington reservoir as 2 cm² on the map.
       Calculate an estimate of the real area of the reservoir.

7 Solve the following quadratic equations using the formula.

(a) $x^2 - 4x - 9 = 0$ (b) $2x^2 + 7x - 1 = 0$ (c) $4x^2 - 5 = 2x$

8 The number of cars crossing the Humber Bridge in the year 1999/2000 was 5.1×10^6, to two significant figures.
The toll for each car was £2.30.

Calculate the minimum possible amount of money collected from car tolls during the year 1999/2000.

9 The cone shown here is to be enlarged so that the volume is increased by 60%.

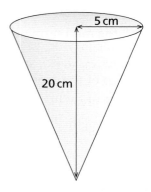

(a) What is the volume factor of the enlargement?

(b) Find the scale factor of the enlargement, correct to three significant figures.

(c) Calculate the radius and height of the enlarged cone.

(d) Calculate the volume of each cone using the formula for the volume of a cone, and check that the volume has been increased by 60%.

10 Rectangle ABCD has length $a + 5$ and width $a - 2$.

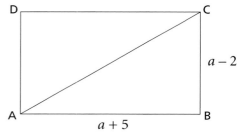

(a) Use Pythagoras' theorem to show that the length of diagonal AC is given by
$$AC^2 = 2a^2 + 6a + 29$$

(b) Given that the length of the diagonal is 13 cm, use your equation in (a) to find the lengths of the sides of the rectangle.

11 Jackie is redesigning her garden.
To help with the design, she has made a scale drawing of the garden.

(a) The length of the real garden is 20 metres.
She uses a scale of $1 : 50$.

What is the length of the garden on the scale drawing?

(b) The real patio has an area of $22 \, \text{m}^2$.
What area will it cover on the scale drawing?

(c) Jackie draws a flower bed with an area of $40 \, \text{cm}^2$ on her scale drawing.
What will the area of the real flower bed be?

(d) She draws a circular pond with radius 5 cm.
What will the area of the real pond be?

12 (a) Find an expression for the surface area of this cuboid.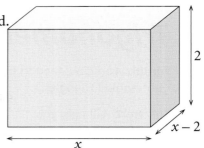
Write the expression in its simplest form.

   (b) The surface area of the cuboid is $62\,\text{cm}^2$.
Form and solve an equation to find
the value of x.

   (c) Calculate the volume of the cuboid.

13 In 1991 the number of staff employed by a company was 1250 to the nearest ten.
By 2001 this number had increased to 1480 to the nearest ten.

   (a) Calculate the maximum possible increase in the number of staff.

   (b) Calculate the maximum possible percentage increase in staff numbers.

14 A circle of radius $9\,\text{cm}$ has its centre at C.
A second circle, of radius $6\,\text{cm}$, has its centre at a
point A on the circumference of the first circle.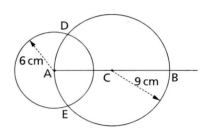

The second circle cuts the first circle at D and E.

The line through A and C meets the first circle
again at B.

Calculate the length BD, showing your method.

*15 The lengths a, b, h, x, y are as shown in this diagram.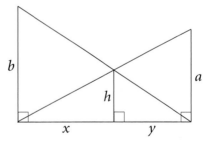

   (a) Explain why $\dfrac{h}{a} = \dfrac{x}{x+y}$.

   (b) Write a similar equation for $\dfrac{h}{b}$.

   (c) Show that $\dfrac{1}{a} + \dfrac{1}{b} = \dfrac{1}{h}$.

   (d) Rearrange to make h the subject.

   (e) Calculate h given that $a = 5$ and $b = 3$.

Challenge

This number is made by using only the four numbers 1, 2, 3, 4 and powers: $3^{2^{1^{4}}}$

- What is the largest number that can be made in this way?
- If you think of 1, 2, 3, 4 as digits, you can make, for example: 2^{134} $4^{2^{31}}$
What is the largest number that can be made in this way?

35 Trigonometric graphs

For this work you need to be familiar with the use of sine, cosine and tangent in right-angled triangles.

This work will help you

◆ use sines, cosines and tangents of angles greater than 90° and negative angles
◆ work with trigonometric graphs
◆ relate trigonometric graphs to real-life situations

A The sine graph

On sheet G175, label each point marked on the circle with an angle like this. Continue up to 360°.

Then take the y-coordinate of each labelled point and plot it against its angle on sheet G176.
Join your points with a smooth curve.

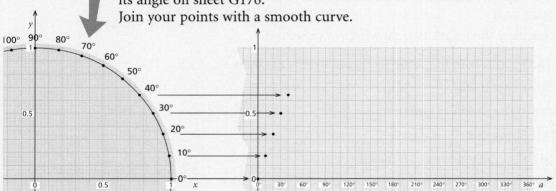

The circle's radius is 1 unit.
So you can see that when the angle (call it a) is between 0° and 90° the y-coordinate of the point on the circle is sin a.

You can think of it this way …

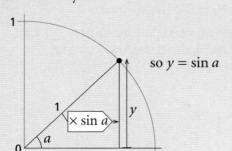

so $y = \sin a$

… or this way …

$$\sin a = \frac{\text{opposite}}{\text{hypotenuse}} = \frac{y}{1}$$

so $y = \sin a$

The sine of an angle outside the range 0° to 90° is also defined as the *y*-coordinate of the corresponding point on the circle.

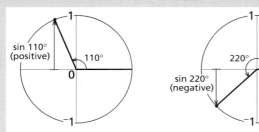

So on sheet G176 label the vertical axis of your graph 'sin *a*'.

Remember: here we are measuring angles anticlockwise, starting from the point (1, 0).

A1 Use your sine graph to get approximate values for these.

(a) sin 65°   (b) sin 105°   (c) sin 153°   (d) sin 212°   (e) sin 285°

A2 Key sin 65° into your calculator to see how close your approximate value was in A1(a). Check your other values the same way.

A3 (a) From your graph, find an angle that has 0.8 as its sine. Can you find more than one angle?

(b) Key sin⁻¹ 0.8 into your calculator (the angle whose sine is 0.8). Does this agree with what you found from your graph?

When you look for sin⁻¹ of a value, there is usually more than one answer between 0° and 360°, but your calculator only gives you one of them.

Making a sketch of a sine graph and using the graph's symmetry can help you find all the angles.

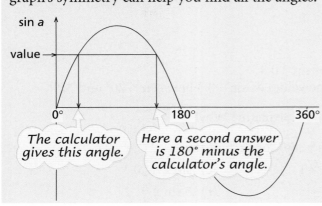

The calculator gives this angle.

Here a second answer is 180° minus the calculator's angle.

Alternatively, sketching a circle like the one on sheet G175 may help.

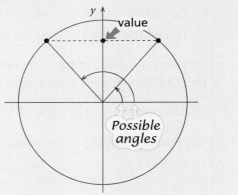

value

Possible angles

A4 What angle between 0° and 360° has **exactly** the same sine as each of these?

(a) 56°   (b) 112°   (c) 299°   (d) 199°   (e) 11°

(f) 251°   (g) 286°   (h) 82°   (i) 101°   (j) 321°

A5 Find these to the nearest 0.1°, using a calculator. Give all possible answers from 0° to 360°.

(a) $\sin^{-1} 0.7$ (b) $\sin^{-1} {}^-0.5$ (c) $\sin^{-1} 0.2$ (d) $\sin^{-1} {}^-0.35$ (e) $\sin^{-1} 0$

(f) $\sin^{-1} 0.95$ (g) $\sin^{-1} {}^-0.1$ (h) $\sin^{-1} 1$ (i) $\sin^{-1} 0.6$ (j) $\sin^{-1} {}^-1$

A6 You are told that $\sin 72° = 0.95$ to 2 d.p.
What angles between 0° and 360° have $^-0.95$ as their sine?

A7 You are told that $\sin 223° = {}^-0.68$ to 2 d.p.
What angles between 0° and 360° have 0.68 as their sine?

- Where should you mark angles 370°, 380°, 390°, ... on the circle on sheet G175?
- Where should you mark angles $^-10°$, $^-20°$, $^-30°$, ... ?

A8 Read y-coordinates from the circle to get approximate values for these.

(a) $\sin 480°$ (b) $\sin 650°$ (c) $\sin 555°$ (d) $\sin 390°$ (e) $\sin 705°$

(f) $\sin {}^-70°$ (g) $\sin {}^-200°$ (h) $\sin {}^-160°$ (i) $\sin {}^-300°$ (j) $\sin {}^-260°$

A9 Key $\sin 480°$ into your calculator to see how close your approximate value was in A8 (a).
Check your other values the same way.

A10 Complete the graph of $\sin a$ on sheet G176 for values of a from $^-360°$ to $0°$.

A11 What angles between 0° and 360° have **exactly** the same sine as

(a) $^-20°$ (b) $^-230°$ (c) $^-290°$ (d) $^-95°$ (e) $^-28°$

A12 With a calculator find these to the nearest 0.1°. Give all possible answers from $^-360°$ to $0°$.

(a) $\sin^{-1} 0.6$ (b) $\sin^{-1} {}^-0.4$ (c) $\sin^{-1} 0.1$ (d) $\sin^{-1} {}^-0.25$ (e) $\sin^{-1} 1$

A13 **Sketch** the graph of $\sin a$, for a between $^-720°$ and 720°.

(a) List all the values of a where $\sin a = 0$.

(b) List all the values of a where $\sin a = 1$.

(c) List all the values of a where $\sin a = {}^-1$.

A14 Use your calculator to find one value of $\sin^{-1} 0.3$.
With the help of your sketch, give all the values of $\sin^{-1} 0.3$ between $^-720°$ and 720°.

A15 What happens if you key $\sin^{-1} 2$ into your calculator? Why?

A16 Find possible values of x between 0° and 360° for each of these.

(a) $\sin x = 0.7$ (b) $2 \sin x = 1$ (c) $\sin x = {}^-0.3$

(d) $\sin x + \frac{1}{4} = 0$ (e) $4 \sin x = 3$ (f) $5 \sin x = {}^-2$

A17 Toni says: 'I'm thinking of an angle between 0° and 720°. Its sine is greater than 0.6.'
What angle could she be thinking of? Describe all the possibilities.

B *The cosine graph*

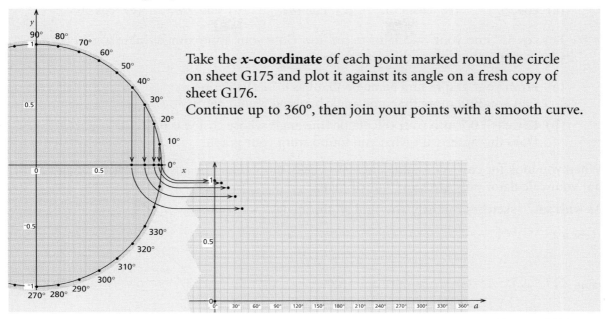

Take the **x-coordinate** of each point marked round the circle on sheet G175 and plot it against its angle on a fresh copy of sheet G176.
Continue up to 360°, then join your points with a smooth curve.

As with the sine, you can see that when the angle a is between 0° and 90° the x-coordinate of the point on the circle is cos a.

You can think of it this way ...

so $x = \cos a$

... or this way ...

$$\cos a = \frac{\text{adjacent}}{\text{hypotenuse}} = \frac{x}{1}$$

so $x = \cos a$

We define the cosine of an angle outside the range 0° to 90° also as the x-coordinate of the corresponding point on the circle.

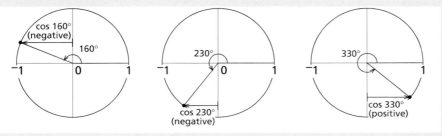

So on your second copy of sheet G176 label the vertical axis of your graph 'cos a'.

B1 Use your cosine graph to get approximate values for these.

(a) cos 55°   (b) cos 95°   (c) cos 147°   (d) cos 208°   (e) cos 305°

B2 Key cos 55° into your calculator to see how close your approximate value was in B1(a).
Check your other values the same way.

B3 (a) From your graph, find an angle that has 0.7 as its cosine.
Can you find more than one angle between 0° and 360°?

(b) Key cos⁻¹ 0.7 into your calculator (the angle whose cosine is 0.7).
Does this agree with what you found from your graph?

When you look for cos⁻¹ of a value, there is usually more than one answer between 0° and 360°, but your calculator only gives you one of them.

As with sin⁻¹, sketches can help you find all the angles.

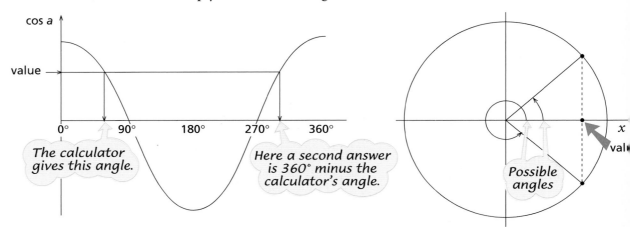

The calculator gives this angle.

Here a second answer is 360° minus the calculator's angle.

Possible angles

B4 What angle between 0° and 360° has **exactly** the same cosine as each of these?

(a) 62°   (b) 108°   (c) 301°   (d) 195°   (e) 2°

B5 Extend your cosine graph to show cosines of angles between ⁻360° and 0°.
What angles between ⁻360° and 360° have exactly the same cosine as each of these?

(a) 272°   (b) 355°   (c) 71°   (d) 99°   (e) 317°
(f) ⁻10°   (g) ⁻250°   (h) ⁻310°   (i) ⁻85°   (j) ⁻32°

B6 Find these to the nearest 0.1°, using a calculator.
Give all possible answers from ⁻360° to 360°.

(a) cos⁻¹ 0.6   (b) cos⁻¹ ⁻0.5   (c) cos⁻¹ 0.1   (d) cos⁻¹ ⁻0.45   (e) cos⁻¹ 1
(f) cos⁻¹ 0.75   (g) cos⁻¹ ⁻0.2   (h) cos⁻¹ ⁻1   (i) cos⁻¹ 0.3   (j) cos⁻¹ 0

B7 You are told that cos 76° = 0.24 to 2 d.p.
What angles between ⁻360° and 360° have ⁻0.24 as their cosine?

B8 You are told that cos 100° = ⁻0.17 to 2 d.p.
What angles between ⁻360° and 360° have 0.17 as their cosine?

B9 **Sketch** the graph of cos a, for a between ⁻720° and 720°.
  (a) List all the values of a where cos a = 0.
  (b) List all the values of a where cos a = 1.
  (c) List all the values of a where cos a = ⁻1.

B10 Use your calculator to find one value of cos⁻¹ 0.4.
With the help of your sketch, give all the values of cos⁻¹ 0.4 between ⁻720° and 720°.

B11 Find possible values of x between 0° and 360° for each of these.
  (a) cos x = 0.1   (b) 7 cos x = 1   (c) cos x = ⁻0.25
  (d) cos $x + \frac{1}{5} = 0$ (e) 3 cos x = 2   (f) 5 cos x = ⁻4

B12 Robert says: 'I'm thinking of an angle between 0° and 720°. Its cosine is less than ⁻0.6.'
What angle could he be thinking of? Describe all the possibilities.

*B13 If sin b = ⁻sin a and cos b = ⁻cos a, what can you say about a and b?

*B14 If cos q = cos p and sin q = ⁻sin p, what can you say about p and q?

C *The tangent graph*

We can also get the tangent of an angle (call it a) by considering
a point moving anticlockwise round a circle of unit radius.

When a is between 0° and 90°, in the normal way:

$$\tan a = \frac{\text{opposite}}{\text{adjacent}} = \frac{y\text{-coordinate of point P}}{x\text{-coordinate of point P}}$$

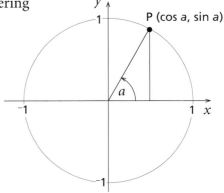

We also regard the tangent of an angle outside the
range 0° to 90° as

$$\frac{y\text{-coordinate of point P}}{x\text{-coordinate of point P}}.$$

Notice that this is $\frac{\sin a}{\cos a}$.

- For what angles between 0° and 360° is the tangent positive?
- For what angles between 0° and 360° is it negative?
- For what angles is it zero?
- What happens around 90°?
- What happens around 270°?

Here again is part of the graph of the sine function ...

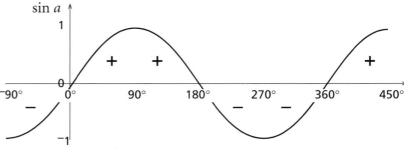

... and here is the cosine function.

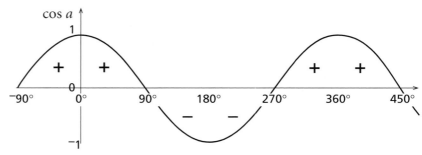

The tangent function is obtained by dividing sine values by cosine values.

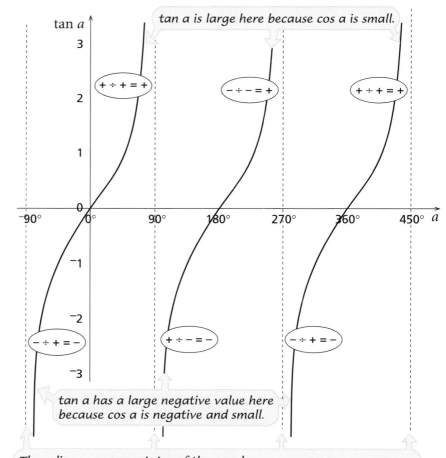

tan a is large here because cos a is small.

$+ \div + = +$ $- \div - = +$ $+ \div + = +$

$- \div + = -$ $+ \div - = -$ $- \div + = -$

tan a has a large negative value here because cos a is negative and small.

These lines are **asymptotes** of the graph.
The graph gets closer and closer to them but never touches them.
They divide the graph into disconnected sections.

C1 (a) Find tan 75° on your calculator to 2 d.p.

   (b) Find two other angles whose tangent is the same as tan 75°.
      Use the tangent graph opposite as a guide.

C2 What angles between $^-90°$ and 450° have exactly the same tangent as each of these?

   (a) 30°   (b) 440°   (c) 200°   (d) $^-80°$ (e) 150°

   (f) 300°   (g) 48°   (h) 399°   (i) 143°   (j) $^-27°$

C3 (a) Find $\tan^{-1} 0.9$ on your calculator.

   (b) Use the tangent graph opposite as a guide to finding two other results for $\tan^{-1} 0.9$.

C4 Find these to the nearest 0.1°, using a calculator.
   Give all the possible answers from $^-90°$ to 450°.

   (a) $\tan^{-1} 0.6$ (b) $\tan^{-1} 1.7$ (c) $\tan^{-1} 45$ (d) $\tan^{-1} 0.01$ (e) $\tan^{-1} {}^-0.4$

   (f) $\tan^{-1} {}^-2.5$ (g) $\tan^{-1} {}^-5$ (h) $\tan^{-1} {}^-0.1$ (i) $\tan^{-1} {}^-0.05$ (j) $\tan^{-1} {}^-3.79$

C5 You are told that tan 63° = 1.96 to 2 d.p.
   What angles between 0° and 360° have $^-1.96$ as their tangent?

C6 You are told that tan 114° = $^-2.25$ to 2 d.p.
   What angles between 0° and 360° have 2.25 as their tangent?

C7 Find possible values of x between 0° and 360° for each of these.

   (a) $\tan x = 90$ (b) $\tan x = {}^-0.5$ (c) $5 \tan x = 8$

   (d) $\frac{1}{2} \tan x = {}^-2$ (e) $\tan x + \frac{1}{5} = 0$ (f) $3 \tan x = {}^-2$

The graphs of sin a and cos a are **periodic graphs**.
They 'repeat themselves' after every 360°.
We say 360° is the **period** of each graph.

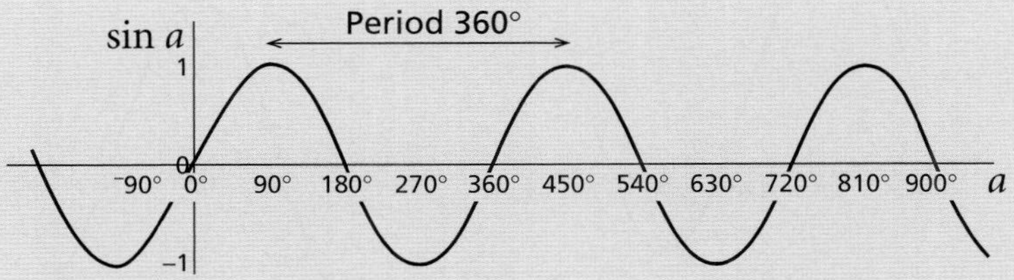

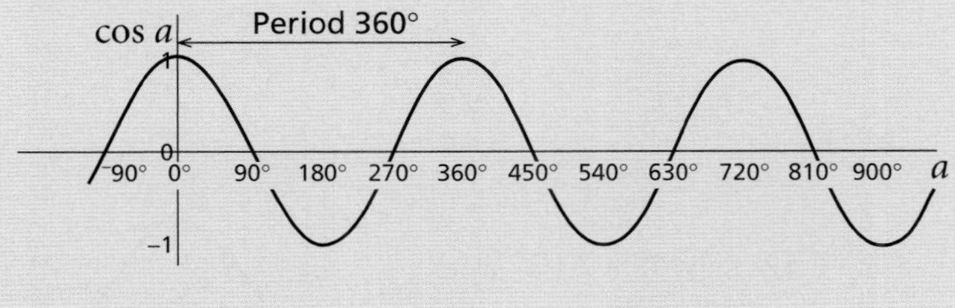

The graph of cos a has the same repeating shape as sin a, but is shifted 90° to the left.

C8 The graph of tan a is also periodic.
   What is the period of the graph of tan a?

D Graphs based on trigonometric functions

D1 (a) Complete this table for the graph of $y = 3 \sin x$.

(b) Draw the graph on graph paper with scales clearly labelled.

(c) How is it different from the graph of $y = \sin x$?

x	0°	45°	90°	360°
$\sin x$	0	0.71	1	
$3 \sin x$	0	2.13	3	

D2 (a) Complete this table for the graph of $y = \cos x + 1$.

(b) Sketch the graph, marking key values on the axes.

(c) How is it different from the graph of $y = \cos x$?

x	0°	90°	180°	360°
$\cos x$	1	0		
$\cos x + 1$	2	1		

D3 (a) Make a table for the graph of $y = {}^-\sin x$.

(b) Sketch the graph, marking key values on the axes.

(c) How is it different from the graph of $y = \sin x$?

D4 Sketch these graphs, first working out key points.
Mark key values on your sketches.

(a) $y = \sin x - 1$ (b) $y = \frac{1}{2} \sin x$ (c) $y = {}^-3 \cos x$ (d) $y = 2 \cos x - 1$

D5 Match the graphs below to these equations.
Remember to check key points,
such as when $x = 0°, 90°, \ldots$

 $y = \cos x - 2$

 $y = {}^-2 \cos x$

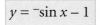 $y = {}^-\sin x - 1$

 $y = {}^-\cos x + 1$

$y = 2 \sin x + 1$

$y = 2 \sin x + 2$

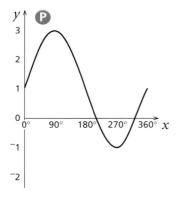

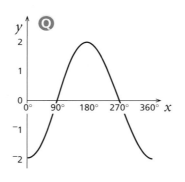

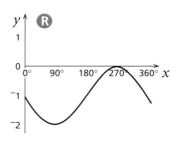

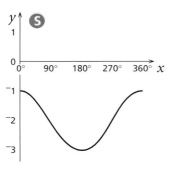

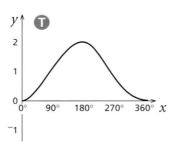

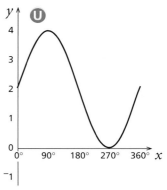

D6

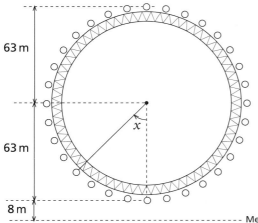

63 m

63 m

8 m

Mean river level

These are the dimensions of the London Eye.

This graph shows your height, h metres above river level after you have travelled through an angle x (starting at the **bottom** of the circle).

(a) What is the equation of this graph? (Test your ideas by trying key points.)

(b) Use your calculator to find your height above river level when you have travelled through

   (i) 100°   (ii) 240°

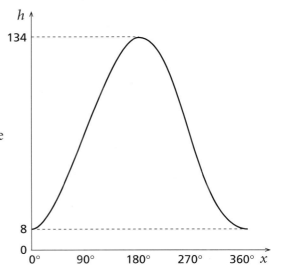

D7 The diagram shows a sketch of part of the curve with equation

$$y = p + q \cos x°$$

where p and q are integers.

(a) Find the values of the integers p and q.

(b) Find the two values of x between 0 and 360 for which

$$p + q \cos x° = 2$$

Edexcel

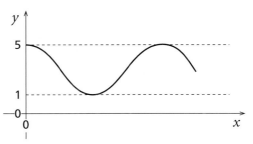

• What happens when you fill in this table for the graph $y = \sin(4x)$?

• Do you have enough information to sketch the graph?

x	0°	90°	180°	270°
$4x$	0°	360°		
$\sin(4x)$	0			

To sketch the graph of $y = \sin(4x)$ it is not enough to mark points every 90°.

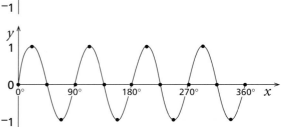

Even marking points every 45° is no use.

Marking every $22\frac{1}{2}°$ gives enough key points, though it is worth checking with more frequent points for part of the graph.

When you have to sketch a graph like $y = \sin(ax)$ or $y = \cos(ax)$, where a is a particular number, you should work out key points every $(90 \div a)°$.

D8 Sketch these graphs, first working out key points. Mark key values on your sketches.

(a) $y = \cos(3x)$ (b) $y = \sin(2x)$ (c) $y = 3\cos(4x)$ (d) $y = \sin(3x) - 1$

D9 Match the graphs below to these equations. $y = \cos(2x)$ $y = 2\sin(2x)$

$y = \sin(2x) - 1$ $y = {}^-2\cos(2x)$

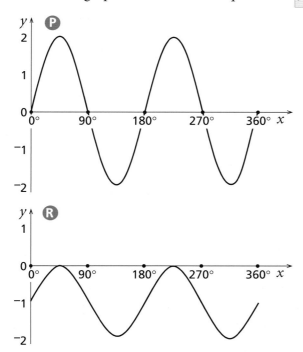

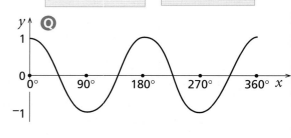

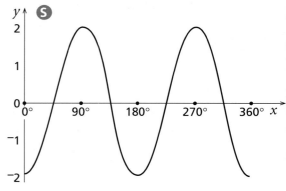

*D10 Sketch these graphs, first working out key points.
Mark key values on your sketches.

(a) $y = \sin(x + 90°)$ (b) $y = 2\sin(x - 60°)$

*D11 This is part of the graph of $y = a\sin(x + b°)$.
What are the values of a and b?

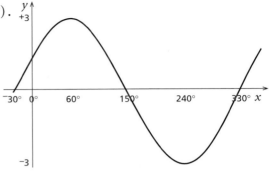

Sine and cosine graphs in the real world

If you hang a weight on a spring, pull down and
let go, the graph of the weight's height against
time is approximately a sine curve.

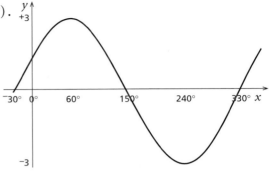

Mains electricity uses alternating current (AC).
The current flows in one direction and then the
other, 50 times a second. We say it has a
frequency of 50 **cycles per second**, or 50 **hertz**.

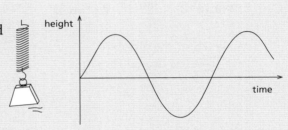

Sounds can be described using sine graphs.
High-pitched sounds (piccolo, triangle) have
high frequencies, while low-pitched sounds
(bass guitar) have low frequencies.

Young people can hear sounds with frequencies
from 20 hertz to about 20 000 hertz.
Dogs and bats can hear higher frequencies
than that.

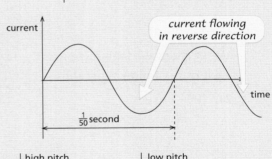

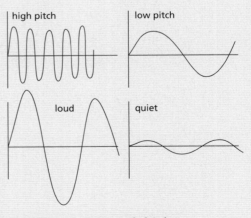

Radio signals also involve the sine graph idea but the frequencies are much higher.
Local FM radio stations work on about 100 megahertz (100 million cycles per second).

Test yourself

T1 Here is a sketch of the graph of
$y = \sin x$ for $0° \leq x \leq 360°$.

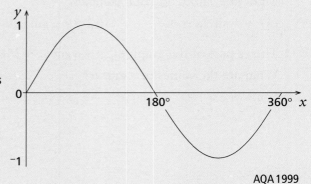

(a) On a copy of the sketch show
the locations of the two solutions
of the equation
$$\sin x = \tfrac{1}{2}$$

(b) Work out accurately the two
solutions of the equation
$$4 \sin x = {}^-3$$

AQA 1999

T2 Arrange these into groups with the same value.

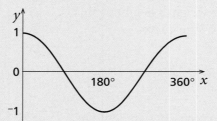

$\boxed{{}^-\tan 145°}$ $\boxed{{}^-\tan 35°}$ $\boxed{\tan 35°}$ $\boxed{\tan 145°}$ $\boxed{\tan 215°}$ $\boxed{\tan 325°}$

T3 A sketch of the graph of
$y = \cos x°$ is drawn for you.

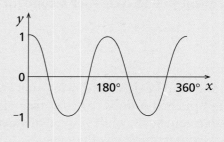

Write down the equations of
each of the following graphs.

(a)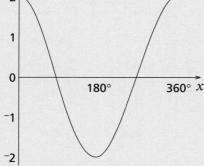

(b)

(c)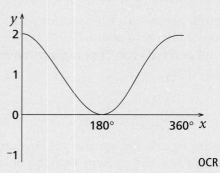

OCR

36 Pattern and proof

This work will help you

◆ find and use a rule for the *n*th term of a linear or quadratic sequence
◆ find a rule for the *n*th term of a sequence generated from a context and prove it
◆ prove a general statement is false using a counter-example
◆ prove a general statement is true using algebra

A Linear sequences

Find an expression for the *n*th term of the linear sequence   7, 11, 15, 19, 23, …

7   11   15   19   23   …

ook at the differences.

+4   +4   +4   +4   …

*Sequences where the differences are constant are called **linear**.*

The differences are all 4 so the rule begins with 4*n*.

Term numbers (n)	1	2	3	4	5	…
Terms of the sequence	7	11	15	19	23	…
4n	4	8	12	16	20	…
Sequence terms – 4n	3	3	3	3	3	…

Subtract 4*n* from each term.

So the nth term of the linear sequence is **4n + 3**.

A1 Which of these sequences are linear?

   A   5, 6, 8, 11, 15, 20   B   6, 10, 14, 18, 22, 26
   C   32, 27, 22, 17, 12, 7   D   3, 4, 5, 7, 9, 12
   E   3, 4.5, 6, 7.5, 9, 10.5   F   2, 4, 8, 16, 32, 64

A2 Copy and complete each linear sequence below, to show the first five terms.

   (a) 2, 9, __ , 23, __ , …   (b) 20, 18, __ , __ , 12, …
   (c) 1, __ , __ , 19, __ , …   (d) 30, __ , __ , __ , 18, …
   (e) k, $k + 1$, $k + 2$, __ , __ , …   (f) $p - 1$, __ , $p + 5$, $p + 8$, __ , …

A3 (a) Find an expression for the *n*th term of each of these linear sequences.

        (i) 5, 7, 9, 11, 13, …   (ii) 1, 4, 7, 10, 13, …
        (iii) −3, −2, −1, 0, 1, …   (iv) 8, 15, 22, 29, 36, …

   (b) Find the 10th term of each sequence in part (a).

A4 Find an expression for the *n*th term of each of these linear sequences.

   (a) 19, 18, 17, 16, 15, …   (b) 23, 21, 19, 17, 15, …

A5 A linear sequence begins   4, 9, 14, 19, …

   (a) Find the next term of this sequence.

   (b) Find an expression, in terms of n, for the nth term of this sequence.

   (c) What is the 20th term of this sequence?

   (d) Show that 1000 cannot be a term in this sequence.

   (e) Calculate the number of terms in the linear sequence   4, 9, 14, 19, … , 499

A6 A sequence of matchstick patterns begins

   Pattern 1   Pattern 2   Pattern 3

   (a) Find an expression for the number of matchsticks in pattern n.

   (b) How many matchsticks will be in pattern 50?

   (c) Which pattern will use 301 matchsticks?

   (d) Show that you cannot make a pattern in this sequence that uses 100 matchsticks.

B *Quadratic sequences*

Find an expression for the nth term of the quadratic sequence   8, 15, 26, 41, 60, …

8   15   26   41   60   …

+7 +11 +15 +19 …

+4   +4   +4   …

> Look at the differences.

> Sequences with constant second differences are called **quadratic**.

> The second differences are all 4 so the rule begins with $2n^2$. (The coefficient of n^2 is half the value of the second differences.)

> Subtract $2n^2$ from each term to obtain a linear sequence.

Term numbers (n)	1	2	3	4	5	…
Terms of the sequence	8	15	26	41	60	…
$2n^2$	2	8	18	32	50	…
Sequence terms – $2n^2$	6	7	8	9	10	…

The nth term of the linear sequence 6, 7, 8, 9, 10, … is $n + 5$.

So the nth term of the quadratic sequence is $2n^2 + n + 5$.

B1 Find the next two terms for each of these quadratic sequences.

   (a) 3, 4, 6, 9, 13, …   (b) 3, 6, 11, 18, 27, …

   (c) 1, 2, 7, 16, 29, …   (d) 50, 49, 47, 44, 40, …

B2 Which of these sequences are quadratic?

A   6, 8, 12, 18, 26, 36   B   6, 8, 12, 20, 36, 68

C   6, 8, 10, 12, 14, 16   D   3, 4, 7, 12, 19, 28

E   1, 6, 12, 19, 27, 36   F   2, 5, 9, 15, 24, 37

B3 Find an expression for the nth term of each of these quadratic sequences.

(a) 5, 8, 13, 20, 29, ...   (b) 0, 3, 8, 15, 24, ...

(c) 2, 8, 18, 32, 50, ...   (d) 1, 10, 25, 46, 73, ...

B4 (a) Find an expression for the nth term of each of these quadratic sequences.

    (i)   4, 10, 18, 28, 40, ...   (ii)   2, 9, 20, 35, 54, ...

    (iii) $1\frac{1}{2}$, 4, $7\frac{1}{2}$, 12, $17\frac{1}{2}$, ...   (iv)   5, 16, 33, 56, 85, ...

(b) Find the 10th term of each sequence in part (a).

B5 A quadratic sequence begins   4, 7, 12, 19, 28, ...

(a) Find the next term of this sequence.

(b) Find an expression, in terms of n, for the nth term of this sequence.

(c) What is the 20th term of this sequence?

(d) Calculate the number of terms in the quadratic sequence   4, 7, 12, 19, ... , 103

(e) Show that 203 cannot be a term in this sequence.

B6 A quadratic sequence begins   3, 8, 15, 24, ...

(a) Find the next term of this sequence.

(b) Find an expression, in terms of n, for the nth term of this sequence.

(c) Calculate the number of terms in the quadratic sequence   3, 8, 15, 24, ... , 675

(d) Is 445 a term in this sequence?

C *All sorts of sequences*

C1 Write down the first five terms of the sequences that have the following nth terms.

(a) $3n - 7$ (b) $n^3 + 1$ (c) $2n^2 - 5n$ (d) $n^3 + n - 2$

(e) $n^4 - n^2$ (f) $\frac{12}{n}$ (g) $\frac{n}{n+1}$ (h) $3^n + 1$

C2 (a) Find the first three terms of the sequence whose nth term is $2n - 1$.

(b) Find the first three terms of the sequence whose nth term is $n^3 - 6n^2 + 13n - 7$.

(c) What do you notice?

(d) Work out the 4th term of each sequence in parts (a) and (b).

C3 (a) Find the first three terms of the sequence whose nth term is n^3.

(b) Find the first three terms of the sequence whose nth term is $\frac{36}{n} + 25n - 60$.

(c) What do you notice?

(d) Work out the 4th term of each sequence in parts (a) and (b).

C4 A **quadratic** sequence begins 1, 8, 27, …

Find an expression for the nth term of this sequence.

When dots are used to show that a sequence continues, it usually means the sequence continues in the simplest way possible.

For example, the sequence 2, 4, 6, 8, …

- is usually the linear sequence that continues 10, 12, … with nth term $2n$
- not the sequence that continues 34, 132, … with nth term $n^4 - 10n^3 + 35n^2 - 48n + 24$.

C5 A sequence begins 5, 9, 13, 17, …

Write down an expression, in terms of n, for the nth term of the sequence.

AQA 2003 Specimen

C6 Give the nth term of each of the following sequences.

(a) 1, 4, 9, 16, 25, 36, …

(b) 2, 5, 10, 17, 26, 37, …

(c) 3, 7, 13, 21, 31, 43, … AQA 2000

C7 The first five terms of a sequence are

$$-2, \ 1, \ 6, \ 13, \ 22$$

Find the nth term of this sequence. OCR

C8 These are the first five terms of a sequence.

$$3, \ 12, \ 27, \ 48, \ 75$$

Find an expression for the nth term of the sequence. OCR

C9 Find the nth term of the sequence

$$4, \ 11, \ 22, \ 37, \ 56, \ 79, \ …$$ OCR

C10 The nth term of a sequence is $\dfrac{5n}{4n + 5}$.

(a) Write down the first two terms of this sequence.

(b) Which term of the sequence has value 1? AQA(NEAB) 1997

C11 Write down the nth term of each of these sequences.

(a) $\frac{1}{3}, \ \frac{2}{4}, \ \frac{3}{5}, \ \frac{4}{6}, \ \frac{5}{7}, \ …$

(b) 4, 16, 36, 64, 100, … OCR

D *Finding a counter-example*

You can prove a statment is false by means of a **counter-example**.

Chords and regions

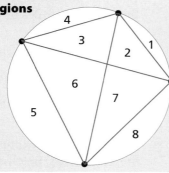

4 points
8 regions

- Is this the maximum number of regions with 4 points on the circumference of a circle?

- Investigate for different numbers of points.

I think that the expression
$n^2 + 23n + 23$
will give a prime number for
any positive integer value of n.

I only need to find one counter-example to prove you wrong.

D1 State which of these statements is false and use a counter-example to show it is false.

    A   The sum of two odd numbers is even.

    B   The sum of two even numbers is even.

    C   The product of two odd numbers is even.

    D   The product of two even numbers is even.

D2 Jim says 'Prime numbers are always odd.'
Explain what a prime number is and prove that Jim is wrong.   OCR

D3 Jagdeep said 'The square root of a number is always smaller than the number itself.'
Is he correct? Give an example to support your answer.   OCR

D4 Show that each statement below is false.

    (a) $3n + 1$ is odd for all integers n.   (b) $2n$ is even for all values of n.

    (c) $2^n \geq 1$ for all values of n.   (d) $2n^2 + 11$ is prime for all integers n.

D5 Show that each statement below is false.

    (a) $(x + y)^2 = x^2 + y^2$ for all values of x and y.

    (b) $\frac{12}{x + y} = \frac{12}{x} + \frac{12}{y}$ for all values of x and y.

D6 Show that each statement below is false.

    (a) If $k^2 > 0$ then $k > 0$.   (b) If k is even then $\frac{k}{2}$ is even.

    (c) If p is prime then $p + 2$ is prime.   (d) If $a < 1$ and $b < 1$ then $ab < 1$.

E *True to form*

The patterns

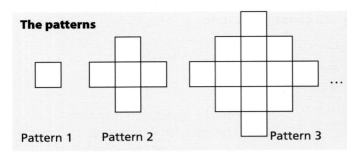

Pattern 1   Pattern 2   Pattern 3

- Can you find an expression for the number of tiles in the *n*th pattern?
- How can you be sure it is correct?
- Can you convince someone else?

Matchstick patterns

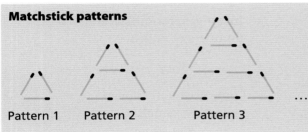

Pattern 1   Pattern 2   Pattern 3

- Can you find an expression for the number of matchsticks in the *n*th pattern?
- How can you be sure it is correct?
- Can you convince someone else?

In each question, explain how you found your rule and how you know it is true.

E1 Each 'hollow square' model is made with cubes.

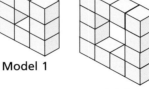

(a) Find a rule for the number of cubes in model *w*.

(b) Some of the cube faces cannot be seen, even if you look all round each model. (For example, there are 16 hidden faces in model 1.) Find a rule for the number of hidden cube faces in model *w*.

Model 1

Model 2

Model 3 ⋰

(c) Can you make a hollow square with 150 cubes? Explain your answer.

E2 (a) Find a rule for the number of cubes in the *n*th model.

(b) How many cubes are in the 10th model?

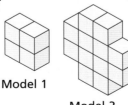

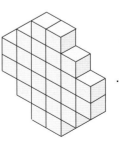

Model 1

Model 2

Model 3

E3 The numbers of cubes in these designs are called 'triangle numbers'.

(a) Work out the first six triangle numbers.

(b) Find a rule for the nth triangle number.

(c) What is the 100th triangle number?

(d) 820 is a triangle number.
How many triangle numbers are less than this?

E4 (a) Find a rule for the number of matchsticks in the nth pattern.

(b) How many of these matchstick designs use fewer than 100 matches?

E5 The numbers of dots in these designs can be called 'pentagon numbers'.

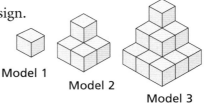

(a) Work out the first five pentagon numbers.

(b) Find a rule for the nth pentagon number.

(c) Show that 100 is not a pentagon number.

*E6 Each layer of cubes in these designs is a square.
Find a rule for the number of cubes in the nth design.

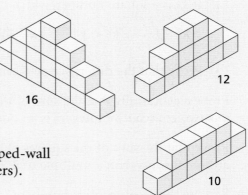

Model 1

Model 2

Model 3

Stepped-wall numbers

Make a 'stepped-wall' like this.

- Start with a row of cubes of any length.

- Build another row on top with two fewer cubes to make steps at the ends.

- Stop at any time after **two or more rows**.

The numbers of cubes in these walls are called stepped-wall numbers, (so 10, 12 and 16 are stepped-wall numbers).

Investigate these numbers.

F *Further proof*

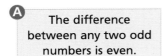

A The difference between any two odd numbers is even.

B The sum of any two consecutive odd numbers is divisible by 4.

C The product of an odd and an even number is even.

D The sum of any two consecutive numbers is odd.

E The sum of any two consecutive triangle numbers is a square number.

F1 Prove the following statements.

    A   The sum of any two odd numbers is even.

    B   The difference between any two even numbers is even.

    C   The product of any two odd numbers is odd.

F2 Prove the following statements.

    A   The sum of any three consecutive numbers is divisible by 3.

    B   The sum of any four consecutive numbers is not divisible by 4.

    C   The sum of any five consecutive numbers is a multiple of 5.

F3 Prove that the sum of any two consecutive even numbers cannot be a multiple of 4.

F4 Choose any three consecutive numbers.
Multiply the first and last number and add 1.
Square the middle number.

Try this with a few sets of numbers. What do you notice? Prove it.

F5 Prove that, for any four consecutive numbers, the product of the middle two numbers is always 2 more than the product of the first and last numbers.

F6 (a) If k^2 is a square number, which of these expressions gives the next square number?

$$k^2 + 1 \qquad k^2 + 3 \qquad (k + 2)^2 \qquad (k + 1)^2 \qquad 2k^2$$

    (b) Prove that the difference between two consecutive square numbers is always odd.

F7 Prove algebraically that the sum of the squares of
any two consecutive integers is an odd number.   Edexcel

F8 Prove that the sum of the squares of any three consecutive numbers is always one less than a multiple of 3.

F9 For any three consecutive numbers …

    (a) Explain why one of the numbers must be even.

    (b) Explain why one of the numbers must be a multiple of 3.

    (c) Explain why the product of the three numbers must be a multiple of 6.

F10 The nth term of a sequence is $n(n + 1)$.

  (a) Work out the first five terms of this sequence.

  (b) Write down an expression for the kth term of this sequence.

  (c) Write down an expression for the $(k + 1)$th term of this sequence.

  (d) Prove that if you add any two consecutive terms of this sequence and then divide by 2, your result will be a square number.

F11 (a) Factorise $p^2 - q^2$.

  (b) Here is a sequence of numbers   0, 3, 8, 15, 24, 35, 48, ...
       Write down an expression for the nth term of this sequence.

  (c) Show algebraically that the product of **any** two consecutive terms of the sequence

         0, 3, 8, 15, 24, 35, 48, ...

       can be written as the product of four consecutive integers.   Edexcel

F12 Prove that there is only one prime number of the form $k^2 - 1$ where k is a whole number.

F13 (a) Choose a triangle number, multiply it by 8 and add 1.
       Is your result a square number?

  (b) Try this with some other triangle numbers.
       Do you think the result will always be a square number?

       Prove it.

  (c) Show that the result has to be an **odd** square number.

F14 The mean of the five consecutive integers 1, 2, 3, 4, 5 is 3.
     So the square of the mean of the consecutive integers 1, 2, 3, 4, 5 is 9.
     The mean of the squares of those five consecutive integers $1^2, 2^2, 3^2, 4^2, 5^2$ is 11.

     Show algebraically that the square of the mean of **any** five consecutive integers is
     always 2 less than the mean of the squares of those five consecutive integers.   Edexcel

---

Consecutive sums

Here are two ways to make 9 by adding consecutive whole numbers.
They are called 'consecutive sums'.

4 + 5

2 + 3 + 4

Investigate consecutive sums.
Is it possible to find a consecutive sum for every number?

Test yourself

T1 Matches are arranged to form a sequence of diagrams as shown.

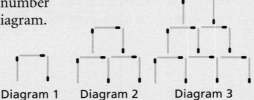

Diagram 1   Diagram 2   Diagram 3

Write an expression, in terms of n, for the number of matches needed to form
the nth diagram.   AQA 2003 Specimen

T2 (a) Write down the 10th term of the sequence which begins

    3, 7, 11, 15, …

(b) Write down an expression for the nth term of this sequence.

(c) Show that 1997 cannot be a term in this sequence.

(d) Calculate the number of terms in the sequence

    3, 7, 11, 15, …, 399   AQA (NEAB) 1997

T3 Write down the nth term of each of the following sequences.

(a) 2, 5, 10, 17, 26, 37, …

(b) 4, 16, 36, 64, 100, 144, …   OCR

T4 Here are the first five terms of a sequence.

    17, 14, 11, 8, 5

Find, in terms of n, an expression for the nth term of the sequence.   Edexcel

T5 Find the nth term of the sequence   2, 7, 14, 23, 34, 47, …

T6 Find an expression for the number
of matchsticks in the nth diagram.

Diagram 1   Diagram 2   Diagram 3

T7 Give an example to show that the following statement is **not true.**

> Any **even** number greater than 2 can be expressed as
> the sum of two prime numbers **in only one way**.

AQA 2003 Specimen

T8 Prove that the sum of any seven consecutive numbers is a multiple of 7.

T9 Prove that, for any three consecutive numbers, the difference between the squares of
the first and last numbers is 4 times the middle number.

Histograms

This work will help you

♦ read and draw histograms

♦ understand frequency density

A *Unequal class intervals*

Mili is in charge of a skiing party.
She draws this frequency chart of
the ages of the people in the party.

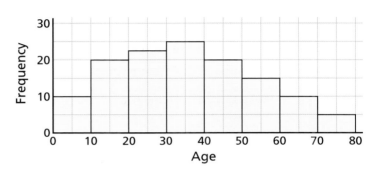

Mili decides to put all the people aged
50 or over into one group.
The total frequency for this new group
is 15 + 10 + 5 = 30.

Here is Mili's new frequency chart.

This chart is misleading. It exaggerates
the number in the 50–80 age group.
This is because our eye is drawn to
the **area**, not the height, of the bars.

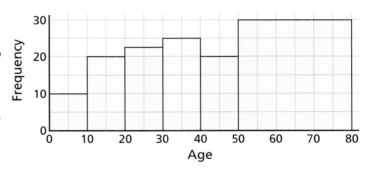

A chart which uses **area** to show frequency
is called a **histogram**.

Here is a histogram for the ages of the
skiing party. It does not exaggerate the
number in the 50–80 group.

We can work out the number of people
in each age group by using the 'area scale'
printed above the chart.

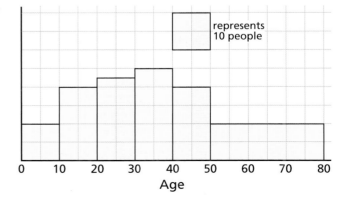

represents
10 people

A1 This histogram shows the distribution of heights in a primary year group.

Copy and complete this frequency table.

Height, h cm	Frequency
95 < h ≤ 100	
100 < h ≤ 110	
110 < h ≤ 115	
115 < h ≤ 120	
120 < h ≤ 130	

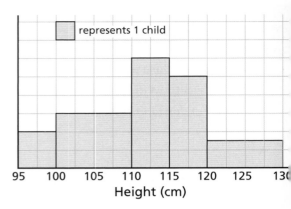

A2 This histogram shows the distribution of weights of a group of babies.

Copy and complete this frequency table.

Weight, w kg	Frequency
1.0 < w ≤ 2.0	
2.0 < w ≤ 2.5	
2.5 < w ≤ 3.0	
3.0 < w ≤ 4.0	
4.0 < w ≤ 4.5	

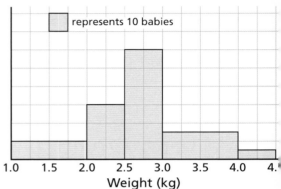

A3 This histogram shows the weights of the eggs collected from a hen-house.

There are 60 eggs in the interval $60 < w \le 65$.

(a) Work out the number of eggs in the interval $45 < w \le 55$.

(b) Work out the total number of eggs collected.

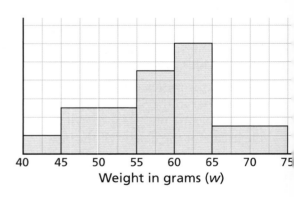

A4 The table below shows the distribution of the prices of houses for sale in a town. ('80k' means 80 000.)

Copy and complete the histogram for the distribution.

Price, £P	Frequency
70k < P ≤ 80k	40
80k < P ≤ 100k	120
100k < P ≤ 120k	200
120k < P ≤ 150k	120
150k < P ≤ 200k	50

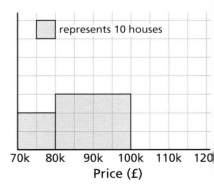

A5 The results of a survey of the lengths of local phone calls are summarised in the table below.

Time, t minutes	Frequency
$0 < t \le 1$	34
$1 < t \le 2$	13
$2 < t \le 5$	12
$5 < t \le 10$	10
$10 < t \le 20$	4
$20 < t$	0

(a) Use mid-interval values to estimate the mean length of a call.

(b) Copy and complete the histogram.

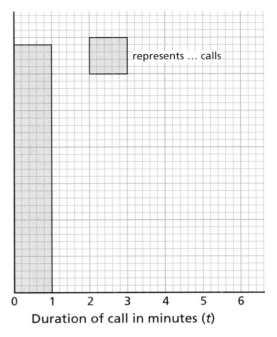

represents ... calls

Duration of call in minutes (t)

B *Frequency density*

So far we have not used a vertical scale.

A vertical scale is useful to help work out the frequency represented by a bar.

Here is a histogram of children's heights. The area of the first bar represents 20 children.

The 'length' of this bar is 10 (from 100 to 110). To make the area 20, the height has to be **2**.

This leads to the vertical scale shown. It is called a **frequency density** scale.

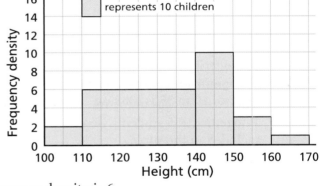

represents 10 children

Frequency density

Height (cm)

For the bar from 110 to 140 (length 30), the frequency density is 6.
So the number of babies represented by this bar is $30 \times 6 = $ **180**. (Check by counting squares.)

frequency = length of interval × frequency density

B1 (a) What are the missing numbers on this frequency density scale?

(b) The bar for the interval 70–75 is missing. If there are 30 eggs in this interval, what is the frequency density for this bar?

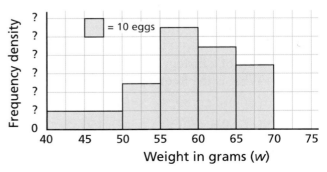

= 10 eggs

Frequency density

Weight in grams (w)

B2 Copy and complete the frequency table below, using the information in the histogram.

Weight, w g	Frequency
100 < w ≤ 150	
150 < w ≤ 250	
250 < w ≤ 300	
300 < w ≤ 450	

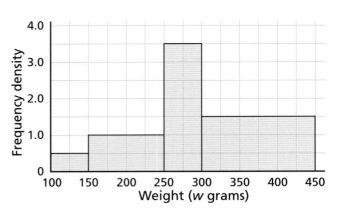

B3 Copy and complete the frequency table below, using the information in the histogram.

Height, h cm	Frequency
120 < h ≤ 130	
130 < h ≤ 155	
155 < h ≤ 175	
175 < h ≤ 190	

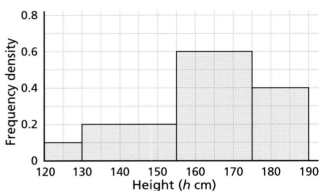

B4 A survey was carried out of the times taken by students to get to school. The distribution is shown in the histogram, except that the bar for the interval 20 to 35 is missing.

Altogether, 255 students were included in the survey.

What is the frequency density for the interval 20 to 35?

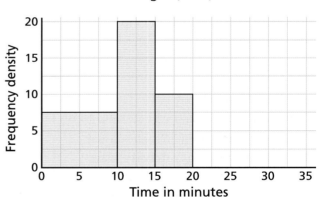

B5 The table below shows the results of a survey of times spent watching TV one evening.

Time, t minutes	0 < t ≤ 30	30 < t ≤ 60	60 < t ≤ 150	150 < t ≤ 300
Frequency	6	15	54	30

Part of a histogram for this data is shown here.
What is the frequency density for the interval
(a) 0 < t ≤ 30   (b) 30 < t ≤ 60
(c) 60 < t ≤ 150   (d) 150 < t ≤ 300

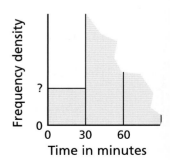

C *Drawing a histogram*

From the fact that **frequency = length of interval × frequency density**

it follows that **frequency density** = $\dfrac{\text{frequency}}{\text{length of interval}}$

To draw a histogram from a frequency table,
first calculate the frequency density for each interval.

Example

Draw a histogram to show
this data.

Weight, w g	Frequency
100 < w ≤ 150	24
150 < w ≤ 250	60
250 < w ≤ 300	15
300 < w ≤ 450	9

First find the frequency densities.

Frequency density
24 ÷ 50 = 0.48
60 ÷ 100 = 0.6
15 ÷ 50 = 0.3
9 ÷ 150 = 0.06

Then choose convenient graph
scales and draw the bars.

C1 Visitors to a museum spend different amounts of time
looking around. This table shows the frequency
distribution of the times spent in the museum by
visitors on a Saturday in summer.

(a) Calculate the frequency density for each interval.

(b) Draw a histogram for the data.

Time, t minutes	Frequency
0 < t ≤ 30	24
30 < t ≤ 45	30
45 < t ≤ 60	15
60 < t ≤ 90	12
90 < t ≤ 180	18

C2 This table shows the distribution of the heights of
trees in a wood.

(a) Draw a histogram for the data.

(b) Use the histogram to estimate the number of trees
whose heights are between 10 m and 20 m.

Height, h m	Frequency
0 < h ≤ 5	8
5 < h ≤ 7.5	12
7.5 < h ≤ 10	18
10 < h ≤ 15	7
15 < h ≤ 25	6

You need sheet G177 for questions C3 and T1.

C3 The table on the right shows the age distributions of the population of the UK in 1901 and in 1998, expressed as percentages.

(a) A histogram for the 1901 distribution is A on sheet G177.
On the same axes, but using a different colour, draw the histogram for the 1998 distribution.

(b) Comment on any differences between the two distributions.

Age A years	Percentage in 1901	Percentage in 1998
A < 5	11.9	6.2
5 ≤ A < 10	10.7	6.6
10 ≤ A < 20	20.3	12.6
20 ≤ A < 30	18.3	13.3
30 ≤ A < 45	19.6	22.5
45 ≤ A < 60	12.1	18.3
60 ≤ A < 65	2.8	4.8
65 ≤ A < 75	3.3	8.4
75 ≤ A < 110	1.3	7.3

Test yourself

T1 The unfinished histogram and the table (B on sheet G177) give information about the heights, in centimetres, of the year 11 students at Mathstown High School.

(a) Use the histogram to complete the table.

(b) Use the table to complete the histogram.

Edexcel

T2 The speeds of 100 cars travelling along a road are shown in this table.

Speed (s km/h)	20 ≤ s < 35	35 ≤ s < 45	45 ≤ s < 55	55 ≤ s < 65	65 ≤ s < 85
Frequency	6	19	34	26	15

(a) Draw a histogram to show this information.

(b) The speed limit along this road is 48 km/h. Estimate the number of cars exceeding the speed limit.

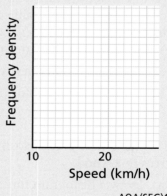

AQA(SEG)1999

38 *Algebraic fractions and equations*

You will revise how to
♦ simplify expressions involving straightforward algebraic fractions
♦ solve equations involving fractions with numerical denominators

This work will help you
♦ simplify more complex expressions with algebraic fractions
♦ solve equations involving fractions with algebraic denominators

A *Review*

A1 Find the value of each of the following when $a = 2$, $b = \frac{1}{2}$, $c = 3$ and $d = {}^-2$.

(a) ab^2 (b) bd^2 (c) $\dfrac{a}{b}$ (d) $\dfrac{ac}{d}$

A2 Simplify these.

(a) $k^3 \times k^2$ (b) $k^3 \div k^2$ (c) $(k^2)^3$ (d) $(k^3)^2$

A3 Simplify these.

(a) $\dfrac{2ab}{b}$ (b) $\dfrac{c^2d}{cd}$ (c) $\dfrac{4e^2f^2}{2e^3f^3}$ (d) $\dfrac{g^3h^5}{g^2h^3}$

A4 Simplify these.

(a) $3a^3 \times 4a^2b$ (b) $pq^{-2} \times p^{-3}q^2$ (c) $(3u^2v^3)^2$ (d) $(x^2y^2)^3 \div (x^3y^3)^2$

(e) $\sqrt{16s^4t^8}$ (f) $\sqrt{\dfrac{4a^4b^2}{c^6}}$ (g) $\sqrt{\dfrac{u^2v^{16}}{25x^2y^8}}$ (h) $\sqrt{k^{-4}l^6} \times \sqrt{k^8l^2}$

A5 Write each of these as simply as possible.

(a) $\dfrac{4a + 8}{2}$ (b) $\dfrac{5b + 15}{5}$ (c) $\dfrac{3c^2 + 12}{3}$ (d) $\dfrac{4ab + 8b}{2b}$

A6 Write each of these as a single fraction.

(a) $\dfrac{a}{2} + \dfrac{a}{6}$ (b) $\dfrac{a}{2} - \dfrac{a}{6}$ (c) $\dfrac{a+1}{2} + \dfrac{a}{6}$ (d) $\frac{1}{2}(a + 1) + \frac{1}{6}(a - 2)$

(e) $\dfrac{2a+1}{3} + \dfrac{3a-2}{6}$ (f) $\dfrac{2a+1}{3} - \dfrac{3a-2}{6}$ (g) $\dfrac{2a+1}{2a} + \dfrac{3a-2}{6a}$ (h) $\dfrac{2a+b}{2a} - \dfrac{3a+2b}{4a}$

A7 Write each of these as simply as possible as a single fraction.

(a) $\dfrac{3e}{2} \times \dfrac{1}{6}$ (b) $\dfrac{2e}{3} \times \dfrac{1}{4e}$ (c) $\dfrac{cd}{2} \times \dfrac{4c}{d}$ (d) $\dfrac{1}{u} \times \dfrac{u}{v}$

(e) $\dfrac{2p}{3} \div \dfrac{p}{2}$ (f) $\dfrac{st}{3} \div \dfrac{6s}{t}$ (g) $\dfrac{de^2}{2} \div \dfrac{3de}{4}$ (h) $\dfrac{2m^2n^2}{3} \div \dfrac{4mn^2}{5}$

B Cancelling

When simplifying a fraction, first factorise as much as you can.

1 Simplify $\dfrac{2x - 6}{x^2 - 5x + 6}$.

$$\frac{2x - 6}{x^2 - 5x + 6} = \frac{2(x-3)}{(x-2)(x-3)} = \frac{2}{x - 2}$$

2 Simplify $\dfrac{2x^2 + 3x - 2}{x^2 + x - 2}$.

$$\frac{2x^2 + 3x - 2}{x^2 + x - 2} = \frac{(2x - 1)(x+2)}{(x+2)(x - 1)} = \frac{2x - 1}{x - 1}$$

B1 Simplify and express each of these as a fraction in its simplest form.

(a) $\dfrac{2x + 6}{3x + 9}$

(b) $\dfrac{6x - 3}{12x - 6}$

(c) $\dfrac{x - 2}{10 - 5x}$

(d) $\dfrac{2x^2 + 10}{3x^2 + 15}$

(e) $\dfrac{bx + b}{cx + c}$

(f) $\dfrac{x + 6}{x^2 + 8x + 12}$

B2 There are four pairs of equivalent expressions below. Find them.

A $\dfrac{x + 2}{x - 2}$ **B** $\dfrac{x + 2}{4 - x^2}$ **C** $\dfrac{1}{x + 2}$ **D** $\dfrac{1}{x - 2}$ **E** $\dfrac{1}{2 - x}$ **F** $\dfrac{x - 2}{x^2 - 4}$ **G** $\dfrac{x + 2}{x^2 - 4}$ **H** $\dfrac{x^2 - 4}{(x-2)^2}$

B3 Simplify these by factorising and then cancelling.

(a) $\dfrac{p^2 - 2p - 3}{p - 3}$

(b) $\dfrac{n + 1}{n^2 + 3n + 2}$

(c) $\dfrac{a^2 + ab}{ac + bc}$

(d) $\dfrac{x^2 - 9}{x^2 + 6x + 9}$

(e) $\dfrac{y^2 + y - 2}{y^2 - 4}$

(f) $\dfrac{p^2 - p - 2}{p^2 - 2p - 3}$

B4 Copy and complete each of these simplifications.

(a) $\dfrac{n^2 + \blacksquare + 2}{n + 1} = n + 2$

(b) $\dfrac{n^2 - \blacksquare + 2}{n - \blacksquare} = n - 1$

(c) $\dfrac{\blacksquare x - 6}{x^2 - 5x + 6} = \dfrac{3}{x - \blacksquare}$

(d) $\dfrac{x^2 - \blacksquare x - \blacksquare}{x^2 - 6x + 8} = \dfrac{x + 1}{x - 2}$

B5 Simplify these where possible.

(a) $\dfrac{a^2 - 1}{a^2 - 2a + 1}$

(b) $\dfrac{(a - 1)^2}{a^2 - 2a + 1}$

(c) $\dfrac{(a + 1)^2}{a^2 - 2a + 1}$

(d) $\dfrac{a^2 + 1}{a^2 - 2a + 1}$

B6 Simplify fully the expression $\dfrac{6x - 18}{x^2 - 5x + 6}$.

AQA(SEG)1998

B7 Simplify fully the expression $\dfrac{7x + 21}{2x^2 + 5x - 3}$.

AQA(SEG)1999

C *More than one fraction*

When adding or subtracting fractions with expressions in the denominator,
first try to find the simplest expression that they will both divide into.
Factorise the expressions first if necessary.

$$\frac{x}{x+1} + \frac{4}{x}$$

$$= \frac{x \times x}{x(x+1)} + \frac{4(x+1)}{x(x+1)}$$

$$= \frac{x^2 + 4(x+1)}{x(x+1)}$$

$$= \frac{x^2 + 4x + 4}{x(x+1)}$$

$$= \frac{(x+2)^2}{x(x+1)}$$

$$\frac{x-1}{6x+3} + \frac{x}{4x+2}$$

$$= \frac{x-1}{3(2x+1)} + \frac{x}{2(2x+1)}$$

$$= \frac{2(x-1)}{6(2x+1)} + \frac{3x}{6(2x+1)}$$

$$= \frac{2(x-1) + 3x}{6(2x+1)}$$

$$= \frac{2x - 2 + 3x}{6(2x+1)}$$

$$= \frac{5x - 2}{6(2x+1)}$$

$$\frac{x+2}{x^2-1} - \frac{1}{x+1}$$

$$= \frac{x+2}{(x+1)(x-1)} - \frac{1}{x+1}$$

$$= \frac{x+2}{(x+1)(x-1)} - \frac{x-1}{(x+1)(x-1)}$$

$$= \frac{x+2 - (x-1)}{(x+1)(x-1)}$$

$$= \frac{x+2 - x+1}{(x+1)(x-1)}$$

$$= \frac{3}{(x+1)(x-1)} \text{ or } \frac{3}{x^2-1}$$

C1 Write each of these as a single fraction. Simplify your answers where possible.

  (a) $\dfrac{2}{a} + \dfrac{3}{2a}$ (b) $\dfrac{5}{2z} - \dfrac{1}{4z}$ (c) $\dfrac{x}{4} + \dfrac{4}{x}$ (d) $z + z^{-1}$ (e) $1 - a^{-2}$

C2 Write each of these as a single fraction.

  (a) $\dfrac{3q}{q+2} + \dfrac{5}{q+2}$ (b) $\dfrac{1}{z+1} + \dfrac{1}{z-1}$ (c) $3 - \dfrac{1}{y-1}$

  (d) $\dfrac{5}{2n-3} - \dfrac{3}{2n-3}$ (e) $1 - (1+s)^{-1}$ (f) $\dfrac{3}{1-x} + \dfrac{5}{(1-x)^2}$

C3 (a) (i) Factorise $x^2 - 4$.   (ii) Factorise $3x - 6$.

  (b) What is the simplest expression that both $x^2 - 4$ and $3x - 6$ will divide into?

  (c) Simplify (i) $\dfrac{x}{x^2-4} + \dfrac{1}{3(x-2)}$ (ii) $\dfrac{1}{3(x-2)} - \dfrac{1}{x^2-4}$

C4 Write each of these as a single fraction.

  (a) $\dfrac{1}{a+2} - \dfrac{1}{a-2}$ (b) $\dfrac{2}{2a-3} - \dfrac{3}{3a-2}$ (c) $x^{-2} - x^{-4}$

  (d) $\dfrac{1}{x^2-1} - \dfrac{1}{x^2+x}$ (e) $\dfrac{x-1}{2x^2} + \dfrac{x}{3x-x^2}$ (f) $1 + \dfrac{1}{a+2} + \dfrac{4a}{a^2-4}$

C5 Simplify $\dfrac{1}{2x+3} + \dfrac{1}{2x-1}$.

                                                     Edexcel

D Solving equations with fractions

When solving equations involving fractions it is usually easiest
to multiply to get rid of all the fractions first.

$$\frac{2x + 1}{x - 2} = 3$$

$$\frac{(2x + 1)(x - 2)}{x - 2} = 3(x - 2) \qquad [\times (x - 2)]$$

$$2x + 1 = 3(x - 2) \qquad [\text{cancel}]$$

$$2x + 1 = 3x - 6 \qquad [\text{expand brackets}]$$

$$\mathbf{7 = x}$$

$$\frac{x}{x + 3} \; + \; \frac{2}{x} \; = 1$$

$$\frac{x \times x(x + 3)}{x + 3} + \frac{2 \times x(x + 3)}{x} = 1 \times x(x + 3) \qquad [\times x(x + 3)]$$

$$x^2 + 2(x + 3) = x(x + 3) \qquad [\text{cancel}]$$

$$x^2 + 2x + 6 = x^2 + 3x \qquad [\text{expand brackets}]$$

$$2x + 6 = 3x \qquad [- x^2]$$

$$6 = x \text{ or } \mathbf{x = 6} \qquad [- 2x]$$

$$\frac{1}{x - 6} - \frac{1}{x - 2} = \frac{1}{8}$$

$$\frac{8(x-6)(x-2)}{x - 6} - \frac{8(x-6)(x-2)}{x - 2} = \frac{8(x-6)(x-2)}{8} \qquad [\times 8(x - 6)(x - 2)]$$

$$8(x - 2) - 8(x - 6) = (x - 6)(x - 2) \qquad [\text{cancel}]$$

$$8x - 16 \; - 8x + 48 = x^2 - 8x + 12 \qquad [\text{expand brackets}]$$

$$0 = x^2 - 8x - 20 \qquad [\text{collect terms}]$$

$$0 = (x - 10)(x + 2) \qquad [\text{factorise}]$$

$$\mathbf{x = 10 \text{ or } ^-2}$$

D1 By multiplying both sides by $x + 1$, solve $\dfrac{x + 5}{x + 1} = 2$.

D2 Solve each of these.

(a) $\dfrac{x + 1}{x - 3} = 3$ (b) $\dfrac{2x + 1}{x - 4} = 5$ (c) $\dfrac{1 - 2x}{2x + 1} = 1$

D3 By multiplying both sides by $(x + 1)(x + 2)$, solve $\dfrac{x}{x + 1} - \dfrac{1}{x + 2} = 1$.

D4 Solve each of these. (Hint: each equation has only one solution.)

(a) $\dfrac{1}{x} + \dfrac{2x}{x + 3} = 2$ (b) $\dfrac{2x}{2x + 1} + \dfrac{1}{x - 1} = 1$ (c) $\dfrac{1}{x + 1} + \dfrac{x}{x - 1} = 1$

D5 By multiplying both sides by $(x + 1)(x + 2)$, solve $\dfrac{x}{x + 1} = \dfrac{x - 1}{x + 2}$.

D6 Solve these.

(a) $\dfrac{x - 1}{x + 3} = \dfrac{x - 2}{x + 1}$ (b) $\dfrac{x + 4}{x + 2} = \dfrac{x + 2}{x - 4}$ (c) $\dfrac{2x - 1}{x + 1} = \dfrac{4x - 3}{2x + 1}$

D7 Solve each of these. In each one, you should obtain a quadratic equation that factorises.

(a) $\dfrac{4}{a} + 4a = 10$ (b) $\dfrac{x}{x - 6} + \dfrac{1}{x} = 0$ (c) $\dfrac{s}{4 + s} - 2s = 3$

(d) $\dfrac{x + 3}{3x + 1} = \dfrac{2x - 1}{3 - 2x}$ (e) $\dfrac{1}{2x + 3} + \dfrac{1}{3x + 4} = 2$ (f) $\dfrac{x - 1}{2x^2 - 2} + \dfrac{1}{3x + 4} + 1 = 0$

D8 Solve these equations. (You may not be able to factorise.)

(a) $\dfrac{x}{2x + 3} + 1 = x$ (b) $\dfrac{1}{x - 3} + \dfrac{2}{x - 1} = \dfrac{1}{3}$ (c) $\dfrac{x + 3}{x + 2} = \dfrac{3x + 1}{x + 4}$

(d) $\dfrac{1 - 2x}{x + 2} = \dfrac{3x + 1}{x + 4}$ (e) $\dfrac{a - 4}{a^2 - 16} - \dfrac{1}{a - 5} = 2$ (f) $\dfrac{x + 1}{2x - 1} - \dfrac{x - 1}{x + 2} = 2$

D9 Solve $\dfrac{2(x - 2)}{x^2 - 4} + \dfrac{3}{2x - 1} = 1$ *Edexcel*

D10 Solve these equations. (a) $x^{-3} = 0.125$ (b) $\dfrac{4}{m} + \dfrac{m}{4} = 2.5$ *AQA(SEG)1998*

D11 I drive 100 miles at x m.p.h. and then 60 miles at $(x + 10)$ m.p.h.

(a) Write down an expression in x for the total time taken.

(b) The total time I take is exactly 3 hours.
Form an equation in x and show that it simplifies to $3x^2 - 130x - 1000 = 0$.

(c) Solve this equation and thus work out my two speeds.

*D12 I always cycle uphill at 3 km/h less than I go on the flat.
I always cycle downhill at 3 km/h more than I go on the flat.

My route to work is 1 km uphill and 1 km downhill.
It takes me a quarter of an hour to cycle.

What is my speed on the flat?

*D13 The sum of the reciprocals of two consecutive even numbers is $\dfrac{9}{40}$.
Find the numbers.

E *Rearranging formulas with fractions*

Rearranging a formula involving fractions is very like solving an equation.

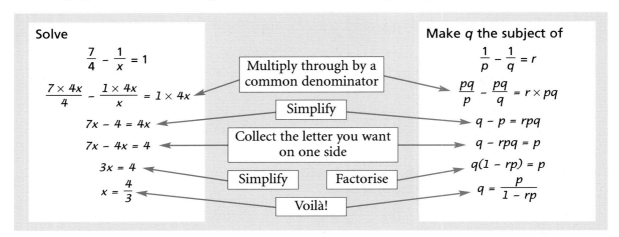

Solve

$$\frac{7}{4} - \frac{1}{x} = 1$$

Multiply through by a common denominator

$$\frac{7 \times 4x}{4} - \frac{1 \times 4x}{x} = 1 \times 4x$$

Simplify

$$7x - 4 = 4x$$

Collect the letter you want on one side

$$7x - 4x = 4$$

Simplify

$$3x = 4$$

$$x = \frac{4}{3}$$

Voilà!

Make q the subject of

$$\frac{1}{p} - \frac{1}{q} = r$$

$$\frac{pq}{p} - \frac{pq}{q} = r \times pq$$

$$q - p = rpq$$

$$q - rpq = p$$

Factorise

$$q(1 - rp) = p$$

$$q = \frac{p}{1 - rp}$$

E1 Rearrange $\dfrac{1}{p} - \dfrac{1}{q} = r$ to make p the subject.

E2 Make v the subject of the formula $\dfrac{1}{f} = \dfrac{1}{u} + \dfrac{1}{v}$.

Give your answer in its simplest form.

AQA(SEG)2000

E3 Rearrange the following to make the letter in brackets the subject.

(a) $P = \dfrac{mu}{t} - \dfrac{mv}{t}$ $[v]$ (b) $\dfrac{v^2}{s} - \dfrac{u^2}{s} = 2f$ $[s]$ (c) $\dfrac{x}{a} - \dfrac{y}{b} = 1$ $[x]$

(d) $y = \dfrac{2(x + a)}{(x - a)}$ $[x]$ (e) $l = \dfrac{a^2}{k} - \dfrac{b^2}{g}$ $[k]$ (f) $\dfrac{1}{a} + \dfrac{1}{b} = c$ $[b]$

E4 When you graph the formula $\dfrac{x^2}{a^2} + \dfrac{y^2}{b^2} = 1$ you get a curve called an ellipse.

Rearrange this formula to make y the subject.

If you have access to a graphical calculator or graphing program, substitute values for a and b and check that the original formula and your rearrangement give the same curve.

E5 The harmonic mean h of two numbers x and y is defined by the formula $\dfrac{1}{h} = \dfrac{1}{2}\left(\dfrac{1}{x} + \dfrac{1}{y}\right)$

(a) (i) Show that this formula can be rearranged to give $h = \dfrac{2xy}{x + y}$.

    (ii) Find the value of h when $x = \frac{2}{5}, y = \frac{2}{7}$.

(b) Rearrange the formula to make y the subject.

OCR(MEG)

F Mixed questions

F1 This formula shows the relationship between x and A.
$$A = \frac{8}{2x-1} - \frac{4}{x}$$

   (a) Show clearly that this formula can be simplified to $A = \frac{4}{2x^2 - x}$.

   (b) Find the values of x when $A = 2$.   AQA(SEG)1998

F2 (a) Factorise $x^2 + 3x + 2$.

   (b) Write as a single fraction in its simplest form $\frac{3}{x+1} - \frac{4}{x}$.   Edexcel

F3 (a) Simplify the following expression. $\frac{x^2 + x - 2}{x^2 - 4}$

   (b) Solve $\frac{2x+1}{x-1} = \frac{7x+3}{4x-3}$.

   (c) Solve $(x-5)(x+1) = 0$.   OCR

F4 (a) Simplify as much as possible $\frac{x}{3} + \frac{2x}{9} - \frac{x}{12}$.

   (b) Solve the equation $\frac{3}{x-4} - \frac{6}{x-2} = 1$.   AQA1999

F5 Make x the subject of the formula $y = \sqrt{\dfrac{3}{x+a}}$.   OCR(MEG)

F6 Solve the following equation. $\frac{3}{x+2} - \frac{2}{2x-3} = \frac{1}{7}$ WJEC

F7 A class takes x weeks to collect £84 for charity.
The mean amount collected per week is, therefore, $£\frac{84}{x}$.

   The following week the class collects £20.

   (a) Write down, in terms of x, the new mean amount collected per week.

   The collection of £20 increases the mean amount collected per week by £1.

   (b) (i) Use your answer to part (a) to write an equation.

       (ii) Show that this equation can be written as $x^2 - 19x + 84 = 0$.

   (c) Solve this equation to calculate x.   AQA(SEG)1998

F8 Sebastian sets out from home to catch the bus to school
He jogs the first 300 metres at a steady speed of x metres per second.
He runs the next 210 metres at a steady speed of $(x+4)$ metres per second.

   (a) Obtain an expression, in terms of x, for the total time he takes.

   (b) (i) The total time Sebastian takes is 130 seconds.
          Form an equation in x and show that it simplifies to $13x^2 + x - 120 = 0$.

       (ii) Solve this equation.   OCR

Test yourself

T1 Simplify these by factorising and then cancelling.

(a) $\dfrac{n^2 + 5n + 4}{n + 1}$
(b) $\dfrac{k - 2}{2k^2 - k - 6}$
(c) $\dfrac{a^2 - b^2}{a + b}$

T2 Write each of these as a single fraction.

(a) $\dfrac{4a}{a + 2} + \dfrac{1}{a + 2}$
(b) $\dfrac{1}{b + 3} + \dfrac{1}{b - 3}$
(c) $1 - \dfrac{2}{c - 1}$

(d) $\dfrac{d}{d^2 - 4} - \dfrac{1}{d + 2}$
(e) $\dfrac{e + 2}{e^2 - 1} - \dfrac{e - 2}{e^2 - e}$
(f) $1 - \dfrac{2f - 1}{f + 3}$

T3 Solve the equation $\dfrac{6}{2x - 1} - \dfrac{3}{x + 1} = 1$. AQA(SEG) 2000

✗ T4 Solve the equation $\dfrac{1}{x} + \dfrac{3x}{x - 1} = 3$. AQA 2003

T5 (a) Make x the subject of the formula $y = \dfrac{x + 3}{x}$. OCR

(b) Simplify $\dfrac{2x}{x + 1} - \dfrac{6}{x - 3}$.

✗ T6 Solve the equation $\dfrac{4}{x + 1} + \dfrac{3}{x - 1} = 1$. OCR

T7 Solve the equation $\dfrac{5}{2x - 1} + \dfrac{6}{x + 1} = 3$.

Give your answer correct to two decimal places. OCR

T8 Rearrange the following to make the letter in brackets the subject.

(a) $\dfrac{a}{x} + \dfrac{b}{y} = 1$ $[x]$
(b) $\dfrac{1}{x^2} - \dfrac{1}{y^2} = 1$ $[x]$
(c) $G = \dfrac{m^2}{a} - \dfrac{m^2}{b}$ $[b]$

T9 Sue travels 70 miles at an average speed of x miles per hour.
She then travels 50 miles at $(x + 10)$ miles per hour.

(a) Write down an expression in terms of x for the total time taken, in hours, for the whole journey.

(b) If the whole journey takes 3 hours, form an equation in x and show that it can be simplified to $3x^2 - 90x - 700 = 0$.

(c) Solve this equation to find the value of x. OCR

Review 9

1 This is a sketch of the graph of $y = \sin x$. Use the graph to help you sketch each of the following graphs.

(a) $y = 2 \sin x$

(b) $y = \sin 2x$

(c) $y = 1 + \sin x$

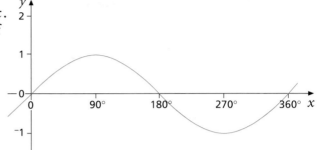

2 Work these out. Give each answer in its simplest form.

(a) $\frac{4}{5} + \frac{3}{4}$

(b) $\frac{2}{3} - \frac{5}{8}$

(c) $2\frac{3}{4} - \frac{7}{10}$

(d) $1\frac{3}{8} + \frac{4}{5}$

(e) $\frac{5}{6} \times \frac{3}{4}$

(f) $1\frac{2}{3} \times 1\frac{4}{5}$

(g) $\frac{9}{10} \div \frac{3}{4}$

(h) $\frac{3}{5} \div \frac{2}{3}$

3 Each of these 'step patterns' is made using square tiles.

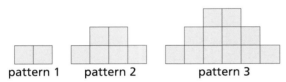

pattern 1   pattern 2   pattern 3

(a) How many tiles would there be in the 10th pattern?

(b) Find a rule for the number of tiles in the nth step pattern.

(c) Show that you could not make a step pattern using exactly 100 tiles.

4 This table shows the distribution of heights of a group of 10-year-old girls.

Draw a histogram for the data.

Height, h cm	Frequency
$120 < h \le 130$	3
$130 < h \le 135$	9
$135 < h \le 140$	12
$140 < h \le 145$	11
$145 < h \le 155$	5

5 Simplify the following.

(a) $\dfrac{a^2 b^3}{ab}$

(b) $(3m)^2$

(c) $r^3 s^{-2} \times r^{-2} s^4$

(d) $\sqrt{\dfrac{9x^2 y^5}{4x^6 y}}$

6 Write each of these as a single fraction.
Simplify your answers where possible.

(a) $\dfrac{1}{z} + \dfrac{2}{3z}$

(b) $\dfrac{1}{p+1} - \dfrac{2}{2p+1}$

(c) $\dfrac{y}{y+1} + \dfrac{2y+1}{y-1}$

7 Find all possible values of x between 0° and 360° for each of these.

(a) $\sin x = 0.5$ (b) $\cos x = 0.2$ (c) $3 \sin x = 1$

(d) $\tan x = 2$ (e) $4 \cos x = {}^-3$ (f) $5 \tan x = {}^-1$

8 Here are the first five terms of a sequence.

      5, 12, 21, 32, 45

(a) Find, in terms of n, the nth term of the sequence.

(b) Calculate the number of terms in the quadratic sequence
      5, 12, 21, 32, 45, …, 725

9 Solve these equations.

(a) $\dfrac{x+1}{x-2} = \dfrac{x-1}{x+4}$ (b) $\dfrac{2x+1}{x-2} = \dfrac{x-1}{x+2}$ (c) $\dfrac{3}{x} + \dfrac{x+3}{3} = x$

10 A survey was carried out to find out the amount of time spent doing homework one week by a group of year 8 pupils. The distribution is shown in this histogram.

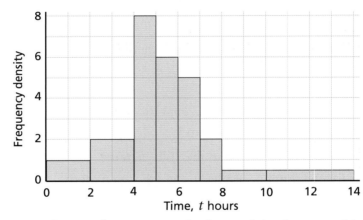

(a) How many pupils spent between 5 and 6 hours doing homework?

(b) How many pupils were sampled altogether?

(c) What percentage of pupils spent over 8 hours doing their homework?

(d) Calculate an estimate of the mean length of time spent doing homework.

11 Solve these equations.

(a) $\dfrac{3x-4}{x-1} = \dfrac{x+4}{x+1}$ (b) $\dfrac{x}{x-3} + \dfrac{2}{x+1} = 4$ (c) $\dfrac{3}{x} + \dfrac{x+1}{2x-3} = 2$

12 Find the value of n for which the nth term of the sequence $\dfrac{n}{n-4}$ is equal to n.

13 (a) Show that the curved surface area of this cone
is given by the formula

$$A = \pi r \sqrt{(r^2 + h^2)}$$

(b) Rearrange the formula to make h the subject.

(c) Calculate h when $A = 360\,\text{cm}^2$ and $r = 5.0\,\text{cm}$.
Give your answer to an appropriate degree of accuracy.

14 (a) Sketch the graph of $y = \cos \frac{x}{2}$ for $0° \le x \le 360°$.

(b) What is the period of the graph of $y = \cos \frac{x}{2}$?

(c) Use your sketch to help you find all of the solutions to $\cos \frac{x}{2} = 0.5$ between $0°$ and $360°$.

15 The frequency of vibration of a stretched string is given by the formula

$$f = \frac{1}{2l} \sqrt{\frac{s}{d}}$$

Rearrange the formula to make s the subject.

16 Below are four graphs.
Match each graph with one of the following equations.

(a) $y = 2 - x^2$ (b) $y = 2x^3$ (c) $y = 2 - x$ (d) $y = \frac{2}{x}$

A
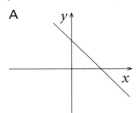

B
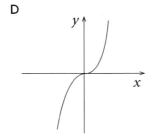

C
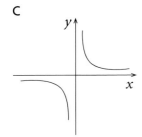

D
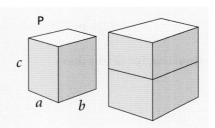

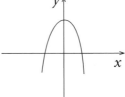

Challenge

Cuboid P has dimensions a, b, c, where $a < b < c$.

Two cuboids identical to P can be stacked on top of each other to make a cuboid which is similar to cuboid P.

- Find the ratio $a : b : c$.

39 Indices 2

This work will help you understand
◆ fractional indices
◆ exponential growth and decay

A *Powers: review*

Rules of indices

$$a^m \times a^n = a^{m+n} \qquad \frac{a^m}{a^n} = a^{m-n} \qquad (a^m)^n = a^{mn}$$

$$a^0 = 1 \qquad a^{-1} = \frac{1}{a} \qquad a^{-n} = \frac{1}{a^n}$$

A1 Write each of these as a single power of 2.

(a) $2^3 \times 2^4$ (b) 2×2^5 (c) $\dfrac{2^8}{2^3}$ (d) $(2^4)^3$ (e) $\dfrac{1}{2^5}$

(f) $2 \times 2^3 \times 2^7$ (g) $\dfrac{2^5 \times 2^8}{2^7}$ (h) $\dfrac{2^4}{2^9}$ (i) $(2^{-4})^2$ (j) $(2^{-5})^{-2}$

A2 (a) Copy and complete this list of the first ten powers of 2: $2^1 = 2$, $2^2 = 4$, $2^3 = 8$, …

(b) Write each of these as a power of 2.

(i) 256×128 (ii) $\dfrac{64}{1024}$ (iii) $\dfrac{1}{64 \times 8}$ (iv) 512^2 (v) 1024^5

A3 Find the value of n in each of these equations.

(a) $2^n = 256 \times 64$ (b) $2^n = \dfrac{1024}{64}$ (c) $2^n = \sqrt{256}$ (d) $2^n = \dfrac{1}{512}$

(e) $2^{n+1} = 256$ (f) $2^{n-1} = 1024$ (g) $2^{2n} = 64$ (h) $2^{3n} = 512$

A4 Find the value of n in each of the following equations.

(a) $(2^n)^2 = 256$ (b) $(2^n)^3 = 64$ (c) $\sqrt{2^n} = 8$ (d) $2^{\frac{1}{2}n} = 512$

A5 Write each of these expressions as a power of a.

(a) $\dfrac{a^2 \times a^3}{a^7}$ (b) $(a^3)^{-1}$ (c) $(a^{-2})^{-4}$ (d) $\left(\dfrac{1}{a^{-2}}\right)^{-3}$

Question for discussion

What do you think $a^{\frac{1}{2}}$ means? Why?

B *Fractional indices*

Waterweed is growing in a lake.
The area covered by the weed doubles
during the course of each year.
It covers 1 km² when measurement starts.

The growth is shown in the table below.
The table can be extended backwards to show
the area at time ⁻1 (one year before 0), ⁻2, ⁻3, and so on.

Time (years)

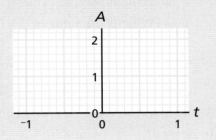

| 0 | 1 | 2 | 3 |

1 km²   2 km²   4 km²   8 km²

Time, t years	⁻3	⁻2	⁻1	0	1	2	3
Area, A km²				1	2	4	8

- Copy and complete the table to show the values of A when $t = $ ⁻1, ⁻2 and ⁻3.
- What is the formula connecting A and t?
- Plot the values from the table on graph paper.
  Suitable scales are shown here.
- Suppose the area grows continuously.
  Draw a smooth curve through the plotted points.
- Assume that the formula connecting A and t is true
  for **all** values of t, including fractions.
  What value does the graph give for $2^{\frac{1}{2}}$?

We can show that $2^{\frac{1}{2}}$ must be $\sqrt{2}$ by using the rules of indices:

$$2^{\frac{1}{2}} \times 2^{\frac{1}{2}} = 2^{\frac{1}{2}+\frac{1}{2}} = 2^1 = 2 \quad \text{so} \quad \mathbf{2^{\frac{1}{2}} = \sqrt{2}}$$

B1 From the rules of indices it follows that $2^{1\frac{1}{2}} = 2^1 \times 2^{\frac{1}{2}} = 2\sqrt{2}$.

Use your graph to find a value for $2^{1\frac{1}{2}}$ and use a calculator to check that
this value is an approximation to $2\sqrt{2}$.

B2 From the rules of indices it follows that $2^{-\frac{1}{2}} = \dfrac{1}{2^{\frac{1}{2}}} = \dfrac{1}{\sqrt{2}}$.

Use your graph to find a value for $2^{-\frac{1}{2}}$ and check that it is an approximation to $\dfrac{1}{\sqrt{2}}$.

⊠ B3 Use the rules of indices to write each of these as a power of 2.

(a) $2^5 \times \sqrt{2}$ (b) $4 \times \sqrt{2}$ (c) $\dfrac{4}{\sqrt{2}}$ (d) $(\sqrt{2})^3$ (e) $\dfrac{\sqrt{2}}{16}$

$$a^{\frac{1}{2}} = \sqrt{a} \qquad\qquad a^{-\frac{1}{2}} = \frac{1}{\sqrt{a}}$$

B4 Find the value of each of these.

(a) $9^{\frac{1}{2}}$ (b) $25^{\frac{1}{2}}$ (c) $(\frac{1}{16})^{\frac{1}{2}}$ (d) $4^{\frac{1}{2}} \times 100^{\frac{1}{2}}$ (e) $36^{\frac{1}{2}} \div 81^{\frac{1}{2}}$

B5 Find the value of each of these.

(a) $(\frac{1}{4})^{\frac{1}{2}}$ (b) $16^{-\frac{1}{2}}$ (c) $(\frac{1}{9})^{-\frac{1}{2}}$ (d) $(0.01)^{\frac{1}{2}}$ (e) $(100)^{-\frac{1}{2}}$

B6 (a) Use the rules of indices to simplify $5^{\frac{1}{3}} \times 5^{\frac{1}{3}} \times 5^{\frac{1}{3}}$.

(b) What does $5^{\frac{1}{3}}$ mean?

B7 (a) Copy and complete this list of cubes of positive integers from 1 to 5.

$1^3 = 1, \ 2^3 = 8, \ 3^3 = \ldots, \ 4^3 = \ldots, \ 5^3 = \ldots$

(b) Write down the value of   (i) $27^{\frac{1}{3}}$ (ii) $125^{\frac{1}{3}}$ (iii) $1000^{\frac{1}{3}}$

(c) What is the value of   (i) $8^{-\frac{1}{3}}$ (ii) $64^{-\frac{1}{3}}$ (iii) $125^{-\frac{1}{3}}$

$3^4 = 3 \times 3 \times 3 \times 3 = 81$

3 is called the 'fourth root' of 81, written $\sqrt[4]{81}$.

Another way to write $\sqrt[4]{a}$ is $a^{\frac{1}{4}}$. This is because $(a^{\frac{1}{4}})^4 = a^{4 \times \frac{1}{4}} = a^1 = a$.

In general, $a^{\frac{1}{n}} = \sqrt[n]{a}$ and $a^{-\frac{1}{n}} = \frac{1}{\sqrt[n]{a}}$.

B8 What is the value of each of these?

(a) $16^{\frac{1}{4}}$ (b) $16^{-\frac{1}{4}}$ (c) $32^{\frac{1}{5}}$ (d) $32^{-\frac{1}{5}}$ (e) $625^{\frac{1}{4}}$

B9 What is the value of each of these?

(a) $1000^{-\frac{1}{3}}$ (b) $(\frac{1}{8})^{\frac{1}{3}}$ (c) $(\frac{1}{100})^{-\frac{1}{2}}$ (d) $(\frac{1}{16})^{\frac{1}{4}}$ (e) $(\frac{1}{32})^{-\frac{1}{5}}$

B10 Solve each of these equations.

(a) $2^n = 8$ (b) $8^n = 2$ (c) $8^n = \frac{1}{2}$ (d) $(\frac{1}{9})^n = \frac{1}{3}$ (e) $(\frac{1}{9})^n = 3$

*B12 Solve these equations.

(a) $(\frac{1}{4})^n = 16$ (b) $16^n = \frac{1}{4}$ (c) $16^n = 2$ (d) $16^n = \frac{1}{2}$ (e) $81^n = 3$

C *More fractional indices*

One of the rules of indices is $(a^p)^q = a^{pq}$.

From this rule it follows that $27^{\frac{2}{3}} = (27^{\frac{1}{3}})^2 = (\sqrt[3]{27})^2 = 3^2 = 9$.

In general, $a^{\frac{p}{q}} = (\sqrt[q]{a})^p$.

Examples

Find the value of $(125)^{-\frac{2}{3}}$.

$$(125)^{-\frac{2}{3}} = \frac{1}{(125)^{\frac{2}{3}}} = \frac{1}{(\sqrt[3]{125})^2} = \frac{1}{5^2} = \frac{1}{25}$$

Write $\dfrac{4}{\sqrt[3]{2}}$ as a power of 2.

$$\frac{4}{\sqrt[3]{2}} = \frac{2^2}{2^{\frac{1}{3}}} = 2^{2-\frac{1}{3}} = 2^{1\frac{2}{3}} = 2^{\frac{5}{3}}$$

C1 Find the value of each of these.
  (a) $8^{\frac{2}{3}}$ (b) $9^{\frac{3}{2}}$ (c) $16^{\frac{3}{4}}$ (d) $1000^{\frac{2}{3}}$ (e) $25^{\frac{3}{2}}$

C2 Find the value of each of these.
  (a) $(27)^{-\frac{2}{3}}$ (b) $(16)^{\frac{3}{2}}$ (c) $(9)^{-\frac{3}{2}}$ (d) $(64)^{\frac{2}{3}}$ (e) $(\frac{1}{9})^{-\frac{3}{2}}$

C3 This is an extract from a table of squares, cubes and fourth powers of numbers.

x	x^2	x^3	x^4
7	49	343	2401
8	64	512	4096
9	81	729	6561

Find the value of
  (a) $729^{\frac{2}{3}}$ (b) $\sqrt{2401}$ (c) $64^{\frac{3}{2}}$
  (d) $2401^{\frac{1}{2}}$ (e) $6561^{\frac{1}{4}}$ (f) $6561^{\frac{1}{8}}$
  (g) $6561^{\frac{3}{8}}$ (h) $343^{\frac{4}{3}}$ (i) $4096^{\frac{1}{12}}$

C4 Write each of these as a power of 2.
  (a) $8\sqrt{2}$ (b) $4\sqrt[3]{2}$ (c) $(\sqrt{2})^4$ (d) $\dfrac{16}{\sqrt{2}}$ (e) $\dfrac{\sqrt[3]{2}}{4}$

C5 Write each of these as a power of 2.
  (a) $(\sqrt[3]{2})^{-2}$ (b) $\dfrac{1}{\sqrt[3]{2}}$ (c) $\left(\dfrac{1}{\sqrt[3]{2}}\right)^{-2}$ (d) $\dfrac{(\sqrt[3]{2})^2}{2}$ (e) $\dfrac{\sqrt{2}}{\sqrt[3]{2}}$

C6 Solve these equations.
  (a) $2^{2x+1} = 32$ (b) $4^x = 32$ (c) $8^x = 16$ (d) $4^x = \sqrt{2}$ (e) $9^x = 27$

D Powers and roots on a calculator

- The key for calculating powers may be labelled $\boxed{x^y}$ or $\boxed{\wedge}$.
  To work out $5^{0.3}$ you do $\boxed{5}\,\boxed{x^y}\,\boxed{0}\,\boxed{.}\,\boxed{3}$ or $\boxed{5}\,\boxed{\wedge}\,\boxed{0}\,\boxed{.}\,\boxed{3}$

- The fifth root of 7 (written $\sqrt[5]{7}$) is the number x such that $x^5 = 7$.
  $\sqrt[5]{7}$ as a power is $7^{\frac{1}{5}}$, because $(7^{\frac{1}{5}})^5 = 7^1 = 7$.
  You use the power key to find $\sqrt[5]{7}$ like this: $\boxed{7}\,\boxed{x^y}\,\boxed{(}\,\boxed{1}\,\boxed{\div}\,\boxed{5}\,\boxed{)}\,\boxed{=}$
  Alternatively, your calculator may have a key labelled $\boxed{x^{\frac{1}{y}}}$ or $\boxed{\sqrt[x]{}}$.

D1 Use a calculator to find the following, to four significant figures.

  (a) 2^{10} (b) $2^{0.5}$ (c) $\sqrt[4]{10}$ (d) $3.5^{2.8}$ (e) $0.8^{\frac{3}{4}}$

*D2 Solve these equations. Give each result to four significant figures.

  (a) $x^4 = 20$ (b) $x^{10} = 100$ (c) $x^7 = 250$ (d) $2x^8 = 100$ (e) $x^{0.2} = 4.5$

E Exponential growth

The population, P, of a colony of animals is increasing by 35% in every year.
When the population is first counted it is 500.

After 1 year, P is 500×1.35, after 2 years $500 \times 1.35 \times 1.35$, and so on.

The equation for P in terms of t is $P = 500 \times 1.35^t$.
A function of t in which t appears as a power is called an **exponential function**.
('Exponent' is another word for 'power'.)

The graph of P against t looks like this.

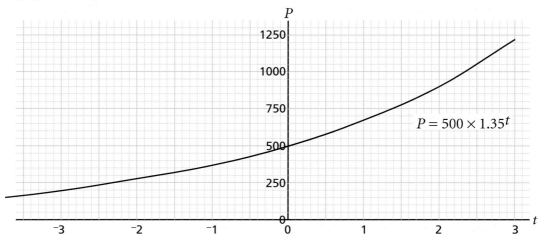

E1 (a) For the function shown in the graph on the previous page,
   calculate the value of P when $t = 2.5$ and check from the graph.

   (b) Calculate P when $t = {}^-1.8$ and check that your result agrees with the graph.

E2 The area, $A\,\mathrm{m}^2$, of a mouldy patch on a wall is given by the formula $A = 6.5 \times 1.05^t$, where t is the time in weeks since the first measurement.

   (a) What was the area of the patch when first measured?

   (b) By what percentage does the area increase each week?

   (c) Calculate, to two significant figures, the area after $4\frac{1}{2}$ weeks.

E3 The population, P, of a city is predicted to rise by 9% each year during the next five years. The population now is 247 000.

   (a) Write down the formula for the value of P after t years.

   (b) Copy and complete this table.

Time, t years	0	1	2	3	4	5
Population, P	247 000					

   (c) Draw a graph of P against t.

   (d) From the graph, estimate the value of t for which $P = 300\,000$.

   (e) Substitute this value of t into the formula for P and check that it is approximately correct.

E4 The population of a colony of micro-organisms grows at a rate of 15% per hour. The population is measured as 2600 to start with.

   (a) Write down an expression for the population after t hours.

   (b) Calculate the population $3\frac{1}{2}$ hours after the first measurement.

E5 In 1990 the population of North America was 300 million and increasing at 1% per year. For South America the corresponding figure was 120 million, with a growth rate of 2.3%.

   If the growth rates remain unchanged, find by trial and improvement how many years it would take the population of South America to be equal to that of North America.

E6 The population of a country grows at a rate of 2% each year.
   Show that it will double in 35 years.

E7 Amy sees an advert in a free newspaper for loans at 1% compound interest a week. She wants to borrow £2000 towards a new car.

   She reckons that she will pay back the loan and the interest in one payment at the end of one year.

   How much will she have to pay back?

E8 For the function shown in the graph on the opposite page, calculate the percentage increase in P during each period of 0.5 year.

F Exponential decay

When a quantity grows exponentially, it is multiplied by the same number in equal periods of time. The multiplier is greater than 1.

When the multiplier is less than 1, we get **exponential decay**.

For example, suppose a population of insects decreases by 6% every year.

The population is multiplied by 0.94 every year. In t years it is multiplied by 0.94^t.

F1 The population of a town is 24 300 now. The population decreases by 4% every year.

(a) Calculate, to three significant figures, the population 4 years from now.

(b) If P is the population after t years, write down the equation connecting P and t.

(c) If the population has been decreasing at the same rate in the past, calculate the population 2 years ago.

F2 Which of these sketch graphs shows exponential decay? Give reasons for your choice.

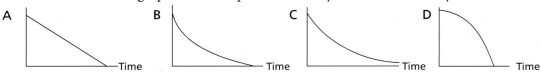

F3 It has been observed that the population of a certain snail is dropping by 8% each year.

At the start of the observation the snail population was 10 000.

(a) How many snails would be left after 1 year?

(b) Copy and complete this table for the population.

Time, t years	0	1	2	3	4	5	6
Population, P	10 000						

(c) Draw the graph of P against t.

(d) From the graph, how long will it be before the snail population has fallen by 30%?

(e) Calculate how many snails there were 2 years before the observation started.

(f) What is the equation connecting P and t?

G Mixed questions

⊠ **G1** Solve these equations.

(a) $2^x = 32$

(b) $10^x = \frac{1}{100}$

(c) $7^x = 1$

(d) $3^x + 4^x = 25$

(e) $64^x = 2$

(f) $\sqrt[4]{x} = 3$

(g) $3^x = \frac{1}{81}$

(h) $5^x = 0.04$

G2 Write each of these as a power of 2.

(a) 128 (b) $\sqrt{128}$ (c) $\sqrt[3]{128}$ (d) $\sqrt{128}$ (e) $\sqrt{\frac{1}{128}}$

G3 The Indian mathematician Ramanujan discovered that $\left(\frac{2143}{22}\right)^{\frac{1}{4}}$ is a good approximation for π. How close is it to your calculator value of π?

G4 Doctors sometimes need to estimate a person's skin area.
One formula they use is
$$S = kh^a w^b$$
where $k = 0.02$, $a = 0.42$, $b = 0.51$,
S is the person's skin area in m^2,
h is their height in cm,
w is their weight in kg.

Anita is 163 cm tall and weighs 55 kg.
Use the formula to estimate her skin area.

G5 Pauline discovered an old bank book in which £1 had been invested 200 years ago at 3% p.a. compound interest.
How much is in the account now?

Test yourself

T1 Find the value of (a) 4^{-2} (b) $36^{\frac{1}{2}}$ (c) $27^{\frac{2}{3}}$ AQA(NEAB) 2000

T2 Simplify each of the following. (i) $16^{\frac{3}{4}}$ (ii) $9^{-\frac{1}{2}}$ WJEC

T3 Work out, as a fraction, the exact value of $\left(\frac{49}{4}\right)^{-\frac{3}{2}}$. OCR

T4 Find the value of $\left(\frac{27}{125}\right)^{-\frac{1}{3}}$. Edexcel

T5 (a) Find the value of w in $2^w = 1$.
(b) Find the value of x in $2^x = \frac{1}{4}$.
(c) Find the value of y in $32^y = 2$.
(d) Find the value of z in $16^{z+3} = 2^z$. AQA(SEG) 1999

T6 (a) Show that $8^4 = 2^{12}$.
(b) Hence, or otherwise, solve the equation $4^x = 8^4$. AQA(NEAB) 2000

T7 (a) Draw a graph of $y = 2^x$ for values of x from $^-3$ to 3.
(b) Use your graph to estimate the positive solution of the equation $2^x - x = 3$, correct to one decimal place. OCR

This work will help you

◆ use vector notation

◆ express positions and lines in terms of a combination of vectors

◆ solve simple geometrical problems using vectors

A *Going places*

A **vector** describes a movement from one position to another.

On this diagram **a** and **b** are vectors.

We can describe them in columns: $\mathbf{a} = \begin{bmatrix} 2 \\ 5 \end{bmatrix}$ $\mathbf{b} = \begin{bmatrix} 3 \\ -4 \end{bmatrix}$

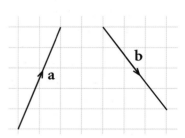

Writing vectors

In this unit, vectors are written in bold type, for example **a**.

Some books and exam papers write them with a wavy line underneath, for example a̰, and this is often done in handwritten work.

The vector $\overrightarrow{CD}$ has a column vector $\begin{bmatrix} 4 \\ 10 \end{bmatrix}$ which is equivalent to 2**a**.

On the diagram above
• which vector is equivalent to **a**?

• which vector is equivalent to 2**b**?

A1 Every vector on the diagram opposite can be described in terms of either **a** or **b**.

Write each of these vectors on the diagram in terms of either **a** or **b**.

(a) $\overrightarrow{EF}$ (b) $\overrightarrow{GH}$ (c) $\overrightarrow{IJ}$ (d) $\overrightarrow{KL}$ (e) $\overrightarrow{MN}$

(f) $\overrightarrow{QR}$ (g) $\overrightarrow{ST}$ (h) $\overrightarrow{UV}$ (i) $\overrightarrow{WX}$ (j) $\overrightarrow{YZ}$

A2 (a) What do you notice about all the vectors which can be described in terms of **a**?

    (b) What do you notice about all the vectors which can be described in terms of **b**?

A3 This diagram shows vectors **c** and **d**.

On squared paper draw and label vectors equivalent to

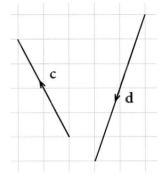

(a) 2**c**   (b) 3**d**   (c) 5**c**   (d) ⁻**c**

(e) ⁻**d**   (f) $\frac{1}{2}$**c**   (g) $1\frac{1}{2}$**d**   (h) ⁻2**c**

A4 This diagram shows a grid and vectors **s** and **t**.

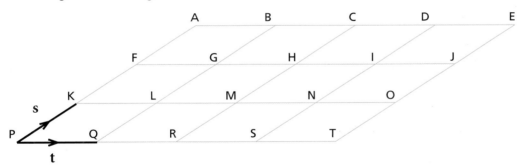

(a) $\overrightarrow{PK}$ = **s**. Give another vector equivalent to **s**.

Write these vectors in terms of **s** or **t**.

(b) $\overrightarrow{LG}$ (c) $\overrightarrow{TE}$ (d) $\overrightarrow{FJ}$ (e) $\overrightarrow{BG}$

(f) $\overrightarrow{LK}$ (g) $\overrightarrow{GI}$ (h) $\overrightarrow{TR}$ (i) $\overrightarrow{DS}$

B *Combining vectors*

Vectors can be combined.

Vector **a** followed by vector **b** is equivalent to the single vector **a** + **b**.
a + **b** is sometimes called the **resultant** of vectors **a** and **b**.

$$\begin{bmatrix} 2 \\ 5 \end{bmatrix} + \begin{bmatrix} 3 \\ -4 \end{bmatrix} = \begin{bmatrix} 5 \\ 1 \end{bmatrix}$$

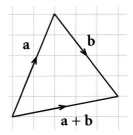

B1 Vectors $\mathbf{c} = \begin{bmatrix} 1 \\ 6 \end{bmatrix}$, $\mathbf{d} = \begin{bmatrix} -2 \\ 3 \end{bmatrix}$, $\mathbf{e} = \begin{bmatrix} -1 \\ -4 \end{bmatrix}$

What single column vector is equivalent to

(a) **c** + **d**   (b) **c** + **e**   (c) **d** + **e**   (d) **c** + **d** + **e**

Draw a diagram to check your answer each time.

B2 Vectors **a**, **b** and **c** are shown on this grid.

(a) Write down the column vectors for

(i) **a**   (ii) **b**   (iii) **c**

(b) Write an equation connecting **a**, **b** and **c**.

(c) Draw a diagram which shows the equation is true.

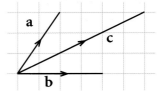

B3 Vectors $\mathbf{f} = \begin{bmatrix} 1 \\ 4 \end{bmatrix}$ and $\mathbf{g} = \begin{bmatrix} -4 \\ 2 \end{bmatrix}$

On squared paper draw vectors equivalent to

(a) **f** + **g**   (b) 2**f** + 2**g**   (c) 3**f** + **g**   (d) $\mathbf{f} + \frac{1}{2}\mathbf{g}$

B4 Vectors $\mathbf{m} = \begin{bmatrix} 2 \\ 5 \end{bmatrix}$, $\mathbf{n} = \begin{bmatrix} -3 \\ 2 \end{bmatrix}$, $\mathbf{o} = \begin{bmatrix} -2 \\ -2 \end{bmatrix}$

(a) Write the vector $\begin{bmatrix} 0 \\ 3 \end{bmatrix}$ in terms of **m**, **n** and **o**.

(b) Write these vectors using **m**, **n** and **o**.   (i) $\begin{bmatrix} -1 \\ 7 \end{bmatrix}$ (ii) $\begin{bmatrix} -5 \\ 0 \end{bmatrix}$ (iii) $\begin{bmatrix} 0 \\ 6 \end{bmatrix}$

B5 On this grid $\overrightarrow{AI} = \mathbf{p}$ and $\overrightarrow{AB} = \mathbf{q}$.

Write these vectors using **p** and **q**.

(a) $\overrightarrow{CL}$ (b) $\overrightarrow{FP}$ (c) $\overrightarrow{AT}$

(d) $\overrightarrow{FK}$ (e) $\overrightarrow{NG}$ (f) $\overrightarrow{OE}$

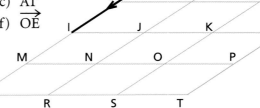

C *Subtracting vectors*

Vectors **m** and **n** are shown on this grid.

- What is $\overrightarrow{CA}$ in terms of vectors **m** and **n**?
- Use this to write $\overrightarrow{BA}$ in terms of **m** and **n**.
- Write down three other vectors on the grid which are equivalent to $\overrightarrow{BA}$.
- What is $\overrightarrow{FD}$ in terms of **m** and **n**?
- What is $\overrightarrow{CD}$ in terms of **m** and **n**?

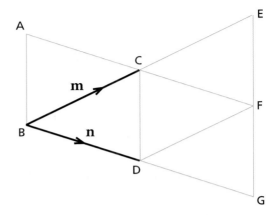

The vector **a** – **b** can be shown in two ways:

$$\mathbf{a} + {}^-\mathbf{b} = \mathbf{a} - \mathbf{b}$$

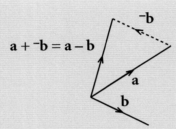

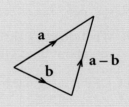

The vectors **a** – **b** are the same length and direction in both diagrams.

The second diagram shows **a** – **b** as going from the end of vector **b** to the end of vector **a**.

C1 (a) If $\mathbf{u} = \begin{bmatrix} 1 \\ 3 \end{bmatrix}$ and $\mathbf{v} = \begin{bmatrix} 4 \\ 2 \end{bmatrix}$, draw two different diagrams on squared paper to show **u** – **v**.

(b) Draw a diagram to show the vector **v** – **u**.

C2 This diagram shows two vectors **j** and **k**.
Express these vectors in terms of **j** and **k**.

(a) $\overrightarrow{HD}$ (b) $\overrightarrow{IE}$

(c) $\overrightarrow{BE}$ (d) $\overrightarrow{HA}$

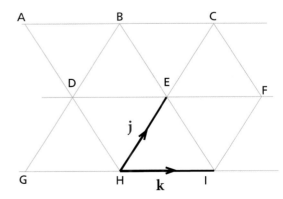

C3 In this parallelogram, express these vectors in terms of **m** and **n**.

(a) $\overrightarrow{AD}$

(b) $\overrightarrow{DA}$

(c) $\overrightarrow{CB}$

(d) $\overrightarrow{BC}$

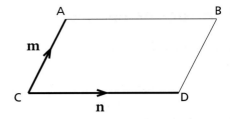

C4 In this diagram lengths AB and BC are equal. Lengths AD and DE are also equal.

$\overrightarrow{AB} = \mathbf{u}$ $\qquad$ $\overrightarrow{AD} = \mathbf{v}$

(a) Express these vectors in terms of **u** and **v**,

(i) $\overrightarrow{AC}$ $\qquad$ (ii) $\overrightarrow{AE}$

(iii) $\overrightarrow{DB}$ $\qquad$ (iv) $\overrightarrow{BE}$

(v) $\overrightarrow{DC}$ $\qquad$ (vi) $\overrightarrow{EC}$

(b) What do your answers to part (a) tell you about lines DB and EC?

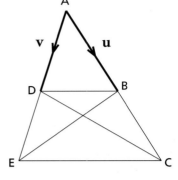

C5 ABCDEF is a regular hexagon.

$\overrightarrow{OA} = \mathbf{m}$ and $\overrightarrow{OF} = \mathbf{n}$

(a) Express these in terms of **m** and **n**.

(i) $\overrightarrow{DA}$ $\qquad$ (ii) $\overrightarrow{OE}$ $\qquad$ (iii) $\overrightarrow{AB}$

(iv) $\overrightarrow{FC}$ $\qquad$ (v) $\overrightarrow{FB}$ $\qquad$ (vi) $\overrightarrow{FD}$

(b) Write each of the vectors $\overrightarrow{AF}, \overrightarrow{FE}, \overrightarrow{ED}, \overrightarrow{DC},$ $\overrightarrow{CB}, \overrightarrow{BA}$ in terms of **m** and **n**.

(c) Add together the vectors for the sides of the hexagon.
Explain why you get this result.

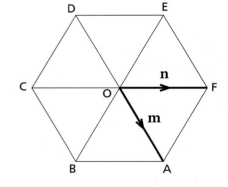

C6 ABCD are the four vertices of a quadrilateral.

$\overrightarrow{BC} = \overrightarrow{AD}$ $\quad$ and $\quad$ $\overrightarrow{AB} = \overrightarrow{DC}$

What can you say about sides BC and AD?
What can you say about the quadrilateral?

C7 A, B and C are three points.

Describe in words as fully as possible the situations described by each of these,

(a) $\overrightarrow{AB} = \overrightarrow{BC}$ $\qquad\qquad$ (b) $\overrightarrow{AB} = 2\overrightarrow{BC}$ $\qquad\qquad$ (c) $\overrightarrow{AB} = {}^{-}\overrightarrow{BC}$

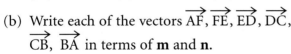

D *Fractions of vectors*

This shows points O, A and B and the vectors **a** and **b**.

When vectors are used to give the positions of points from a given origin O, they are sometimes referred to as **position vectors**.

The position vector of point C from O is $\frac{3}{2}\mathbf{a} + \mathbf{b}$.

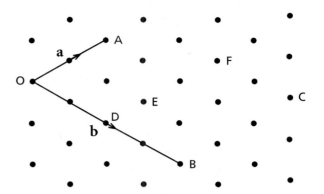

D1 What are the position vectors from O of
  (a) point D   (b) point E   (c) point F

D2 Copy the vectors **a** and **b** and the point O on to triangular dotty paper.
  Mark points G, H, I and J whose position vectors are given by
  (a) $\overrightarrow{OG} = \mathbf{b} - \mathbf{a}$ (b) $\overrightarrow{OH} = 2\mathbf{a} + \frac{1}{2}\mathbf{b}$ (c) $\overrightarrow{OI} = \mathbf{a} + \frac{1}{4}\mathbf{b}$ (d) $\overrightarrow{OJ} = -\frac{1}{2}\mathbf{a} + \frac{3}{4}\mathbf{b}$

D3 PQRS is a parallelogram.
  T is the midpoint of QR.
  U is the point on SR for which SU : UR = 1 : 2.
  $\overrightarrow{PQ} = \mathbf{a}$ and $\overrightarrow{PS} = \mathbf{b}$.
  Write down, in terms of **a** and **b**, expressions for
  (a) $\overrightarrow{PT}$ (b) $\overrightarrow{TU}$

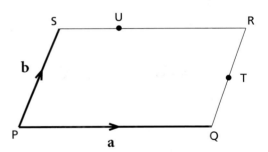

Edexcel

D4 In triangle RST, A and B are the midpoints of RS and RT respectively.
  $\overrightarrow{RS} = \mathbf{x}$ and $\overrightarrow{RT} = \mathbf{y}$.

  Find in terms of **x** and **y**

  (a) $\overrightarrow{ST}$

  (b) $\overrightarrow{RA}$

  (c) $\overrightarrow{AB}$

  (d) What conclusions can you draw about the lines AB and ST?

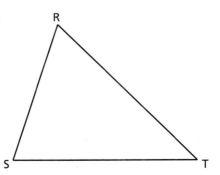

OCR

E *Vector algebra*

This diagram shows

$(\mathbf{a} + \mathbf{b}) + (\mathbf{a} + \mathbf{b}) = 2(\mathbf{a} + \mathbf{b})$

and the effect of

$2\mathbf{a} + 2\mathbf{b}$

- What does this say about the vectors $2(\mathbf{a} + \mathbf{b})$ and $2\mathbf{a} + 2\mathbf{b}$?

- Investigate whether this is true for all vectors **a** and **b**.

- Find some other equivalent expressions to $2(\mathbf{a} + \mathbf{b})$.
  Draw a diagram to show these equivalences are true.

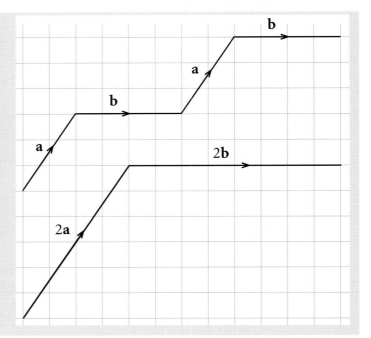

E1 Draw diagrams to illustrate whether these statements are **true** or **false**.

    (a) $\mathbf{a} + \mathbf{b} = \mathbf{b} + \mathbf{a}$ (b) $\mathbf{a} + \mathbf{b} + \mathbf{a} + 2\mathbf{b} = 2\mathbf{a} + 3\mathbf{b}$ (c) $2(\mathbf{a} + \mathbf{b}) + \mathbf{a} = 3\mathbf{a} + \mathbf{b}$

    (d) $\frac{1}{2}\mathbf{a} + \frac{1}{2}\mathbf{b} = \frac{1}{2}(\mathbf{a} + \mathbf{b})$ (e) $\mathbf{a} - \mathbf{b} = \mathbf{b} - \mathbf{a}$ (f) $\mathbf{a} + 2\mathbf{b} = 2\mathbf{a} + \mathbf{b}$

E2 Write these vector expressions in a simpler form.

    (a) $2\mathbf{a} + 3\mathbf{b} + \mathbf{a} + 2\mathbf{b}$ (b) $\mathbf{a} - 2\mathbf{b} + 3\mathbf{a} + \mathbf{b}$ (c) $3(\mathbf{a} + 2\mathbf{b})$

    (d) $4\mathbf{a} - 2(\mathbf{a} + \mathbf{b})$ (e) $\frac{1}{2}(\mathbf{a} + 3\mathbf{b})$ (f) $\frac{1}{2}\mathbf{a} + \frac{2}{3}\mathbf{b} + \frac{1}{3}\mathbf{a} - \frac{1}{2}\mathbf{b}$

E3 ABCD is a trapezium.

$\overrightarrow{DC} = 2\overrightarrow{AB}$

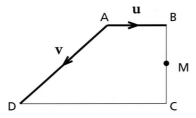

    (a) Write these as simply as possible in terms of **u** and **v**.

        (i) $\overrightarrow{DC}$ (ii) $\overrightarrow{BD}$ (iii) $\overrightarrow{BC}$ (iv) $\overrightarrow{AC}$

    (b) M is the point halfway along BC.
        Write these as simply as possible in terms of **u** and **v**.

        (i) $\overrightarrow{BM}$ (ii) $\overrightarrow{AM}$ (iii) $\overrightarrow{DM}$

E4 This shows an equilateral triangle drawn on a triangular grid.

$\overrightarrow{OA} = \mathbf{a}$ and $\overrightarrow{OB} = \mathbf{b}$

(a) Write these as simply as possible in terms of **a** and **b**.

(i) $\overrightarrow{AB}$ (ii) $\overrightarrow{BA}$ (iii) $\overrightarrow{OP}$

(iv) $\overrightarrow{BP}$ (v) $\overrightarrow{AP}$

(b) M is the midpoint of $\overrightarrow{AB}$.
Write these as simply as possible in terms of **a** and **b**.

(i) $\overrightarrow{MB}$ (ii) $\overrightarrow{MP}$

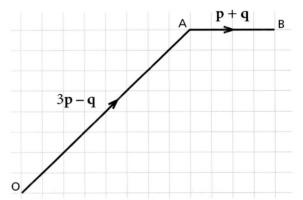

E5 In this diagram OA = $3\mathbf{p} - \mathbf{q}$ and AB = $\mathbf{p} + \mathbf{q}$.

(a) Write the vector OB in terms of **p** and **q**. Write this in its simplest form.

(b) From the diagram write OB in the form $\begin{bmatrix} m \\ n \end{bmatrix}$ where m and n are whole numbers.

(c) Use your answers to (a) and (b) to write **p** in the form $\begin{bmatrix} m \\ n \end{bmatrix}$.

(d) Write **q** in the form $\begin{bmatrix} m \\ n \end{bmatrix}$.

E6 (a) Write down, in terms of **x** and **y**, the vector $\overrightarrow{RQ}$.

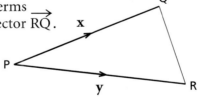

(b) The vectors $\overrightarrow{LM} = \mathbf{a} + 2\mathbf{b}$ and $\overrightarrow{LN} = \mathbf{a} - \mathbf{b}$ are drawn on the grid.

(i) Write down, in terms of **b**, the vector $\overrightarrow{NM}$.

(ii) Hence find **b** in the form $\begin{bmatrix} p \\ q \end{bmatrix}$, where p and q are whole numbers.

(iii) Find **a** in the form $\begin{bmatrix} r \\ s \end{bmatrix}$, where r and s are whole numbers.

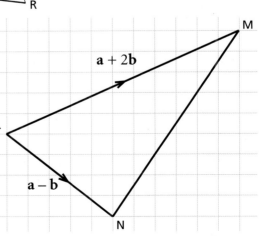

AQA(NEAB) 1997

F *Proof by vectors*

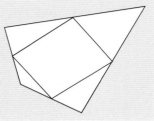

Investigation

- Draw any quadrilateral.
- Join up the midpoints of the sides to make a new quadrilateral.
- What type of quadrilateral do you get?

The result cannot be proved by drawing and measuring quadrilaterals. But it can be proved by using vectors.

Here is a quadrilateral PQRS.

K, L, M and N are the midpoints of the sides.

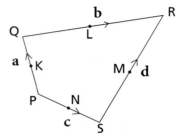

Define these four vectors:

$$\mathbf{a} = \overrightarrow{PQ} \quad \mathbf{b} = \overrightarrow{QR} \quad \mathbf{c} = \overrightarrow{PS} \quad \mathbf{d} = \overrightarrow{SR}$$

The vector $\overrightarrow{QS}$ is the sum of $\overrightarrow{QP}$ and $\overrightarrow{PS}$, so $\overrightarrow{QS} = {}^{-}\mathbf{a} + \mathbf{c} = \mathbf{c} - \mathbf{a}$.

$\overrightarrow{QS}$ is also the sum of $\overrightarrow{QR}$ and $\overrightarrow{RS}$, so $\overrightarrow{QS} = \mathbf{b} + {}^{-}\mathbf{d} = \mathbf{b} - \mathbf{d}$.

So it follows that $\mathbf{c} - \mathbf{a} = \mathbf{b} - \mathbf{d}$.

The vector $\overrightarrow{KN}$ is the sum of $\overrightarrow{KP}$ and $\overrightarrow{PN}$, so $\overrightarrow{KN} = -\frac{1}{2}\mathbf{a} + \frac{1}{2}\mathbf{c} = \frac{1}{2}(\mathbf{c} - \mathbf{a})$.

The vector $\overrightarrow{LM}$ is the sum of $\overrightarrow{LR}$ and $\overrightarrow{RM}$, so $\overrightarrow{LM} = \frac{1}{2}\mathbf{b} + {}^{-}\frac{1}{2}\mathbf{d} = \frac{1}{2}(\mathbf{b} - \mathbf{d})$.

From the fact that $\mathbf{c} - \mathbf{a} = \mathbf{b} - \mathbf{d}$ it follows that $\overrightarrow{KN} = \overrightarrow{LM}$.

So the line segments $\overrightarrow{KN}$ and $\overrightarrow{LM}$ are parallel and equal in length.

- Find expressions for $\overrightarrow{KL}$ and $\overrightarrow{NM}$ in terms of **a**, **b**, **c**, **d** and show that $\overrightarrow{KL} = \overrightarrow{NM}$.

The quadrilateral KLMN has both pairs of opposite sides parallel, so it is a parallelogram.

Useful facts

- If a vector $\overrightarrow{AB}$ is equal to a vector $\overrightarrow{CD}$, then the line segments AB, CD are parallel and equal in length.
- If $\overrightarrow{AB} = k\overrightarrow{CD}$ (where k is a number), then the line segment AB is parallel to CD and the length AB is k times the length CD.

  For example, if you have shown that $\overrightarrow{CD} = \mathbf{p} - \mathbf{q}$ and that $\overrightarrow{AB} = 3(\mathbf{p} - \mathbf{q})$, it follows that CD is parallel to AB, and 3CD = AB.

F1 Describe in words what these facts tell you about the geometrical relations in these situations.

(a) $\overrightarrow{FG} = \mathbf{a} + 2\mathbf{b}$ and $\overrightarrow{JK} = 2\mathbf{a} + 4\mathbf{b}$

(b) $\overrightarrow{MN} = \frac{1}{2}(\mathbf{p} + \mathbf{q})$ and $\overrightarrow{RT} = \frac{1}{2}\mathbf{p} + \frac{1}{2}\mathbf{q}$

(c) The position vector $\overrightarrow{OA} = 2\mathbf{a}$ and position vector $\overrightarrow{OB} = \frac{3}{4}\mathbf{a}$

(d) The position vector of point P is $4(\mathbf{m} + \frac{1}{2}\mathbf{n})$ and the position vector of Q is $2\mathbf{n} + 4\mathbf{m}$.

F2 In this diagram C is $\frac{2}{3}$ of the way along OA and D is $\frac{2}{3}$ of the way along OB.

(a) If $\overrightarrow{OA} = \mathbf{a}$ and $\overrightarrow{OB} = \mathbf{b}$, find expressions for $\overrightarrow{AB}$ and $\overrightarrow{CD}$ in terms of $\mathbf{a}$ and $\mathbf{b}$.

(b) What does this tell you about the lines AB and CD?

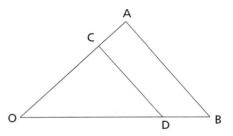

F3 This diagram shows a triangle ABC. Points X, Y and Z are the midpoints of BC, CA and AB respectively.

Vector $\overrightarrow{AZ} = \mathbf{a}$ and vector $\overrightarrow{AY} = \mathbf{b}$.

(a) Express in terms of $\mathbf{a}$ and $\mathbf{b}$

(i) the vector $\overrightarrow{YZ}$

(ii) the vector $\overrightarrow{CB}$

(b) Use your answers to write down **two** facts about the relationship between lines YZ and CB.

(c) Express, in terms of $\mathbf{a}$ and $\mathbf{b}$, the vector $\overrightarrow{AX}$. Simplify your answer.

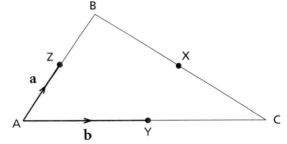

AQA(SEG) 1998

F4 ABCD is a quadrilateral.

$\overrightarrow{AD} = 3\mathbf{v}$, $\overrightarrow{AB} = \mathbf{u}$ and $\overrightarrow{BC} = 2\mathbf{v}$.

(a) Express in terms of $\mathbf{v}$ and $\mathbf{u}$

(i) $\overrightarrow{CD}$ (ii) $\overrightarrow{BD}$ (iii) $\overrightarrow{CA}$

(b) What type of quadrilateral is ABCD?

F5 In this diagram $\overrightarrow{OA} = \mathbf{a}$ and $\overrightarrow{OB} = \mathbf{b}$.

Also $\overrightarrow{AC} = 2\mathbf{a}$ and $\overrightarrow{AD} = 3\mathbf{b} - \mathbf{a}$.

(a) Write $\overrightarrow{AB}$ in terms of $\mathbf{a}$ and $\mathbf{b}$.

(b) $\overrightarrow{OD} = n\mathbf{b}$ where n is a whole number. Find n.

(c) Prove that OAB and ODC are similar triangles.

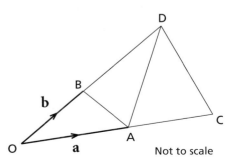

Not to scale

F6 ABCD is a parallelogram.

M is the midpoint of the line AC.

N is a point $\frac{2}{3}$ of the way along the line AD.

$\overrightarrow{AB} = \mathbf{u}$ and $\overrightarrow{AD} = 3\mathbf{v}$.

AB is extended to a new point O so that BO = AB.

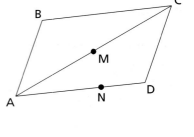

(a) Write $\overrightarrow{AM}$ in terms of $\mathbf{u}$ and $\mathbf{v}$.

Hence write $\overrightarrow{NM}$ in terms of $\mathbf{u}$ and $\mathbf{v}$.

(b) Write $\overrightarrow{NO}$ in terms of $\mathbf{u}$ and $\mathbf{v}$.

(c) Use your answers to (a) and (b) to prove that

(i) N, M and O lie on a straight line

(ii) NM is a quarter of NO

F7 A triangle OAB has midpoints on its sides Q, R and S as shown in this diagram.

$\overrightarrow{OA} = \mathbf{a}$ and $\overrightarrow{OB} = \mathbf{b}$.

P is a point $\frac{2}{3}$ of the way along OQ.

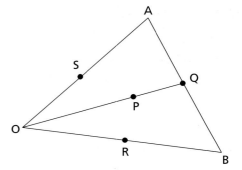

(a) Express $\overrightarrow{AQ}$ in terms of $\mathbf{a}$ and $\mathbf{b}$.

(b) Express $\overrightarrow{OQ}$ in terms of $\mathbf{a}$ and $\mathbf{b}$. Hence, or otherwise, express OP in terms of $\mathbf{a}$ and $\mathbf{b}$.

(c) Express $\overrightarrow{AR}$ in terms of $\mathbf{a}$ and $\mathbf{b}$.

(d) T is a point $\frac{2}{3}$ along AR.

Express $\overrightarrow{AT}$ in terms of $\mathbf{a}$ and $\mathbf{b}$.

(e) Using your answer to (d) express $\overrightarrow{OT}$ in terms of $\mathbf{a}$ and $\mathbf{b}$.

(f) What does this tell you about the points P and T?

(g) If X is a point $\frac{2}{3}$ along BS, what do you think the vector $\overrightarrow{OX}$ will be in terms of $\mathbf{a}$ and $\mathbf{b}$? Prove your result.

F8 ABCDEFGH is a cuboid.

AB = $\mathbf{i}$, AD = $\mathbf{j}$, DH = $\mathbf{k}$

(a) Express these vectors in terms of $\mathbf{i}$, $\mathbf{j}$ and $\mathbf{k}$.

(i) $\overrightarrow{EG}$ (ii) $\overrightarrow{HB}$ (iii) $\overrightarrow{FC}$

(b) Prove that the diagonals AG and FD intersect at a point.

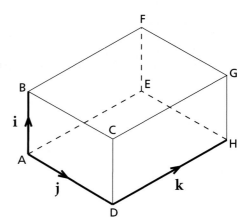

Test yourself

T1 The diagram shows two sets of parallel lines.

Vector $\overrightarrow{AB} = \mathbf{x}$ and vector $\overrightarrow{AH} = \mathbf{y}$.
AC = 4AB and AG = 2AH.

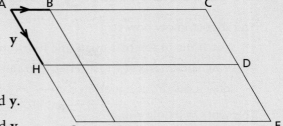

(a) Write the vector $\overrightarrow{AF}$ in terms of **x** and **y**.

(b) Write the vector $\overrightarrow{BG}$ in terms of **x** and **y**.

(c) Find **two** different vectors that can be written as **y** + 4**x**.

(d) Write down **two** statements about line segments which have the same vector.

AQA(SEG) 1998

T2 OABC is a trapezium.

$\overrightarrow{OA} = 2\mathbf{a}$, $\overrightarrow{OC} = 8\mathbf{b}$ and $AB = \frac{1}{4}OC$

Find in terms of **a** and **b**

(a) $\overrightarrow{BA}$ (b) $\overrightarrow{BC}$

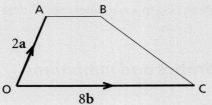

AQA(SEG) 2000

T3 OABC is a parallelogram.
M is the midpoint of BC.

$\overrightarrow{OA} = \mathbf{a}$ and $\overrightarrow{OC} = \mathbf{c}$.

(a) Find in terms of **a** and **c** expressions for the following vectors.

(i) $\overrightarrow{OB}$ (ii) $\overrightarrow{CM}$ (iii) $\overrightarrow{AM}$

P is a point on AM such that AP is $\frac{2}{3}$AM.

(b) Find in terms of **a** and **c** expressions for

(i) $\overrightarrow{AP}$ (ii) $\overrightarrow{OP}$

(c) Describe as fully as possible what your answer to (b)(ii) tells you about the position of P.

AQA(NEAB) 1998

Transforming graphs

You should know how to

◆ sketch the graph of a quadratic function

◆ draw and interpret trigonometric graphs

◆ solve a quadratic equation such as $x^2 + 8x + 13 = 0$ by writing it as $(x + 4)^2 = 3$

This work will help you

◆ find and use the completed-square form of a quadratic expression

◆ transform graphs and find their equations

◆ use function notation

◆ sketch graphs of $y = a\,f(x)$, $f(ax)$, $f(x) + a$ and $f(x + a)$ given the graph of $y = f(x)$

◆ fit a function to a non-linear set of data

A *Quadratics and minimums*

- What is the value of $(x - 2)^2 + 1$ when $x = 5$?
- Try other values of x.
- What is the minimum value of $(x - 2)^2 + 1$?

- What does the graph of $y = (x - 2)^2 + 1$ look like?
- What is its minimum point?
- What is its equation in the form $y = x^2 + px + q$?
- Where does it cut the y-axis?

- What does the graph of $y = (x - 3)^2 - 4$ look like?
- What is its minimum point?
- Where does it cut the x-axis?

A1 (a) What is the value of $(x - 6)^2$ when

    (i) $x = 8$ (ii) $x = 10$ (iii) $x = 3$ (iv) $x = {}^-1$ (v) $x = 6$

(b) What is the minimum value of $(x - 6)^2$?

(c) What is the minimum value of

    (i) $(x - 6)^2 + 7$ (ii) $(x - 6)^2 - 5$

A2 Write down the minimum value of each expression and the value of x that gives the minimum.

    (a) $(x - 4)^2 + 3$ (b) $(x + 2)^2 + 5$ (c) $(x - 3)^2 - 6$ (d) $(x + 1)^2$

A3 These equations match the graphs below.
Match each one with its appropriate graph.

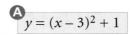
A $y = (x - 3)^2 + 1$

B $y = (x + 3)^2$

C $y = (x - 1)^2 + 3$

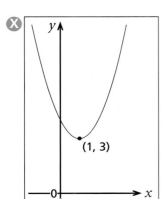

X with point $(1, 3)$

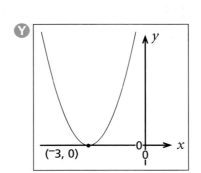

Y with point $(^-3, 0)$

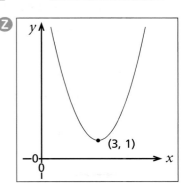

Z with point $(3, 1)$

A4 (a) Which equation below is equivalent to $y = x^2 - 10x + 26$?

$y = (x - 10)^2 + 16$ $y = (x - 5)^2 + 26$ $y = (x - 10)^2 - 74$ $y = (x - 5)^2 + 1$

(b) Hence find the minimum point on the graph of $y = x^2 - 10x + 26$.

A5 (a) Which equation below matches the graph on the right?

$y = (x - 2)^2 + 4$ $y = (x + 4)^2 + 2$

$y = (x + 2)^2 + 4$ $y = (x - 4)^2 + 2$

$y = (x + 4)^2 - 2$ $y = (x - 2)^2 - 4$

graph with point $(4, 2)$

(b) Write the equation in the form $y = x^2 + px + q$.

(c) Where will the graph cut the y-axis?

A6 Solve these equations.

(a) $(x - 4)^2 - 9 = 0$ (b) $(x - 2)^2 - 16 = 0$

(c) $(x + 3)^2 - 25 = 0$ (d) $(x + 5)^2 - 1 = 0$

A7 Work out where the graph of $y = (x + 1)^2 - 16$ cuts the x-axis.

A8 A graph has equation $y = (x - 3)^2 - 4$.

(a) What is the minimum point on this graph?

(b) Where does the graph cut the x-axis?

(c) Where does the graph cut the y-axis?

(d) Sketch the graph of $y = (x - 3)^2 - 4$, marking all the points you have found.

B Completing the square

Example

Write $y = x^2 - 6x + 10$ in completed-square form.

$$y = x^2 - 6x + 10$$

The coefficient of x is $^-6$.
Half of $^-6$ is $^-3$.
So the square is $(x - 3)^2$.

We know that
$$(x - 3)^2 = x^2 - 6x + 9$$

Expand $(x - 3)^2$.

So $(x - 3)^2 + 1 = x^2 - 6x + 10$

Add 1 to each side to give the correct expression on the right.

So $y = (x - 3)^2 + 1$ is the completed-square form.

B1 Which of these expressions are in completed-square form?

A $(x - 3)^2 + 5$ **B** $(x + 4)^2 + x$ **C** $(x - 1)^2 - 3x$ **D** $(x + 6)^2 - 1$

B2 Match each expression with its completed-square form.

A $x^2 + 6x + 13$ **B** $x^2 - 4x + 7$ **P** $(x + 3)^2 - 4$ **Q** $(x + 3)^2 + 4$

C $x^2 + 6x + 5$ **D** $x^2 - 4x + 1$ **R** $(x - 2)^2 - 3$ **S** $(x - 2)^2 + 3$

B3 Copy and complete each identity.
  (a) $x^2 + 4x + 8 = (x + \blacksquare)^2 + 4$ (b) $x^2 - 8x + 5 = (x - 4)^2 - \blacksquare$
  (c) $x^2 + 10x + 28 = (x + \blacksquare)^2 + \blacksquare$ (d) $x^2 - 6x + 2 = (x - \blacksquare)^2 - \blacksquare$

B4 Complete the square for each expression.
  (a) $x^2 + 6x + 15$ (b) $x^2 - 2x + 5$ (c) $x^2 - 8x + 10$

B5 (a) Write the expression $x^2 + 6x + 12$ in the form $(x + p)^2 + q$, where p and q are constants.
  (b) Hence find the minimum value of the expression $x^2 + 6x + 12$.

B6 Calculate the values of p and q in the identity
$$x^2 - 6x + 14 = (x - p)^2 + q$$

AQA 2003 Specimen

B7 The expression $x^2 - 4x + 7$ can be written in the form $(x - a)^2 + b$, where a and b are constants.
Calculate the values of a and b.

B8 The expression $x^2 - 12x + 20$ can be written in the form $(x - a)^2 + b$, where a and b are integers.

    (a) Calculate the values of a and b.

    (b) Hence find the minimum point on the graph of $y = x^2 - 12x + 20$.

    (c) Sketch the graph of $y = x^2 - 12x + 20$, showing the minimum point and y-intercept.

B9 (a) Write the expression $x^2 - 10x + 7$ in the form $(x - a)^2 + b$, where a and b are integers.

    (b) Solve the equation $x^2 - 10x + 7 = 0$.   AQA(SEG) 1998

B10 The expression $x^2 - 6x + 7$ can be written in the form $(x + a)^2 + b$, where a and b are constants.
Calculate the values of a and b and hence find
the minimum value of the expression.   AQA(SEG) 2003

B11 Copy and complete each identity.

    (a) $x^2 + 6x - 1 = (x + \blacksquare)^2 - \blacksquare$ (b) $x^2 - 8x - 3 = (x - \blacksquare)^2 - \blacksquare$

B12 (a) Write $y = x^2 - 4x - 3$ in completed-square form.

    (b) Sketch the graph of $y = x^2 - 4x - 3$, showing the minimum point and y-intercept.

B13 Copy and complete each identity.

    (a) $x^2 + 8x = (x + \blacksquare)^2 - \blacksquare$ (b) $x^2 - 14x = (x - \blacksquare)^2 - \blacksquare$

B14 The expression $x^2 + 6x$ can be written in the form $(x + a)^2 + b$, where a and b are constants.
Find the values of a and b and hence find
the minimum value of the expression.   AQA(SEG) 1998 Specimen

B15 (a) The expression $x^2 - 10x + a$ can be written in the form $(x + b)^2$. Find the values of a and b.

    (b) Solve the equation $x^2 - 10x + 20 = 0$.
Give your answers to two decimal places.

    (c) State the minimum value of y if $y = x^2 - 10x + 20$.   AQA 2003 Specimen

B16 (a) Expand $(3x + 2)^2$.

    (b) Write the expression $9x^2 + 12x + 5$ in the form $(ax + b)^2 + c$, where a, b and c are constants.

B17 The expression $4x^2 - 12x - 2$ can be written in the form $(ax + b)^2 - 11$.
Find the values of a and b.   AQA(SEG) 2000

*B18 Copy and complete each identity.

    (a) $x^2 + 3x + 2 = (x + \blacksquare)^2 - \blacksquare$ (b) $x^2 - 5x + 7 = (x - \blacksquare)^2 + \blacksquare$

*B19 Sketch the graph of $y = x^2 - 7x + 2$, showing clearly the minimum point.

C Transforming linear and quadratic graphs

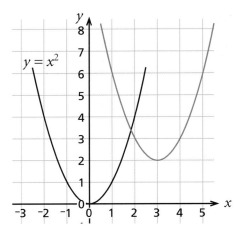

The curve with equation $y = x^2$ is translated $\begin{bmatrix} 3 \\ 2 \end{bmatrix}$ on to the red curve.

- What is the minimum point on the red curve?
- What is the equation of the red curve?

A graphical calculator is useful for checking your solutions.

C1 Each of the following sketches shows the graph of $y = x^2$ after a translation.
Find the equation of each curve.

(a)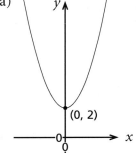

(b)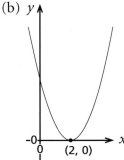

(c)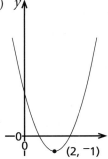

(d)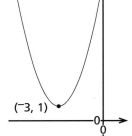

C2 The graph of $y = x^2$ is translated 4 units to the left.
What is the equation of the transformed curve?

C3 What translation will map the graph of $y = x^2$ on to the graph of $y = (x + 5)^2 + 1$?

C4 The graph of $y = x^2$ is translated $\begin{bmatrix} 3 \\ -4 \end{bmatrix}$. What is the equation of the image?

C5 What is the equation of the image of $y = x^2$ after

(a) reflection in the x-axis   (b) reflection in the y-axis

C6 What is the equation of the image of the straight line $y = x$ after

(a) reflection in the x-axis   (b) reflection in the y-axis

C7 (a) Draw a sketch of the image of $y = x^2$ after a reflection in the x-axis followed by a translation of 5 units vertically.

(b) Find the equation of the transformed curve.

Stretching

For example,
to stretch a shape from the x-axis
by a scale factor of 2 in the y-direction,
multiply all y-coordinate values by 2.

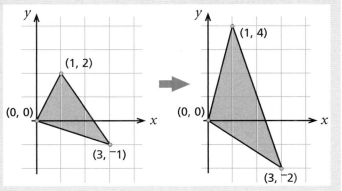

C8 (a) Write down the coordinates of five points on the graph of $y = x^2$.

(b) Double the y-values to find the image of each point under a stretch of scale factor 2 in the y-direction.

(c) On the same set of axes, draw and label graphs of $y = x^2$ and its image after a stretch of scale factor 2 in the y-direction.

(d) What is the equation of the transformed curve?

C9 (a) On the same set of axes, draw and label graphs of $y = x^2$ and its image after a stretch of scale factor $\frac{1}{2}$ in the y-direction

(b) What is the equation of the transformed curve?

C10 Find the image of the **line** $y = x$ after a stretch of scale factor 5 in the y-direction.

C11 Describe the transformation that maps $y = x^2$ to $y = 3x^2$.

C12 What is the equation of the image of $y = x$ after a stretch of scale factor $\frac{1}{2}$ in the y-direction followed by a translation of $\begin{bmatrix} 0 \\ 3 \end{bmatrix}$?

C13 What is the equation of the image of $y = x^2$ after a stretch of scale factor 4 in the y-direction followed by a translation of $\begin{bmatrix} 0 \\ 5 \end{bmatrix}$?

C14 (a) Write down the coordinates of five points on the graph of $y = x^2$.

(b) Double the x-values to find the image of each point under a stretch of scale factor 2 in the x-direction.

(c) On the same set of axes, draw and label graphs of $y = x^2$ and its image after a stretch of scale factor 2 in the x-direction.

(d) What is the equation of the transformed curve?

C15 (a) On the same set of axes, draw and label graphs of $y = x^2$ and its image after a stretch of scale factor $\frac{1}{2}$ in the x-direction.

(b) What is the equation of the transformed curve?

C16 The diagram shows a sketch of the graph $y = x^2$.

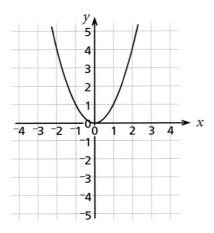

(a) Each of the graphs below is a transformation of this graph. Write down the equation of each graph.

(i)

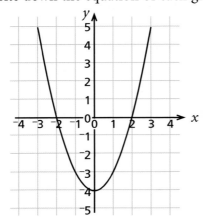

(ii)

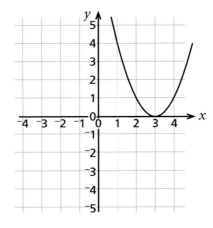

(iii)

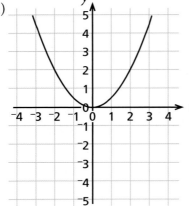

(iv)

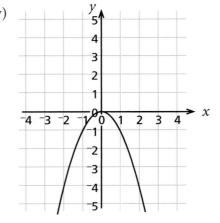

(b) Draw a set of axes, both numbered from $^-5$ to 5, and sketch the graph of $y = 3 - x^2$.

AQA 2003 Specimen

D *Trigonometric graphs*

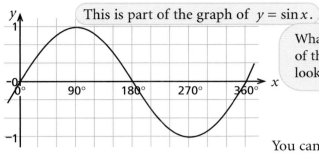

This is part of the graph of $y = \sin x$.

What will the graphs of these equations look like?

- $y = \sin x + 2$
- $y = 2 \sin x$
- $y = {}^-2 \sin x$
- $y = \sin (x + 2°)$
- $y = \sin (2x)$

You can use sheet G178.

D1 Each graph below is the image of $y = \sin x$ after a transformation.
Write the equation of each graph.

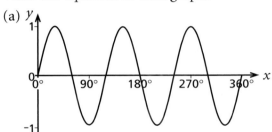

D2 Each graph below is the image of $y = \cos x$ after a transformation.
What is the equation of each graph?

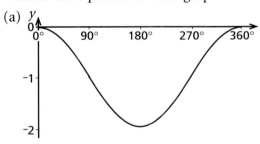

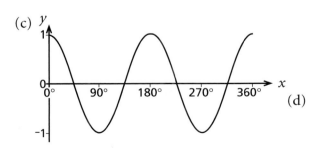

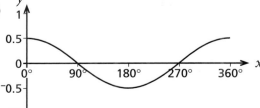

D3 Draw a sketch of the graph $y = \sin (x - 30°)$.

E *Function notation*

A **function** is another name for a rule.

We often use a letter to label a function.
For example, we could use f to mean 'square and add 1'.

We write this as $f(x) = x^2 + 1$.   We say 'f of $x = x^2 + 1$.'

$f(2)$ means the value of $x^2 + 1$ when $x = 2$.
So $f(2) = 2^2 + 1 = 5$.

E1 $f(x) = 2x - 3$
 (a) Evaluate   (i) $f(2)$ (ii) $f(5)$ (iii) $f(0)$
 (b) If $f(x) = 0$, what is the value of x?

E2 $g(x) = 2x + 1$ and $h(x) = 3x - 4$
 (a) Evaluate   (i) $g(3)$ (ii) $h(3)$ (iii) $g(6)$ (iv) $h(6)$
 (b) If $g(x) = h(x)$, find the value of x.

E3 $f(t) = t^2 - 9$
 (a) Evaluate   (i) $f(4)$ (ii) $f(^-4)$ (iii) $f(0)$
 (b) If $f(t) = 0$, what are the possible values of t?

E4 $k(x) = \cos x$ $(^-360° \leq x \leq 360°)$
 (a) Evaluate   (i) $k(0°)$ (ii) $k(180°)$ (iii) $k(^-90°)$
 (b) If $k(x) = 0$, what are all the possible values of x in the range $^-360° \leq x \leq 360°$?

Writing one rule in terms of another

$f(x) = x^2$ is one function.
Let $y = f(x) + 2$.   For example, when $x = 3$,
$$y = f(3) + 2$$
$$= 3^2 + 2$$
$$= 11$$
We could write the rule for y in terms of x as $y = x^2 + 2$.

Let $y = f(x + 2)$.   For example, when $x = 3$,
$$y = f(3 + 2)$$
$$= f(5)$$
$$= 5^2$$
$$= 25$$
We could write the rule for y in terms of x as $y = (x + 2)^2$.

E5 (a) When $f(x) = 2x + 4$, evaluate
  (i) $f(1)$ (ii) $f(3)$
  (b) If $y = f(x - 2)$, find the value of y when
  (i) $x = 5$ (ii) $x = 2$
  (c) Which of these gives the rule for y in terms of x?

  | $y = 2x - 4$ | $y = 2x$ | $y = x + 5$ | $y = 2x + 2$ |

E6 (a) When $f(x) = x^2$, evaluate
  (i) $f(^-4)$ (ii) $f(0)$
  (b) If $y = f(x + 5)$, find the value of y when
  (i) $x = ^-5$ (ii) $x = 2$
  (c) Which of these gives the rule for y in terms of x?

  | $y = x^2 + 5$ | $y = (x + 5)^2$ | $y = x^2 - 5$ | $y = (x - 5)^2$ |

E7 $f(x) = x^2$ and $y = 2f(x)$
  (a) Find the value of y when
  (i) $x = 4$ (ii) $x = ^-1$ (iii) $x = 5$
  (b) Write the rule for y in terms of x.

E8 $f(x) = \sin x$ and $y = f(2x)$
  (a) Evaluate $f(90°)$.
  (b) Find the value of y when   (i) $x = 45°$ (ii) $x = 15°$
  (c) Write the rule for y in terms of x.

E9 If $f(x) = x^2 + 1$, write each of the following rules for y in terms of x.
  Expand any brackets and write each rule in its simplest form.
  (a) $y = f(x) + 2$ (b) $y = f(x - 3)$ (c) $y = f(2x)$

E10 $f(x) = x^2$ and $y = f(x + 2)$
  (a) Find the value of y when   (i) $x = 3$ (ii) $x = ^-2$
  (b) Write the rule for y in terms of x.
  (c) On the same set of axes, sketch the graphs of $y = f(x)$ and $y = f(x + 2)$.

E11 $f(x) = \cos x$ and $y = 3f(x)$
  (a) Write the rule for y in terms of x.
  (b) Sketch the graph of $y = 3f(x)$ for $0° \le x \le 360°$.

F Transforming functions

This is part of the graph of $y = f(x)$.

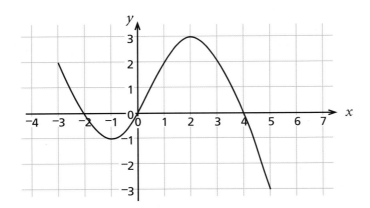

What will the graphs of these equations look like?

- $y = f(x) + 2$
- $y = 2f(x)$
- $y = \frac{1}{2}f(x)$
- $y = {}^{-}f(x)$
- $y = f(x + 2)$
- $y = f(2x)$
- $y = f(\frac{1}{2}x)$
- $y = f({}^{-}x)$

You can use sheet G179.

F1 The graph of $y = f(x)$ is shown on the grid.

(a) Let $y = f(x) + 1$.

  (i) What is the value of y when $x = 1$?

  (ii) What is the value of y when $x = 3$?

  (iii) What transformation will map the graph of $y = f(x)$ to the graph of $y = f(x) + 1$?

(b) Let $y = f(x + 1)$.

  (i) What is the value of y when $x = 1$?

  (ii) What is the value of y when $x = 3$?

  (iii) What transformation will map the graph of $y = f(x)$ to the graph of $y = f(x + 1)$?

  (iv) Sketch the graph of $y = f(x + 1)$.

(c) Let $y = f(2x)$.

  (i) What is the value of y when $x = 1$?

  (ii) What is the value of y when $x = 0.5$?

  (iii) What is the value of y when $x = 2$?

  (iv) Sketch the graph of $y = f(2x)$.

(d) Sketch the graphs of

  (i) $y = 2f(x)$ (ii) $y = {}^{-}f(x)$ (iii) $y = f({}^{-}x)$

(e) Find a value of x so that $f(4x) = 5$

(f) Find **all** the values of x so that

  (i) $f(3x) = 0$ (ii) $f(x + 2) = 0$

F2 The graph of the function $y = f(x)$ is shown on the grid.

The point P $(^-1, 5)$ lies on the curve.

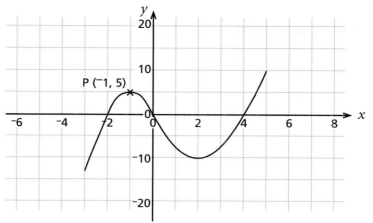

Draw three grids like the one in the diagram.

On each of the grids draw the graph of the transformed function. In each case write down the coordinates of the transformed point P.

(a) $y = f(x + 3)$ (b) $y = 2f(x)$ (c) $y = {}^-f(x)$ AQA 2000

F3 This question is on sheet G180.

F4 This is the sketch of the curve with equation $y = f(x)$.

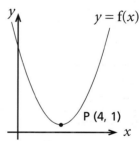

The minimum point of the curve is P $(4, 1)$.

Write down the coordinates of the minimum point for each of the curves having the following equations.

(a) $y = 3f(x)$ (b) $y = f(x + 1)$ (c) $y = f(4x)$ (d) $y = f(\frac{1}{4}x)$

F5 This is the sketch of the curve with equation $y = f(x)$.

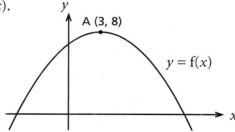

The maximum point of the curve is A $(3, 8)$.

Write down the coordinates of the maximum point for each of the curves having the following equations.

(a) $y = f(x) + 2$ (b) $y = f(x - 3)$ (c) $y = f(-x)$ (d) $y = f(3x)$

Edexcel

Transforming functions: a summary

From the function $y = \mathrm{f}(x)$,

- $y = \mathrm{f}(x) + a$ is a translation of $\begin{bmatrix} 0 \\ a \end{bmatrix}$.

- $y = \mathrm{f}(x + a)$ is a translation of $\begin{bmatrix} {}^-a \\ 0 \end{bmatrix}$.

- $y = a\mathrm{f}(x)$ is a stretch in the y-direction (leaving the x-axis unchanged) with scale factor a.
- $y = \mathrm{f}(ax)$ is a stretch in the x-direction (leaving the y-axis unchanged) with scale factor $\frac{1}{a}$.
- $y = \mathrm{f}\left(\frac{x}{a}\right)$ is a stretch in the x-direction (leaving the y-axis unchanged) with scale factor a.
- $y = {}^-\mathrm{f}(x)$ is a reflection in the x-axis.
- $y = \mathrm{f}({}^-x)$ is a reflection in the y-axis.

G *Finding the connection*

My local aquatic centre sells Koi carp.
This table shows the length, l inches, and the cost, £c, of carp available.

Length of carp in inches (l)	5	6	8	10	12	14	16
Cost of carp in £ (c)	10.00	15.30	28.70	46.00	67.10	92.10	120.90

There is a relationship between the length and the cost.
Use a graphical calculator or graphing software to find it.

G1 The table shows pairs of values of p and q.

p	3.0	4.6	5.2	6.0	7.4
q	24.5	30.6	33.5	38.0	47.4

It is suspected that there is a relationship between p and q of the form $q = ap^2 + b$.
Make a table of values of (p^2, q), graph the values and thus find a and b.

G2 The costs (£ C) of some circular tablecloths are shown in this table.

Diameter of tablecloth (D m)	0.60	0.90	1.50	2.00
Cost (£ C)	2.40	3.50	7.10	11.50

(a) Draw the graph of C against D^2.
(b) It is thought that C and D are connected by a formula of the type $C = pD^2 + q$.
    (i) Does your answer to part (a) support this?
    (ii) If it does, estimate the values of p and q. If not, explain why.

OCR

G3 In an experiment, values of R and S are noted.

There is good reason to suspect that R and S are connected by a formula of the form $S = \dfrac{a}{R} + b$.

R	1.2	1.5	2.0	2.5	3.0
S	6.4	4.4	2.4	1.2	0.4

(a) Tabulate values of $\dfrac{1}{R}$ and S, and graph S against $\dfrac{1}{R}$.

(b) Use your graph to find the values of a and b.

(c) Thus find the value of S when $R = 0.5$.

G4 In an experiment, it is thought that x and y are connected by a formula of the type $y = \dfrac{a}{x^2} + b$.

Some values of x and y are given in the table.

By drawing a suitable graph, find the values of a and b.

x	1	2	3
y	21	$7\frac{1}{2}$	5

G5 In an experiment the following values of x and y were measured.

x	2	6	10	14
y	5.8	7.9	9.3	10.5

It is thought that an approximate relationship between x and y is of the form

$$y = a\sqrt{x} + b.$$

By drawing a suitable graph find the values of a and b.   OCR

Example

In an experiment, these values of p and q were found.

It is suspected that a relationship of the form $q = ap^2 + b$ exists.

Draw a suitable graph to check, and thus find a and b.

p	0.8	1.4	2.0	2.2	2.6
q	3.46	5.44	8.5	9.76	12.64

We first draw up a table showing values of p^2 and q.

p^2	0.64	1.96	4.0	4.84	6.76
q	3.46	5.44	8.5	9.76	12.64

Think of
$q = ap^2 + b$ *as*
$y = mx + c$

Now plot the graph of q against p^2, and see if it is a straight line.

It is a straight line, so find the gradient $\left(\dfrac{7.5}{5} = 1.5\right)$ and intercept on the q-axis (2.5).

So the relationship must be $q = 1.5p^2 + 2.5$.

It is sensible to use a couple of the original values to check your answer.

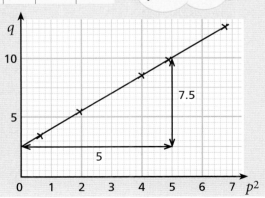

Test yourself

T1 The equation of a curve is $y = f(x)$ where

$$f(x) = x^2 - 10x + 32$$

(a) Complete the square for $f(x)$.

(b) Hence, sketch the graph of

  (i) $y = f(x)$ (ii) $y = f(x + 5)$

$y = f(x)$ where $f(x) = \sin x$ where $0 \le x \le 180°$.

(c) By considering the function $f(ax)$, sketch the graph of $y = \sin 2x$.   Edexcel

T2 What is the equation of the image of $y = x^2$ after

(a) a translation of $\begin{bmatrix} 2 \\ -3 \end{bmatrix}$ (b) a reflection in the line $y = 1$

T3 The diagram shows a sketch of the function $y = f(x)$.

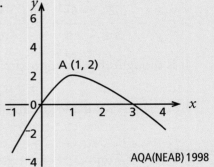

The curve cuts the x-axis at $(0, 0)$ and $(3, 0)$.
A $(1, 2)$ is a point on the curve.

The function is transformed.

On three identical grids, sketch
the following transformed functions.
State the coordinates of A on each new curve.

(a) $y = f(x) - 2$ (b) $y = 2f(x)$
(c) $y = f(2x)$

AQA(NEAB) 1998

T4 (a) This is the graph of
a function $y = f(x)$.

On an identical grid, draw
the graph of $y = f(\frac{1}{2}x)$.

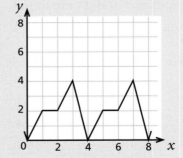

(b) $g(x) = 2x^2 - 3$

Find, in the form $ax^2 + bx + c$, an expression for $g(x - 5)$ OCR

T5 In an experiment, it is thought that x and y are connected
by a formula of the type $y = \frac{a}{x} + b$.

Some values of x and y are given
in the following table.

x	2	4	6	8	10
y	33	18	14	11	9

By drawing a suitable graph,
find the values of a and b.   OCR

1 (a) Find, as a fraction, the exact value of $\left(\frac{25}{16}\right)^{-\frac{1}{2}}$.

(b) Find the value of t if

(i) $2^t = \frac{1}{16}$ (ii) $4^t = \frac{1}{2}$ (iii) $27^t = 9$

2 The centre of the square PQRS is at O.
T, U, V, W are the midpoints of the sides of the square.
$\overrightarrow{OP} = \mathbf{p}$ and $\overrightarrow{OT} = \mathbf{t}$.

Express each of these vectors in terms of $\mathbf{p}$ and $\mathbf{t}$.

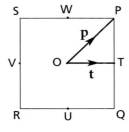

(a) $\overrightarrow{TP}$ (b) $\overrightarrow{VP}$ (c) $\overrightarrow{OQ}$ (d) $\overrightarrow{US}$

(e) $\overrightarrow{RT}$ (f) $\overrightarrow{SQ}$ (g) $\overrightarrow{WR}$ (h) $\overrightarrow{PU}$

3 The function $f(x)$ is defined for
values of x in the interval $^-6 \le x \le 6$.
The graph of $y = f(x)$ is shown on
the right.

Copy the grid and draw on it as much as
you can of each of these graphs.
(To avoid confusion you may need to make
more than one copy of the grid.)

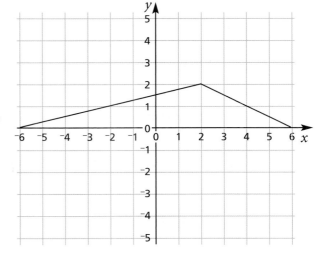

(a) $y = 2f(x)$

(b) $y = -\frac{1}{2}f(x)$

(c) $y = f(^-x)$

(d) $y = f(2x)$

(e) $y = f(\frac{1}{2}x)$

4 Calculate, correct to 2 d.p., the coordinates of the points where the
graph of $y = 2x^2 + 3x - 4$ crosses the x-axis.

5 The depth of the water in this cone is $\frac{2}{3}$ of the depth of the cone.

What percentage of the volume of the cone does the water occupy?
Give your answer correct to the nearest 1%.

6 Rearrange each of these formulas to make x the subject.

(a) $y = \frac{x+3}{x}$ (b) $y = \frac{x}{x+3}$ (c) $y = \sqrt{\frac{x}{x+3}}$ (d) $y = \frac{\sqrt{x}}{\sqrt{x+3}}$

7 OACB is a parallelogram.

$\overrightarrow{OA} = \mathbf{a}$ and $\overrightarrow{OB} = \mathbf{b}$.

D is the midpoint of OA.
E is the midpoint of AC.

Point F is on BC such that $BF = \frac{1}{6}BC$.

Point G is on BC such that $BG = \frac{5}{6}BC$.

(a) Write, in terms of **a** and **b**,

  (i) $\overrightarrow{EC}$ (ii) $\overrightarrow{CG}$ (iii) $\overrightarrow{EG}$ (iv) $\overrightarrow{DF}$

(b) What do your results for $\overrightarrow{EG}$ and $\overrightarrow{DF}$ tell you about the line segments EG and DF?

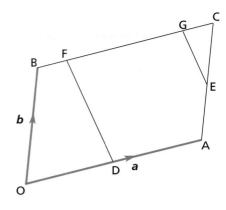

8 This is a sketch of the graph of the function $y = f(x)$.

The graph has a maximum point at $(^-2, 1)$ and a minimum point at $(4, ^-1)$.

What are the coordinates of the maximum and minimum points of the graphs with the following equations?

(a) $y = f(x) + 2$ (b) $y = f(x + 1)$

(c) $y = 2f(x)$ (d) $y = f(2x)$

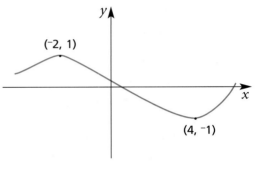

9 Calculate, correct to 2 d.p., the coordinates of the points where

(a) the line $y = 2x - 1$ crosses the curve $y = x^2 - 3x + 5$

(b) the curve $y = x^2 - 3x + 1$ crosses the curve $y = 3x^2 + 4x + 4$

10 This diagram shows parts of the graphs of the functions

$$y = x^2$$
and
$$y = 2^x$$

From the diagram you can see that $x = 2$ is one solution of the equation $x^2 = 2^x$.

(a) There is another solution greater than 2. What is this solution?

(b) There is also a solution between $^-1$ and 0. Use trial and improvement to find this solution, correct to 1 d.p.

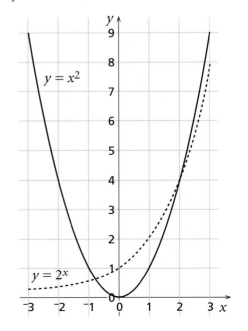

11 The triangle ABC is first translated with vector $\begin{bmatrix} -2 \\ -2 \end{bmatrix}$ and

then rotated through 90° clockwise about (−2, 0).

(a) Copy the diagram and draw the image of ABC, labelling it A′B′C′.

(b) Describe the single transformation that maps ABC on to A′B′C′.

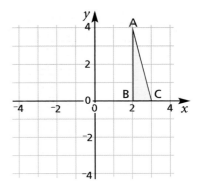

12 North Rustleton has a population of 2845 with a mean age of 35.4.
South Rustleton has a population of 1225 with a mean age of 42.6.

Calculate the mean age of the combined population of the two places, to a reasonable degree of accuracy.

13 The diagram shows a sketch of the graph of
$$y = x^3 - 2x^2$$

(a) What is the equation of the straight line which crosses the curve at the points whose x-coordinates are the solutions of the equation
$$x^3 - 2x^2 - x + 1 = 0$$

(b) Draw the graph of $y = x^3 - 2x^2$ and the graph of the straight line and use them to find approximate values of the solutions of $x^3 - 2x^2 - x + 1 = 0$.

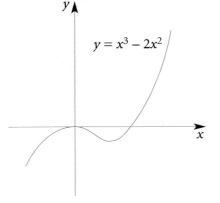

14 Kay takes two tests, whose outcomes are independent of each other.
The probability that she succeeds in the first test is p.
The probability that she succeeds in the second test is $\frac{1}{2}p$.

(a) Find an expression in terms of p for the probability that Kay does not succeed in either of the two tests.

(b) If the probability in (a) has the value $\frac{1}{4}$, what is the value of p, to 2 d.p.?

Challenge

An engineer has three pumps, A, B and C.

Pumps A and B working together will empty a 1000-litre tank in 3 minutes.
Pumps B and C will do it in 4 minutes.
Pumps A and C will do it in 6 minutes.

How long will it take to empty the tank if all three pumps are used?

This work will help you

◆ decide from given information whether two triangles are congruent

◆ use congruent triangles to prove geometrical statements

A *'Fixing' a triangle*

- You are told that the angles of a triangle are 40°, 60° and 80°.
  Is this information enough to fix both the size and shape of the triangle?

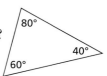

- Is the information in each sketch below enough to fix the size and shape of the triangle?

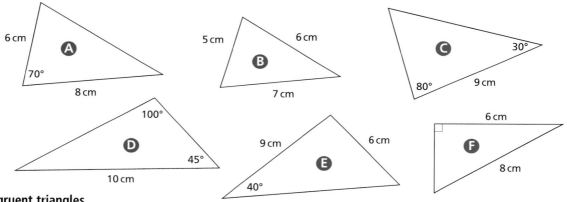

Congruent triangles

Triangles are **congruent** to each other if they are the same size and shape.

Suppose in a diagram there are two triangles and that we can prove that
the three sides of one triangle are equal to the three sides of the other.
As the lengths of the three sides are enough to fix the size and shape of a triangle,
it follows that the two triangles must be congruent.

The same is true for each of the other sets of information that 'fix' a triangle.
So altogether there are four conditions for triangles to be congruent.

- The three sides of one are equal to the three sides of the other (**SSS**).
- Two sides and the angle between them are the same in both triangles (**SAS**).
- Two angles and the side between them are the same in both triangles (**ASA**).
- Both triangles are right-angled, the hypotenuse is the same length in both,
  and so is one other side (**RHS**).

A1 The sketches below are deliberately distorted, so you can't tell by looking whether any of the triangles are congruent to each other.

Use the information about sides and and angles to find as many pairs of congruent triangles as you can. Give the reason (SSS, ...) in each case.

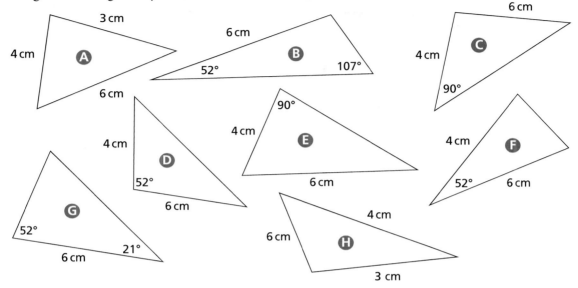

B *Proving that two triangles are congruent*

These facts about angles can be assumed to be true.

- Vertically opposite angles are equal.
- Corresponding angles (made with parallel lines) are equal.
- Alternate angles (made with parallel lines) are equal.

These facts about a circle can be assumed to be true.

- All radii are equal. (This comes from the definition of a circle.)
- A tangent is perpendicular to the radius at the point of contact.

To prove that two triangles are congruent, you have to use one of the four reasons. **SSS   SAS   ASA   RHS**

Example

ABC is a straight line and so is DBE.
AB = BE and DB = BC.
Prove that triangles ABD and EBC are congruent.

AB = EB (given)
DB = CB (given)
angle ABD = angle EBC (vertically opposite)
So triangles $\begin{smallmatrix}ABD\\EBC\end{smallmatrix}$ are congruent (**SAS**).

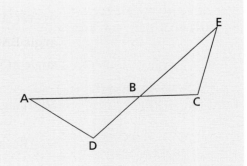

B1 In the diagram on the right, O is the centre of the circle.
M is the midpoint of the chord AB.

(a) Prove that the triangles OAM and OBM are congruent.

(b) What can you deduce about angles OMA and OMB?

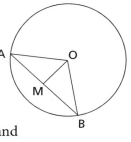

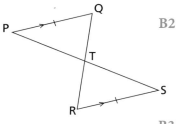

B2 In the diagram on the left, PQ is equal and
parallel to RS.
The lines PS and QR cross at T.

Prove that triangles PTQ and STR are
congruent.

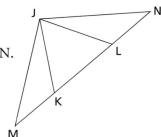

B3 JKL is an equilateral triangle and KM = LN.
Prove that triangles JKM and JLN are
congruent.

C *Proving by congruent triangles*

Congruent triangles are used to help prove geometrical statements.
The process is this.

- First find a pair of triangles that you think may be congruent.
  You may need to draw an extra line ('construction line') to make the triangles.

- Use what you know about these triangles to prove they are congruent.
  Give the reason (SSS, …).

- Then use the fact that all the corresponding pairs of sides and angles must be equal.

Example

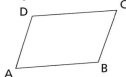

ABCD is a parallelogram.
Prove that AB = CD.

The definition of a parallelogram tells us
that AB is parallel to DC
and AD is parallel to BC.

Draw the line AC to make two triangles.

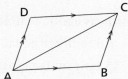

In the triangles ABC
                   CDA

AC = CA (common side)

angle BAC = angle DCA (alternate)

angle ACB = angle CAD (alternate)

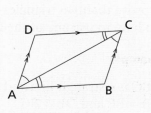

So triangles ABC are congruent. (**ASA**)
              CDA

So AB = CD.

C1 **Proving that the base angles of an isosceles triangle are equal**

ABC is an isosceles triangle, with AB equal to AC.
M is the midpoint of BC.

Explain why triangles ABM and ACM are congruent.

From this it follows that angle ABM = angle ACM.

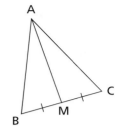

C2 PQRS is a quadrilateral in which PQ = PS and QR = SR (a kite).

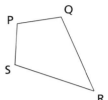

  (a) Draw the diagram and join PR.
     Prove that triangles PQR and PSR are congruent.

  (b) It now follows that angle PQR is equal to another angle.
     Which angle?

C3 *l* is a straight line and X is a point on it.

XA is perpendicular to *l*.

The lines AP and AQ are equal in length.

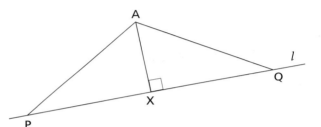

  (a) Explain why triangles XAP and XAQ
     are congruent.

  (b) It follows that another pair of lengths
     are equal. Which are they?

C4 The two lines *p* and *q* cross at V.

VA = VB.

AK is drawn parallel to BL.

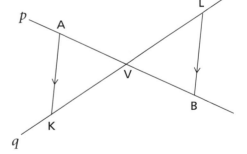

  (a) Explain why triangle VAK is congruent
     to triangle VBL.

  (b) What further deductions can you make
     about other lines and angles in the diagram?

C5 The lines PQ and PR are equal in length.

Line *b* bisects angle QPR.

S is a point on *b*.

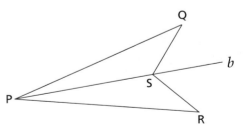

  (a) Explain why triangles PQS and PRS
     are congruent.

  (b) What further deductions can you make
     about other lines and angles in the diagram?

C6 A and B are the centres of two circles that intersect at P and Q.

By using congruent triangles, prove that angle APB = angle AQB.

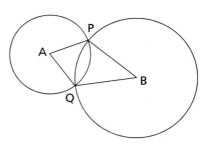

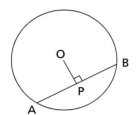

C7 O is the centre of a circle and AB is a chord.
P is on AB and OP is perpendicular to AB.

Prove by congruent triangles that P is the mid-point of AB.

D Justifying ruler-and-compasses constructions

Congruent triangles can be used to prove that ruler-and-compasses constructions are correct.

Example: bisecting an angle

The construction for bisecting an angle is as follows.

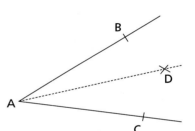

- With centre at the vertex A of the angle, draw arcs with the same radius, cutting the arms of the angle at B, C.

- With centre B draw an arc.
  With centre C and the same radius draw an arc to cut the previous arc at D.

- AD bisects the angle BAC.

Complete the missing parts of this proof of the construction.

Join ...... and ....

In the triangles $\underset{........}{ABD}$

     AD = ... (common side)

    ...... = AC (same radius)

    ...... = ... (.....................)

So triangles $\underset{........}{ABD}$ are congruent (reason: ....).

So angle DAB = angle ....

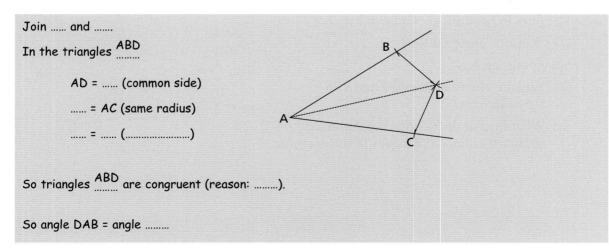

D1 A construction for drawing a perpendicular to a line *l* from a point P on it is as follows.

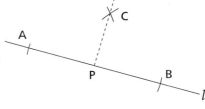

- With centre P, draw arcs with the same radius to cut the line *l* at A and B.
- With centre A, draw an arc. With centre B, draw an arc with the same radius to cut the previous arc at C.
- PC is perpendicular to *l*.

(a) Use congruent triangles to prove that angle APC = angle BPC.

(b) Explain why it follows that each of these angles is a right angle.

D2 A construction for drawing a perpendicular from a point P to a line *l* (when P is not on *l*) is as follows.

- With centre P, draw arcs with the same radius to cut the line *l* at A and B.
- With centre A, draw an arc. With centre B, draw an arc with the same radius to cut the previous arc at D.
- PD is perpendicular to *l*.

The proof of this construction is in two stages.
The point where PD cuts the line has been labelled E.

(a) Stage 1

   (i) Prove that triangles APD and BPD are congruent.

   (ii) Deduce that angle APD = angle BPD.

(b) Stage 2

   (i) Prove that triangles PEA and PEB are congruent.

   (ii) Deduce that angles PEA and PEB are equal, and that each is a right angle.

D3 The standard construction for drawing the perpendicular bisector of a line segment AB is illustrated in this diagram.

Draw and label the diagram and prove that the construction is correct. As in D2, the proof is in two stages.

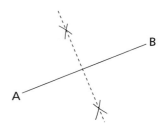

Test yourself

T1 O is the centre of a circle.
AB and CD are two chords of equal length.

   (a) Draw a diagram for this information.

   (b) Prove by using congruent triangles
that angle AOB = angle COD.

T2 ABCD is a parallelogram.
Prove that triangles ABD and
CDB are congruent.

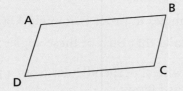

Diagram **NOT**
accurately drawn

Edexcel

T3

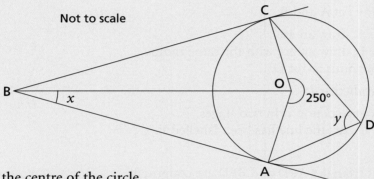

Not to scale

O is the centre of the circle.
A, C and D are points on the circle.
The tangents at A and C meet at B.

   (a) Show that AB = CB by proving that triangles OAB and OCB are congruent.

   (b) (i) Calculate angle x.

      (ii) Calculate angle y, giving your reasons.

OCR

T4 ABCD and BEFG are squares.

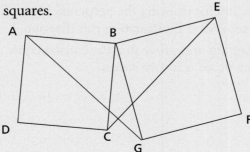

   (a) Explain why angle ABG = angle CBE.

   (b) Show that AG = CE by proving that triangles ABG and CBE are congruent.

43 Circles and equations

You should know how to

◆ write quadratic expressions in completed square form
◆ solve quadratic equations
◆ sketch the graph of a quadratic equation
◆ solve simultaneous equations such as $y = x^2$ and $y = 5x - 6$

This work will help you

◆ find the equation of a circle
◆ sketch a circle given its equation
◆ solve simultaneous equations such as $x^2 + y^2 = 4$ and $y = 3x + 1$

A Review

A1 (a) Factorise $x^2 + 6x - 7$.
  (b) Hence solve the equation $x^2 + 6x = 7$.

A2 (a) Factorise $4x^2 + 19x - 5$.
  (b) Hence solve the equation $4x^2 + 19x - 5 = 0$.

A3 Solve $2x^2 + 3x - 7 = 0$, giving your answers correct to 2 d.p.

A4 (a) Write $x^2 - 8x + 3$ in completed square form.
  (b) Hence solve $x^2 - 8x + 3 = 0$.

A5 Solve the following quadratic equations.
  Give your answers correct to two decimal places where appropriate.
  (a) $x^2 + 3x - 28 = 0$ (b) $x^2 + x - 5 = 0$ (c) $2x^2 - 5x = 0$
  (d) $2x^2 + 17x - 9 = 0$ (e) $x^2 + x = 13$ (f) $3x^2 + 6x + 2 = 0$
  (g) $2x^2 + 3 = 5x$ (h) $x^2 + 7x = 3$ (i) $2x^2 + 15x + 30 = x + 6$

A6 (a) Write $x^2 - 4x + 9$ in completed square form.
  (b) Hence sketch the graph of $y = x^2 - 4x + 9$.
    Show clearly the coordinates of the minimum point and the y-intercept.
  (c) Add a sketch of the line with equation $y = x + 5$ to your graph.
  (d) Find the coordinates of the points of intersection between the line and the parabola.

A7 Solve the simultaneous equations
$$y = 3x - 7$$
$$y = x^2 - 4x + 5$$

B *Circles*

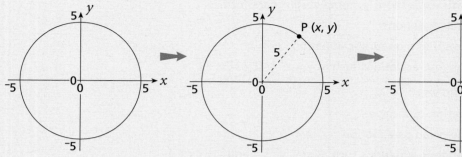

What is the equation of this circle?

Let P (x, y) be a point on the circle.

We can draw a right-angled triangle.

Using Pythagoras we have $x^2 + y^2 = 5^2$

So the equation is $x^2 + y^2 = 25$.

- Which of these points are on this circle?

  $(3, 4)$ $(1, 4)$ $(^-3, 4)$ $(\sqrt{21}, ^-2)$ $(2, ^-3)$ $(1, \sqrt{24})$ $(\sqrt{3}, \sqrt{2})$

- Can you draw this circle using a computer or graphical calculator?

B1 (a) Which of these circles has the equation $x^2 + y^2 = 4$?

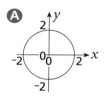

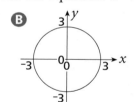

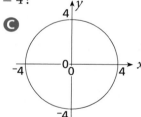

(b) Write down the equations of the other two circles.

B2 Sketch the circle with equation $x^2 + y^2 = 49$.

B3 (a) Give the coordinates of the points where the circle $x^2 + y^2 = 100$ cuts the x-axis.

(b) Which of these points are on the circle $x^2 + y^2 = 100$?

$(0, ^-10)$ $(^-6, ^-8)$ $(20, 30)$ $(1, ^-\sqrt{99})$ $(^-5, \sqrt{75})$

B4 Write down the radius of each of these circles.

(a) $x^2 + y^2 = 64$ (b) $x^2 + y^2 = 400$ (c) $x^2 + y^2 = 11$

B5 Write down the equations of the circles with centre $(0, 0)$ and the following radii.

(a) 6   (b) 1   (c) $\frac{1}{2}$ (d) $\sqrt{7}$ (e) $\sqrt{18}$

B6 What is the radius of the circle with equation $x^2 + y^2 = \frac{1}{16}$?

C Lines and circles

What are the coordinates of the points where
the line $y = 3x - 1$ meets the circle $x^2 + y^2 = 5$?

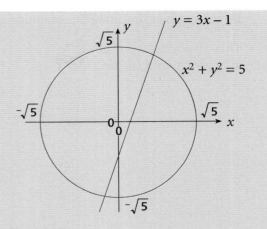

To find the coordinates we need to
solve the two simultaneous equations

$$y = 3x - 1$$
$$x^2 + y^2 = 5$$

Substitute $y = 3x - 1$ into $x^2 + y^2 = 5$ to give

$$x^2 + (3x - 1)^2 = 5$$
$$x^2 + 9x^2 - 6x + 1 = 5$$
$$10x^2 - 6x - 4 = 0$$
$$5x^2 - 3x - 2 = 0$$
$$(5x + 2)(x - 1) = 0$$

$$\text{So } x = {}^-0.4 \text{ or } x = 1$$

$x = {}^-0.4$ gives $y = 3 \times {}^-0.4 - 1 = {}^-2.2$

$x = 1$ gives $y = 3 \times 1 - 1 = 2$

So the solutions are $x = {}^-0.4, y = {}^-2.2$ and $x = 1, y = 2$.

Hence the coordinates are $({}^-0.4, {}^-2.2)$ and $(1, 2)$.

C1 (a) Sketch the circle with equation $x^2 + y^2 = 9$.

    (b) Add a sketch of the line with equation $y = 2x - 3$ to your diagram.

    (c) Find the coordinates of the points where
the line $y = 2x - 3$ meets the circle $x^2 + y^2 = 9$.

C2 The diagram shows a sketch of the
circle $x^2 + y^2 = 17$ and the line $y = x + 5$.

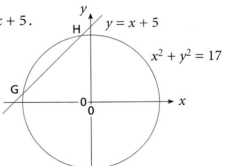

    (a) For the points of intersection,
show that $x^2 + 5x + 4 = 0$.

    (b) Hence find the coordinates of G and H.

C3 The straight line with equation $y = x + 6$ meets the circle with
equation $x^2 + y^2 = 50$ at two points P and Q.
By solving two simultaneous equations, find the coordinates of P and Q.   OCR

C4 This diagram shows the circle
$x^2 + y^2 = 4$ and the line $y = x + 1$.

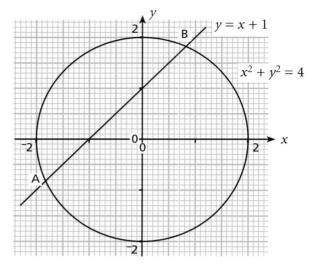

(a) Use the graph to estimate the coordinates of points A and B, correct to 1 d.p.

(b) Solve two simultaneous equations to find the coordinates of A and B, correct to 2 d.p.

C5 The straight line $y = 2x + 1$ meets the circle $x^2 + y^2 = 7$ at two points P and Q.
Find the coordinates of P and Q, correct to 2 d.p.

C6 (a) Try to solve the simultaneous equations

$$x^2 + y^2 = 2$$
$$y = x - 3$$

(b) What does this tell you about the line $y = x - 3$ and the circle $x^2 + y^2 = 2$?
Confirm your answer by sketching the line and circle on the same set of axes.

C7 (a) Solve the simultaneous equations

$$x^2 + y^2 = 5$$
$$y = 5 - 2x$$

(b) What does this tell you about the line $y = 5 - 2x$ and the circle $x^2 + y^2 = 5$?

C8 Solve these pairs of simultaneous equations.

(a) $x^2 + y^2 = 25$ (b) $y = 7 - 3x$ (c) $6x^2 + y^2 = 22$
$\quad y = x - 1$ $\quad x^2 + y^2 = 5$ $\quad x = 2y + 5$

C9 Solve these pairs of simultaneous equations.

(a) $x^2 + y^2 = 10$ (b) $x - 3y = 2$ (c) $x^2 + 2y^2 = 2$
$\quad x + y = 4$ $\quad x^2 + y^2 = 2$ $\quad x - y = 1$

D *Off centre*

What is the equation of a circle with centre (2, 3) and radius 5?

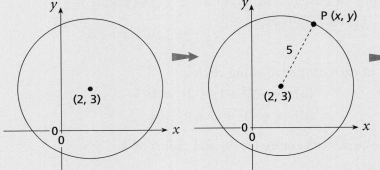

(2, 3)

Let P (x, y) be a point on the circle.

P (x, y)

5

(2, 3)

We can draw a right-angled triangle.

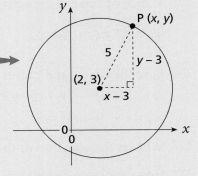

P (x, y)

5

$y - 3$

(2, 3)

$x - 3$

Using Pythagoras we have $(x - 2)^2 + (y - 3)^2 = 5^2$.

So the equation of the circle is $(x - 2)^2 + (y - 3)^2 = 25$.

• Where does this circle cut the x- and y-axes?

D1 Which of these circles could have the equation $(x - 4)^2 + (y - 1)^2 = 4$?

A

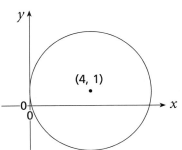

(4, 1)

B

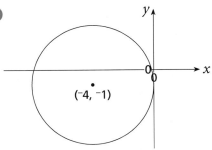

(⁻4, ⁻1)

C

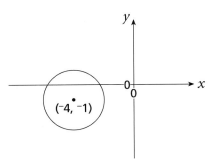

(⁻4, ⁻1)

D

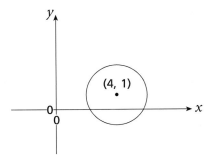

(4, 1)

D2 Sketch the circle with equation $(x-3)^2 + (y-2)^2 = 36$.

D3 Write down the equation of the circle with radius 7 and centre $(4, 5)$.

D4 Where does the circle $(x-1)^2 + (y-6)^2 = 17$ cut the y-axis?

D5 Which of these is the equation of the circle with centre $(^-2, 3)$ and radius 3?

$(x-2)^2 + (y+3)^2 = 9$ $(x+2)^2 + (y-3)^2 = 9$ $(x-2)^2 + (y-3)^2 = 9$

D6 Give the centre and radius of each of the following circles.
  (a) $(x+4)^2 + (y-1)^2 = 9$ (b) $(x-5)^2 + (y+3)^2 = 49$
  (c) $(x+7)^2 + y^2 = 81$ (d) $(x+\frac{1}{2})^2 + (y+9)^2 = \frac{1}{4}$

D7 What is the equation of the circle with centre $(1, ^-9)$ and radius 8?

D8 What is the equation of each circle?

(a)

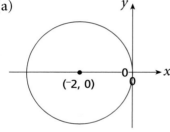

(b)

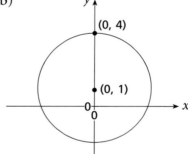

D9 (a) What is the equation of the circle with centre $(0, 3)$ that goes through the origin?
  (b) What is the equation of the circle with centre $(3, 4)$ that goes through the origin?

*__D10__ (a) Write $x^2 + 4x$ in completed square form.
  (b) Hence write $x^2 + 4x + y^2 = 5$ in the form $(x+a)^2 + y^2 = c$, where a and c are constants.
  (c) Write down the centre and radius of the circle $x^2 + 4x + y^2 = 5$.

*__D11__ Find the centre and radius of each of the following circles.
  (a) $x^2 + 6x + y^2 = 0$ (b) $x^2 - 10x + y^2 = 11$

*__D12__ (a) Write $x^2 + 8x$ in completed square form.
  (b) Write $y^2 - 2y$ in completed square form.
  (c) Hence write $x^2 + 8x + y^2 - 2y = 8$ in the form $(x+a)^2 + (y+b)^2 = c$, where a, b and c are constants.
  (d) Sketch the graph of $x^2 + 8x + y^2 - 2y = 8$.

Test yourself

T1 (a) Sketch the graph of $x^2 + y^2 = 121$.

(b) What are the coordinates of the points where the graph cuts the y-axis?

T2 What is the equation of a circle with centre $(0, 0)$ and radius 3?

T3 The diagram shows the circle $x^2 + y^2 = 25$ and the line $y = x + 7$.
The line and the circle intersect at points A and B.

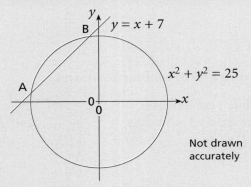

Not drawn
accurately

(a) By substituting the equation $y = x + 7$ into the equation $x^2 + y^2 = 25$, show that
$$x^2 + 7x + 12 = 0.$$

(b) Hence find the coordinates of A and B.   *AQA Specimen 2003*

T4 The line $y = x + 4$ meets the circle $x^2 + y^2 = 26$ at two points P and Q.
By solving two simultaneous equations, find the coordinates of P and Q.

T5 Solve the simultaneous equations
$$y = 2x - 7$$
$$x^2 + y^2 = 61$$
Edexcel

T6 Solve the simultaneous equations
$$x^2 + y^2 = 100$$
$$x - y = 2$$
AQA Specimen 2003

T7 Solve algebraically these simultaneous equations. Show your method clearly.
$$x + y = 5$$
$$x^2 + 3y^2 = 49$$
OCR

T8 (a) Find integers p and q such that such that $x^2 - 8x = (x - p)^2 + q$ for all x.

(b) (i) Using your answer to (a), express $x^2 + y^2 - 8x = 0$ in the form
$(x + u)^2 + y^2 = v$, where u and v are constants.

(ii) Describe the shape and other important features of
the graph of $x^2 + y^2 - 8x = 0$.   *OCR*

44 Exactly so

You will revise
- finding the circumference and area of a circle
- using Pythagoras

This work will help you understand why and how to leave exact values in answers instead of numerical approximations.

A Keeping sight of π

The shapes on the opposite page are drawn on a grid of centimetre squares.

Shape A is made of two semicircles with radius 1 cm and one semicircle with radius 2 cm. Their centres are marked with dots.

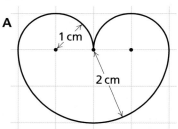

The length in centimetres of the large semicircle is

$$\tfrac{1}{2} \times 2\pi r = \tfrac{1}{2} \times 2 \times \pi \times 2 = 2\pi$$

Leaving π in the result keeps it **exact**. (Writing it to a number of decimal places would involve approximation, since π is irrational and cannot be written as a terminating decimal.)

A1 (a) Find the exact length of one small semicircle in shape A.

(b) Find the exact value of the whole perimeter of shape A.

(c) Show that this perimeter is **exactly the same** as the perimeter of shape B on the opposite page.

A2 Do shapes A and B have exactly the same **area**?
Give a reason.

A3 Find the exact perimeters of all the shapes A to R on the opposite page, and list groups of shapes with exactly the same perimeter.

A4 List A to R in groups with exactly the same area, giving the exact area for each group.

*A5 Sketch shapes with these values.

	Perimeter (cm)	Area (cm²)
(a)	2π	2
(b)	$\pi + 4$	$4 - \pi$
(c)	4π	$\pi + 4$

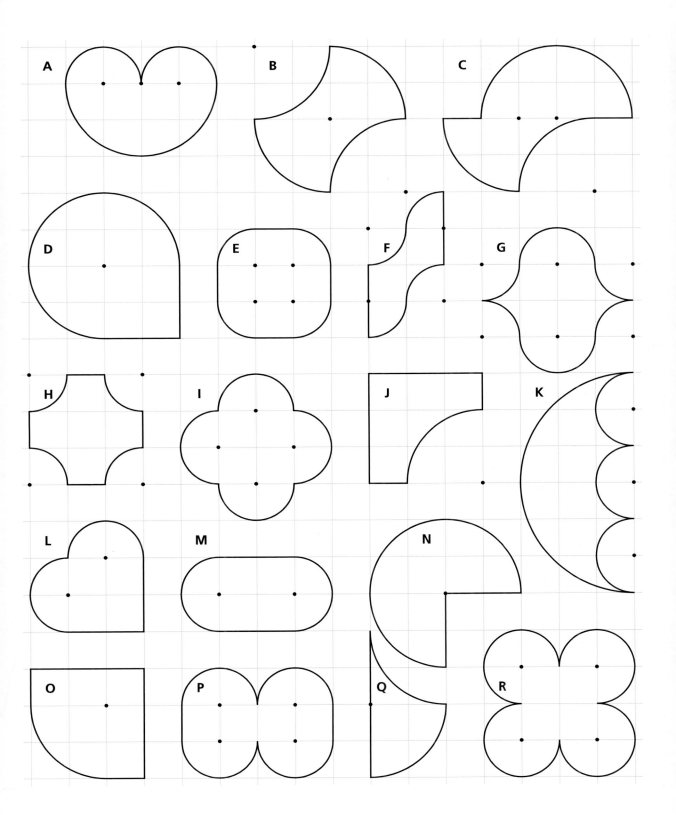

B *Keeping sight of surds*

This triangle is drawn on centimetre squares.

Using Pythagoras' theorem you can see that
its perimeter in centimetres is $1 + 3 + \sqrt{10}$.

Worked out to 2 d.p. this is 7.16.

But sometimes it is better to leave a result like this as an **exact** value, in this case $4 + \sqrt{10}$.

An expression with an exact square root, like $4 + \sqrt{10}$, is called a **surd**.
Another example of a surd is $4 + 3\sqrt{10}$ or just $\sqrt{10}$.

> **B1** Shapes P to W are drawn on centimetre squares.
> By working out exact perimeters (keeping your answers as surds),
> sort them into pairs with the same perimeter.

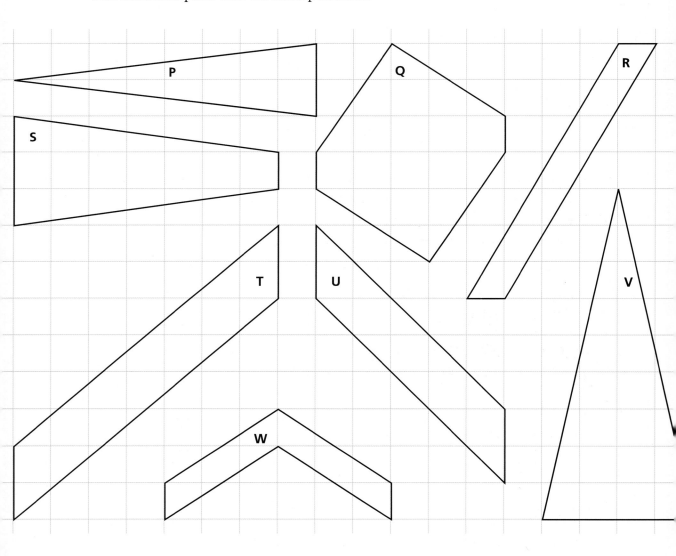

Using Pythagoras, the length of this hypotenuse is $\sqrt{3^2 + 3^2} = \sqrt{18}$ units.

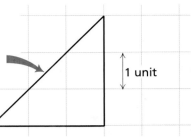

1 unit

However this hypotenuse is $\sqrt{1^2 + 1^2} = \sqrt{2}$ units long.

The large triangle is an enlargement with scale factor 3 of the small one, so $\sqrt{18}$ must be equal to $3\sqrt{2}$.

We can also show that $3\sqrt{2}$ is the square root of 18 by squaring it to get 18:

$(3\sqrt{2})^2 = (3\sqrt{2}) \times (3\sqrt{2}) = 3 \times \sqrt{2} \times 3 \times \sqrt{2} = 3 \times 3 \times \sqrt{2} \times \sqrt{2} = 9 \times 2 = 18$

We can also start with $\sqrt{18}$ and get $3\sqrt{2}$ like this: $\sqrt{18} = \sqrt{9 \times 2} = \sqrt{9} \times \sqrt{2} = 3\sqrt{2}$.

Another example would be: $\sqrt{20} = \sqrt{4 \times 5} = \sqrt{4} \times \sqrt{5} = 2\sqrt{5}$.

B2 Use this triangle and a suitable enlargement of it to demonstrate that $\sqrt{20} = 2\sqrt{5}$.

B3 Find a suitable triangle and an enlargement of it to demonstrate that $\sqrt{160} = 4\sqrt{10}$.

B4 The length of the hypotenuse of this right-angled triangle is $\sqrt{2}$. So the sine of 45° is **exactly** $\dfrac{1}{\sqrt{2}}$.

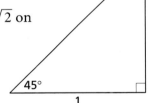

As an approximate check, make sure you can get 0.707... for $1 \div \sqrt{2}$ on a calculator, and that the calculator also gives 0.707... for sin 45°.

(a) Use the triangle to give an exact value for cos 45°.

Do an approximate calculator check for your answer.

(b) Give exact values for these. (i) sin 135° (ii) cos 135°

B5 This equilateral triangle has sides 2 units long. M is the midpoint of side BC.

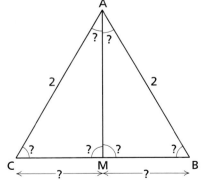

(a) Make a large sketch of the triangle but replace the question marks by values.

(b) How long is the line AM **exactly** (left as a surd)? Mark this on your sketch.

(c) Use the diagram to give these exactly.

(i) sin 60° (ii) cos 60° (iii) tan 60°

(iv) sin 30° (v) cos 30° (vi) tan 30°

Do a calculator check as in B4 for these answers.

(d) Give these exactly.

(i) sin 120° (ii) cos 120° (iii) tan 120° (iv) sin 150° (v) cos 150° (vi) tan 150°

(e) Calculate the area of the equilateral triangle, leaving any surds in your answer.

Pythagoras spiral

How long **exactly** is the hypotenuse of the first triangle?

How long exactly is the hypotenuse of the fourth triangle?

If triangles continue to be added in the same way,
how long exactly will be the hypotenuse of the nth triangle?

Which triangle will have a hypotenuse exactly 3 units long?

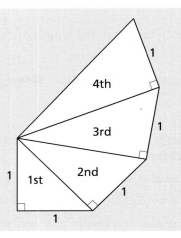

C *Mixed questions*

C1 The shaded area is formed by
two circles with the same centre.
Find the shaded area exactly, in terms of a.

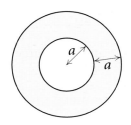

C2 A solid cone has a height of 12 cm and a slant height of 13 cm.
Calculate the total surface area of the cone.
Give your answer in terms of π.

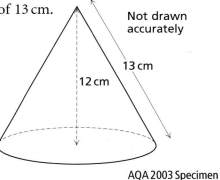

Not drawn
accurately

13 cm

12 cm

AQA 2003 Specimen

C3 Find the area of the shaded
semicircle in terms of π.

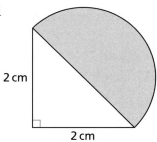

2 cm

2 cm

C4 A circle has an area of 10 square units.
Find its radius in terms of π.

C5 In this right-angled triangle $\cos a = \frac{3}{4}$.

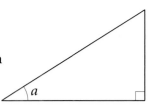

   (a) Make a sketch of the triangle and label all its sides with possible exact lengths.

   (b) Write as surds   (i) $\sin a$ (ii) $\tan a$

Exact expressions for quantities can be used to find relationships between variables.

Example

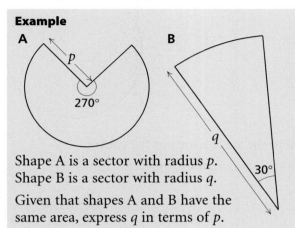

Shape A is a sector with radius p.
Shape B is a sector with radius q.

Given that shapes A and B have the same area, express q in terms of p.

Area of shape A $= \frac{3}{4}\pi p^2$

Area of shape B $= \frac{1}{12}\pi q^2$

Since the areas of the two shapes are equal,

$$\frac{3}{4}\pi p^2 = \frac{1}{12}\pi q^2$$

Multiplying both sides by 12,

$$9\pi p^2 = \pi q^2$$

Dividing both sides by π,

$$9p^2 = q^2$$

Taking the positive square root of both sides,

$$3p = q, \quad so \quad q = 3p$$

C6 (a) A circle of radius r fits exactly inside a square of side $2r$. Exactly what fraction of the square does the circle occupy?

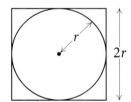

   (b) A square fits exactly inside a circle of radius r. Exactly what fraction of the circle does the square occupy?

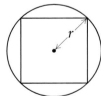

C7 The cone and cylinder have the same radius, r, and the same volume. How are the heights a and b related?

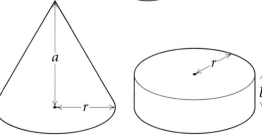

C8 A sphere has a surface area of 25π square units.

   (a) What is its radius?   (b) What is its volume, exactly?

C9 A solid sphere of radius r fits inside a cylindrical space that has radius r and height $2r$.
Exactly what fraction of the cylinder's volume does the sphere occupy?

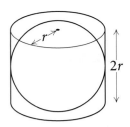

C10 A cylinder has radius x and height $4x$.

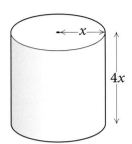

A sphere has radius r.

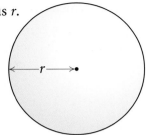

The total surface area of the cylinder is equal to the surface area of the sphere.

(a) Show that $r^2 = \frac{5}{2}x^2$.

(b) When $r = 10$, find the value of x.
Give your answer in the form $a\sqrt{b}$.

AQA 2003 Specimen

Test yourself

T1 These shapes are drawn on centimetre squares with centres of arcs shown by dots.
Find their areas and perimeters, giving exact values.

(a)   (b)   (c)   (d)

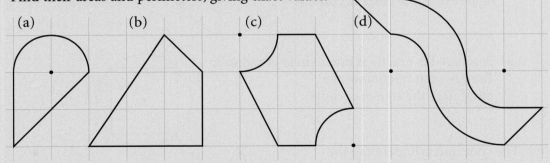

T2 Given that $\tan x = 3$, write exact values for $\sin x$ and $\cos x$.

T3 A semicircle of radius 3 cm has the same area as a circle of radius a cm.
Find the exact value of a.

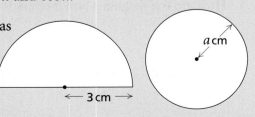

T4 A cone has height 12 cm and volume 196 cm³.
Find, in terms of π, the diameter of its base.

Review 11

1 A kite can be defined as a quadrilateral ABCD in which AB = AD and CB = CD.

   By using congruent triangles, prove that angle ABC = angle ADC.

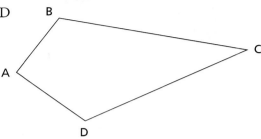

2 A circle has centre (0, 0) and radius 10.

   (a) Write down the equation of the circle.

   (b) By solving a pair of simultaneous equations, find the coordinates of the points where the circle intersects the line $y = x + 2$.

3 This shape consists of a semicircle drawn on the hypotenuse of a right-angled triangle.

   The lengths of the two shorter sides of the triangle are 6 cm and 4 cm.

   (a) Explain why the radius of the semicircle is $\sqrt{13}$ cm.

   (b) Find an expression for the exact value of

      (i) the perimeter of the combined shape

      (ii) the area of the combined shape

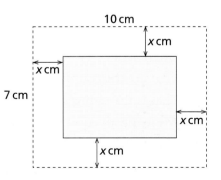

4 A strip of width x cm is cut from each edge of a rectangle whose sides were originally of length 10 cm and 7 cm.

   (a) Write down an expression for the area of the remaining rectangle.

   (b) Given that the area remaining is 50 cm², calculate the value of x, correct to 1 d.p.

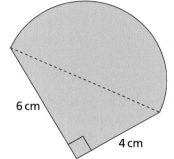

5 A car travels 18 miles in 13 minutes.
   The distance is correct to the nearest mile and the time to the nearest minute.

   Calculate, correct to 2 d.p., the upper and lower bounds of the average speed of the car in m.p.h.

6 Sam has two cubical dice. On one dice the faces are numbered 1, 2, 2, 3, 3, 3.
On the other dice the faces are numbered 1, 1, 1, 2, 2, 3.
She rolls both dice.

   (a) What is the probability that the total score on the two dice is 6?

   (b) What is the probability that exactly one of the dice shows '3'?

   (c) What is the most likely total score on the two dice and what is the
       probability of getting this score?

7 The expression $x^2 + 10x - 3$ can be written in the form $(x + a)^2 + b$.

   (a) Find the values of a and b.

   (b) Find the minimum value of the expression $x^2 + 10x - 3$ and the
       value of x for which it is a minimum.

8 In triangle ABC, M is the midpoint of BC.
BX and CY are perpendicular to AM.

Prove that BX = CY.

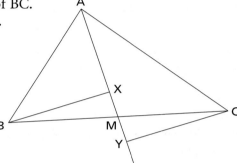

9 Triangle ABC is right-angled at C.
AB = 7 cm and AC = 6 cm.
The size of angle A is x.

Find the exact value of

   (a) $\cos x$ (b) $\tan x$

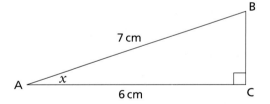

10 The equation $x^2 + (y - 8)^2 = 16$ represents a circle.

   (a) What are the coordinates of the centre of the circle?

   (b) What is the radius of the circle?

   (c) By solving an equation, find the coordinates of the points where
       the line $y = 2x$ crosses the circle.

11 The probability that a spinner shows 'red' is $\frac{2}{5}$.
The spinner is spun twice.

Calculate the probability that it shows 'red' at least once.

12 The distance between two ports is 165 miles.
The passenger ferry between them travels 8 m.p.h. faster than the cargo ferry
and takes 2 hours less for the journey.

By forming and solving an equation, calculate the speed of each ferry.

13 A, B, C and D are points on a circle.
AB and CD meet at point X.

AX = 4 cm, XB = 3 cm and CX = 5 cm.

(a) Explain why triangles AXC and DXB are similar.

(b) Calculate XD.

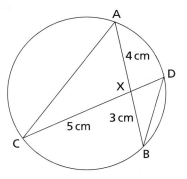

14 Solve these pairs of simultaneous equations.

(a) $y = 2x^2 - 4$
$2x + y = 20$

(b) $y = x^2 + 2$
$2x + 3y = 7$

15 Pritti weighed a potato and then dried it in an oven.
Before drying it weighed 84 g and afterwards 55 g, both correct to
the nearest gram.

Calculate, to the nearest 0.1%, the upper and lower bounds of the
percentage weight loss.

Challenges

1 What fraction of this regular octagon is green?

There are several methods of working this out.
Explain your method.

Try to find another – possibly simpler – method.

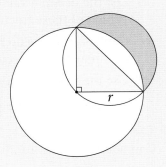

2 The larger circle has radius r.

The blue 'crescent' is part of a semicircle whose diameter
is the hypotenuse of the yellow triangle.

Find exact expressions for the area of the triangle and for
the area of the crescent.

Comment on your answer.

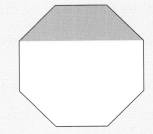

45 Rational and irrational numbers

This work will help you

◆ write recurring decimals as fractions

◆ learn about rational and irrational numbers

◆ manipulate irrational numbers in surd form

A Fractions to decimals: review

Rational numbers are numbers that can be written in the form $\frac{a}{b}$ (a and b are integers). Their decimal equivalents either **terminate** or **recur**.

$$\frac{7}{8} = 7 \div 8 = 0.875 \qquad \begin{array}{r} 0.875 \\ 7\overline{)8.000} \end{array}$$

$$\frac{5}{12} = 5 \div 12 = 0.41666... \qquad \begin{array}{r} 0.41666... \\ 12\overline{)5.00000...} \end{array}$$
$$= 0.41\dot{6}$$

$$\frac{3}{11} = 3 \div 11 = 0.272727... \qquad \begin{array}{r} 0.2727... \\ 11\overline{)3.0000...} \end{array}$$
$$= 0.\dot{2}\dot{7}$$

In the dot notation, a dot is written above the first and last digit of the recurring group.

$$0.2454545... = 0.2\dot{4}\dot{5} \qquad 0.7184184184... = 0.7\dot{1}8\dot{4}$$

A1 Write these fractions as decimals.   (a) $\frac{3}{5}$ (b) $\frac{7}{25}$ (c) $\frac{39}{50}$ (d) $\frac{5}{8}$ (e) $\frac{9}{40}$

A2 Write each of these recurring decimals using the dot notation.

(a) 0.33333...   (b) 0.232323...   (c) 0.233333...   (d) 0.321321321...   (e) 0.3212121...

A3 Write these fractions as decimals.   (a) $\frac{5}{6}$ (b) $\frac{2}{11}$ (c) $\frac{1}{7}$ (d) $\frac{6}{11}$ (e) $\frac{7}{30}$

A4 (a) Write the following fractions as decimals.

$$\frac{1}{9} \qquad \frac{3}{9} \qquad \frac{5}{9} \qquad \frac{7}{9}$$

(b) What do you think is the value of $0.\dot{9}$?

A5 Unit fractions are fractions with numerator 1.

(a) List the decimal values of all the unit fractions from $\frac{1}{2}$ to $\frac{1}{20}$.
   Which of these fractions are equivalent to terminating decimals?

(b) Can you find a rule for deciding whether a fraction is equivalent to a terminating or a recurring decimal?

(c) Use the rule to decide which fractions below are equivalent to terminating decimals.

$$\frac{9}{20} \qquad \frac{7}{8} \qquad \frac{1}{21} \qquad \frac{3}{47} \qquad \frac{1}{25} \qquad \frac{4}{30} \qquad \frac{1}{80} \qquad \frac{3}{64}$$

A6 The number 0.25225222522225222225... continues to follow the pattern.
Can you write this number using recurring decimal notation? Explain your answer.

B Decimals to fractions

Terminating decimals can be converted to fractions by thinking about place values.

$$0.45 = \frac{45}{100} = \frac{9}{20} \qquad 0.384 = \frac{348}{1000} = \frac{48}{125}$$

Recurring decimals need a different approach.

Let f stand for $0.\dot{5}$
Multiply by 10.
Subtract.

$f = 0.5555\ldots$
$10f = 5.5555\ldots$
$9f = 5$
$f = \frac{5}{9}$

Let g stand for $0.9\dot{6}\dot{3}$
Get the non-recurring part before the decimal point, by multiplying by 10.
Multiply both sides by 100 to get the 6s and 3s aligned.
Subtract $10g$ from $1000g$.

$g = 0.96363\ldots$
$10g = 9.6363\ldots$
$1000g = 963.6363\ldots$
$990g = 954$
$g = \frac{954}{990} = \frac{53}{55}$

B1 Write the following decimals as fractions, in their simplest form.

(a) 0.7   (b) 0.85   (c) 0.03   (d) 0.924   (e) 0.025

B2 Write the following decimals as fractions, in their simplest form.

(a) 0.888888...   (b) 0.151515...   (c) 0.272727...   (d) 0.147147...

(e) $0.\dot{2}$ (f) $0.\dot{3}\dot{6}$ (g) $0.\dot{0}0\dot{3}$ (h) $0.\dot{3}2\dot{6}$

B3 (a) Copy and complete the working on the right.

(b) Hence write 0.1555... as a fraction in its lowest terms.

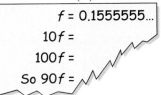

$f = 0.1555555\ldots$
$10f =$
$100f =$
So $90f =$

B4 Write these as fractions in their lowest terms.

(a) 0.1777777...   (b) 0.0444444...   (c) 0.254444444...   (d) 0.312121212...

(e) $0.0\dot{3}$ (f) $0.4\dot{2}\dot{7}$ (g) $0.0\dot{1}\dot{6}$ (h) $0.9\dot{8}9\dot{7}$

B5 (a) Write 0.3333... as a fraction.   (b) Write 0.1 as a fraction.

(c) Hence write 0.43333... as a fraction.

B6 (a) Write 0.666... as a fraction.   (b) Write 0.4 as a fraction.

(c) Hence write 0.2666... as a fraction.

*B7 The sum of the reciprocals of two consecutive numbers is 0.36666666...
Find the numbers.

C Irrational numbers

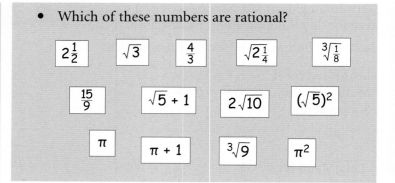

- Can you find two square numbers so that one square number is double the other?

1	4	9	16	25	36	49	6
121	144	169	196	225	256	289	3
441	484	529	576	625	676	729	7

- Which of these numbers are rational?

$2\frac{1}{2}$ $\sqrt{3}$ $\frac{4}{3}$ $\sqrt{2\frac{1}{4}}$ $\sqrt[3]{\frac{1}{8}}$

$\frac{15}{9}$ $\sqrt{5}+1$ $2\sqrt{10}$ $(\sqrt{5})^2$

π $\pi+1$ $\sqrt[3]{9}$ π^2

C1 Which of these numbers are irrational?

$\sqrt{8}$ $\sqrt{16}$ $4\sqrt{7}$ $\frac{\pi}{2}$ $1.\dot{7}$ 5.1879 $\sqrt{125}$ $\sqrt[3]{25}$ $\sqrt{3}+4$ π^2+1

C2 Sort these into four matching pairs of equivalent numbers.

$\sqrt{5}+\sqrt{5}$ $6\sqrt{5}-2\sqrt{5}$ $2(\sqrt{5}+2)$ $2\sqrt{5}$ $(2\sqrt{5})^2$ $2\sqrt{5}+4$ 20 $4\sqrt{5}$

C3 Simplify these.
   (a) $\sqrt{7}+\sqrt{7}$ (b) $(\sqrt{7})^2$ (c) $5\sqrt{7}-2\sqrt{7}$
   (d) $3(\sqrt{7}+2)$ (e) $(2+\sqrt{7})+(5-\sqrt{7})$ (f) $(3\sqrt{7}+6)-(\sqrt{7}-2)$

C4 Which of these numbers are between 2 and 3?

$\pi-1$ $\sqrt{5}$ $\sqrt{3}$ $\sqrt{8}$ $\sqrt{11}$ $\sqrt{12}-1$ $\sqrt{17}-1$

C5 Write down an irrational number between
   (a) 4 and 5   (b) 5 and 6   (c) 0 and 1   (d) 10 and 11

C6 Is the number 0.747744777444777744447777744444… rational or irrational? Explain your answer carefully.

C7 (a) For each of these decimals, state, with reasons, whether it is a rational or an irrational number.
      (i) 0.1   (ii) 0.12112111211112…   (iii) $0.\dot{1}2\dot{3}$
   (b) Write, where possible, the decimals in (a) as fractions in their simplest form.   OCR

C8 Write down the **exact** lengths of the sides of squares with these areas.
   (a) $10\,\text{cm}^2$ (b) $3\,\text{cm}^2$ (c) $16\,\text{m}^2$ (d) $101\,\text{mm}^2$ (e) $81\,\text{cm}^2$

C9 What is the exact length of AC, where ABCD is a square?

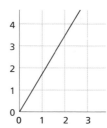

*C10 The sloping line in this diagram has a gradient of $\sqrt{3}$.

Prove that, no matter how far it is extended, it will never go through an intersection of two grid lines.

*C11 Two tiny insects move at the same speed.

One walks round and round the perimeter of a square; the other walks backwards and forwards across a diagonal of the same square.

They start together at one corner of the square.
Prove that their positions never coincide again.

Irrational numbers

Irrational numbers are numbers that cannot be written in the form $\frac{a}{b}$ (a and b are integers).

Their decimal equivalents never terminate or recur.

Examples of irrational numbers are $\quad\sqrt{3},\quad \sqrt[3]{6},\quad \pi,\quad \pi+5,\quad \pi^2$

Any number that involves an irrational root, such as $\sqrt{7}$ or $\sqrt{5}+1$, is said to be in **surd** form.

D **Simplifying**

• Can you sort these into matching pairs of equivalent numbers?

• Which is the odd one out?

A $\sqrt{4}\times\sqrt{9}$ **B** $3\sqrt{2}$ **C** $\sqrt{4}\times\sqrt{6}$ **D** $\sqrt{8}+\sqrt{10}$ **E** $\sqrt{20}$

F $\sqrt{2}\times\sqrt{5}$ **G** $\sqrt{10}$ **H** $\sqrt{18}$ **I** $\sqrt{24}$ **J** $2\sqrt{5}$ **K** $\sqrt{36}$

D1 Write each of these in the form $\sqrt{n}$, where n is an integer.

(a) $\sqrt{3}\times\sqrt{5}$ (b) $\sqrt{2}\times\sqrt{7}$ (c) $\sqrt{3}\times\sqrt{8}$ (d) $\sqrt{5}\times\sqrt{10}$

D2 Simplify these.

(a) $\sqrt{5}\times\sqrt{5}$ (b) $(\sqrt{3})^2$ (c) $\sqrt{2}\times\sqrt{8}$ (d) $\sqrt{5}\times\sqrt{20}$

D3 Simplify these.

(a) $2\sqrt{3} \times \sqrt{3}$ (b) $3\sqrt{7} \times 2\sqrt{7}$ (c) $3\sqrt{2} \times \sqrt{5}$ (d) $2\sqrt{5} \times 4\sqrt{2}$

D4 Work out the value of the letter in each statement.

(a) $\sqrt{3} \times \sqrt{a} = \sqrt{33}$ (b) $\sqrt{21} = \sqrt{b} \times \sqrt{7}$ (c) $\sqrt{30} = \sqrt{3} \times \sqrt{c}$

A root can sometimes be written in the form $p\sqrt{q}$, where p and q are both integers.

Example

$$\sqrt{28}$$
$$= \sqrt{4 \times 7}$$
$$= \sqrt{4} \times \sqrt{7}$$
$$= 2 \times \sqrt{7} = 2\sqrt{7}$$

D5 Write each number below in the form $p\sqrt{q}$, where q is the smallest integer possible.

(a) $\sqrt{27}$ (b) $\sqrt{50}$ (c) $\sqrt{80}$ (d) $\sqrt{98}$ (e) $\sqrt{300}$

D6 Write each of these in the form $p\sqrt{q}$, where q is the smallest integer possible.

(a) $3\sqrt{8}$ (b) $2\sqrt{32}$ (c) $2\sqrt{45}$ (d) $3\sqrt{72}$ (e) $5\sqrt{80}$

D7 Jake thinks that $\sqrt{9} + \sqrt{9} = \sqrt{18}$.
How would you convince him that he is wrong?

D8 Write each of these in the form $p\sqrt{q}$, where q is the smallest integer possible.

(a) $\sqrt{2} \times \sqrt{10}$ (b) $\sqrt{3} \times \sqrt{18}$ (c) $\sqrt{15} \times \sqrt{5}$ (d) $\sqrt{18} \times \sqrt{2}$

D9 (a) Write $\sqrt{12}$ in the form $a\sqrt{b}$, where a and b are both prime numbers.

(b) Hence simplify $\sqrt{12} + 5\sqrt{3}$.

D10 Simplify $4\sqrt{12} - \sqrt{3}$, giving your answer in the form $p\sqrt{q}$.   OCR

D11 Simplify these.

(a) $\sqrt{8} + \sqrt{2}$ (b) $\sqrt{27} - \sqrt{12}$ (c) $\sqrt{20} + 4\sqrt{5}$ (d) $\sqrt{200} - \sqrt{32}$

D12 (a) Write $\sqrt{75}$ in the form $a\sqrt{b}$, where b is a prime number.

(b) Hence simplify $\dfrac{\sqrt{75}}{5}$.

D13 (a) Write $\sqrt{108}$ in the form $a\sqrt{b}$, where b is a prime number.

(b) Hence simplify $\dfrac{\sqrt{108}}{2}$.

D14 Simplify these.

(a) $\dfrac{\sqrt{12}}{2}$ (b) $\dfrac{\sqrt{18}}{\sqrt{2}}$ (c) $\dfrac{5}{\sqrt{5}}$ (d) $\dfrac{\sqrt{72}}{3}$ (e) $\dfrac{\sqrt{200}}{5}$

E Brackets

After multiplying out brackets, collect like terms and simplify where possible.

Examples

$$\sqrt{2}(5 + \sqrt{18})$$
$$= (\sqrt{2} \times 5) + (\sqrt{2} \times \sqrt{18})$$
$$= 5\sqrt{2} + \sqrt{36}$$
$$= 5\sqrt{2} + 6$$

$$(\sqrt{3} + 4)(\sqrt{3} - 1)$$
$$= (\sqrt{3} \times \sqrt{3}) - (\sqrt{3} \times 1) + (4 \times \sqrt{3}) - (4 \times 1)$$
$$= 3 - \sqrt{3} + 4\sqrt{3} - 4$$
$$= 3\sqrt{3} - 1$$

E1 Multiply out the brackets and simplify where possible.

(a) $\sqrt{3}(4 + \sqrt{2})$ (b) $\sqrt{2}(\sqrt{5} + \sqrt{2})$ (c) $\sqrt{5}(\sqrt{20} + 1)$ (d) $\sqrt{2}(\sqrt{8} + \sqrt{32})$

E2 Multiply out the brackets and simplify where possible.

(a) $(3 + \sqrt{2})(2 + \sqrt{2})$ (b) $(\sqrt{5} - 4)(1 + \sqrt{5})$ (c) $(5 + \sqrt{3})(5 - \sqrt{3})$

(d) $(1 - \sqrt{7})(5 - \sqrt{7})$ (e) $(2 + \sqrt{5})^2$ (f) $(1 - \sqrt{3})^2$

E3 What is the area of a square of side $(1 + \sqrt{2})$ units?
Give your answer in the form $a + b\sqrt{2}$.   *AQA 2003 Specimen*

E4 (a) Is $(3 + \sqrt{2})$ a rational or an irrational number?

(b) (i) Simplify $(3 + \sqrt{2})^2$.

    (ii) Is your answer to part (b)(i) a rational or an irrational number?   *AQA(NEAB) 1998*

E5 Multiply out the brackets and simplify where possible.

(a) $(1 + \sqrt{2})(2 + \sqrt{3})$ (b) $(6 - \sqrt{3})(1 + \sqrt{5})$ (c) $(\sqrt{5} + \sqrt{3})(\sqrt{5} - \sqrt{3})$

(d) $(\sqrt{2} + \sqrt{18})^2$ (e) $(\sqrt{5} - \sqrt{3})^2$ (f) $(\sqrt{50} - \sqrt{2})^2$

E6 k is an integer.

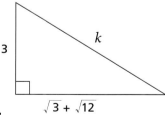

Find the value of k.   *Edexcel*

E7 (a) Expand and simplify $(1 + \sqrt{5})^2$.

(b) Hence, or otherwise, show that $x = 1 + \sqrt{5}$ is a solution to
the equation $x^2 - 2x - 4 = 0$.

E8 (a) Write $\sqrt{48}$ in the form $a\sqrt{b}$, where b is prime.

(b) Write $(\sqrt{24} - \sqrt{2})^2$ in the form $p - q\sqrt{3}$.

E9 Write $(\sqrt{5} + \sqrt{10})^2$ in the form $p + q\sqrt{2}$.

E10 Write $(\sqrt{15} + \sqrt{3})^2$ in the form $p + q\sqrt{r}$, where r is as small as possible.

F Further simplifying

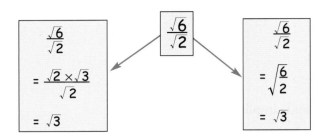

F1 Simplify (a) $\dfrac{\sqrt{12}}{\sqrt{2}}$ (b) $\dfrac{\sqrt{8}}{\sqrt{2}}$ (c) $\dfrac{\sqrt{15}}{\sqrt{3}}$

F2 Evaluate (a) $\sqrt{\dfrac{9}{25}}$ (b) $\sqrt{\dfrac{81}{144}}$ (c) $\sqrt{\dfrac{16}{121}}$

F3 Write $\sqrt{\dfrac{3}{16}}$ in the form $\dfrac{\sqrt{a}}{b}$, where a and b are integers.

F4 Write $\sqrt{\dfrac{25}{2}}$ in the form $\dfrac{a}{\sqrt{b}}$, where a and b are integers.

Sometimes it is useful to write an expression with an integer denominator.
This is called **rationalising the denominator**.

Example $\dfrac{\sqrt{7}}{\sqrt{2}} = \dfrac{\sqrt{7} \times \sqrt{2}}{\sqrt{2} \times \sqrt{2}}$

Multiplying both the numerator and denominator by √2 does not change the value of the expression.

$= \dfrac{\sqrt{14}}{2}$

F5 For each expression, rationalise the denominator and write the result in its simplest form.

(a) $\dfrac{7}{\sqrt{5}}$ (b) $\dfrac{1}{\sqrt{3}}$ (c) $\dfrac{6}{\sqrt{2}}$ (d) $\dfrac{7}{\sqrt{7}}$ (e) $\dfrac{\sqrt{5}}{\sqrt{2}}$ (f) $\dfrac{\sqrt{10}}{\sqrt{3}}$

F6 Write $\dfrac{9}{\sqrt{3}}$ in the form $p\sqrt{q}$.

F7 Show that $\dfrac{1}{\sqrt{2}} + \dfrac{\sqrt{2}}{6} = \dfrac{2\sqrt{2}}{3}$

F8 Simplify $\dfrac{\sqrt{18} \times \sqrt{75}}{\sqrt{50}}$.

Simplifying roots

For all values of a and b it is true that $\sqrt{a}\sqrt{b} = \sqrt{ab}$.

- $\sqrt{2} \times \sqrt{18} = \sqrt{36} = 6$

- $\sqrt{75} = \sqrt{25} \times \sqrt{3} = 5\sqrt{3}$

For all values of a and b it is true that $\dfrac{\sqrt{a}}{\sqrt{b}} = \sqrt{\dfrac{a}{b}}$.

- $\dfrac{\sqrt{18}}{\sqrt{2}} = \sqrt{\dfrac{18}{2}} = \sqrt{9} = 3$

- $\sqrt{\dfrac{2}{25}} = \dfrac{\sqrt{2}}{\sqrt{25}} = \dfrac{\sqrt{2}}{5}$

G Quadratic equations

Example

Solve the equation $x^2 + 10x + 7 = 0$, giving the solution in the form $x = p \pm q\sqrt{r}$, where r is a positive integer and as small as possible.

First method

'Complete the square' by writing
$x^2 + 10x$ as $(x^2 + 10x + 5^2) - 5^2$,
which is $(x + 5)^2 - 25$.

So $\quad (x + 5)^2 - 25 + 7 = 0$
$$(x + 5)^2 = 18$$
$$x + 5 = \pm\sqrt{18}$$
$$= \pm\sqrt{9 \times 2}$$
$$= \pm 3\sqrt{2}$$
So $\qquad\qquad x = -5 \pm 3\sqrt{2}$

Second method

Using the quadratic formula with $a = 1$, $b = 10$ and $c = 7$,
we get $x = \dfrac{-10 \pm \sqrt{100 - 28}}{2}$

so $\quad x = \dfrac{-10 \pm \sqrt{72}}{2}$
$$= \dfrac{-10 \pm \sqrt{36 \times 2}}{2}$$
$$= \dfrac{-10 \pm 6\sqrt{2}}{2}$$
$$= -5 \pm 3\sqrt{2}$$

G1 Solve the equation $x^2 - 8x + 3 = 0$, giving the solution in the form $x = p \pm \sqrt{q}$.

G2 Solve each of these equations, giving the solution in the form $x = p \pm \sqrt{q}$.

(a) $x^2 - 4x + 2 = 0$ (b) $x^2 - 6x + 3 = 0$ (c) $x^2 - 2x - 2 = 0$ (d) $x^2 + 8x + 3 = 0$

G3 (a) Write the expression $x^2 - 6x + 2$ in the form $(x - p)^2 - q$.

(b) Hence solve the equation $x^2 - 6x + 2 = 0$, giving the solution in the form $x = a \pm \sqrt{b}$.

G4 (a) Write the expression $x^2 + 12x + 4$ in the form $(x + p)^2 - q$.

(b) Hence solve the equation $x^2 + 12x + 4 = 0$, giving the solution in the form $x = a \pm b\sqrt{c}$, where c is as small as possible.

G5 Solve each of these equations by 'completing the square', giving the solution in surd form.

(a) $x^2 + 4x + 1 = 0$ (b) $x^2 + 6x + 4 = 0$ (c) $x^2 - 10x + 2 = 0$ (d) $x^2 - 3x - 1 = 0$

G6 Use the quadratic formula to solve the equation $2x^2 - 8x + 3 = 0$, giving the solution in the form $x = a \pm b\sqrt{c}$.

G7 Use the quadratic formula to solve each of these equations, giving the solution in the form $x = a \pm b\sqrt{c}$.

(a) $2x^2 - 4x + 1 = 0$ (b) $2x^2 - 3x - 3 = 0$ (c) $3x^2 - 4x - 1 = 0$

*G8 (a) Write the expression $2x^2 + 8x + 5$ in the form $p(x + q)^2 - r$.

(b) Hence show that the solution of the equation $2x^2 + 8x + 5 = 0$ can be written in the form $x = -2 \pm \frac{1}{2}\sqrt{6}$

H *Mixed questions*

H1 A square has area 24 cm^2. What is the exact length of one edge in the form $p\sqrt{q}$, where p and q are integers?

H2 The shaded areas in each diagram are squares.
Find the area and perimeter of each white rectangle in surd form, simplifying your answers as far as you can.   (a)   (b)

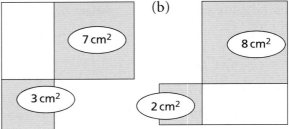

H3 Express $0.\overset{..}{4}\overset{.}{8}$ as a fraction in its simplest terms.   AQA 2000

H4 Solve the equation $x^2 - 4x + 1 = 0$, giving your solutions in the form $a \pm \sqrt{b}$.

H5 Write the perimeter of this triangle in the form $a + b\sqrt{c}$, where the integer c is as small as possible.

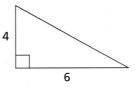

H6 The table shows three pairs of expressions.

For each expression in column A, state whether or not it is exactly equal to the corresponding expression in column B, justifying your answer.

	A	B
(a)	$\sqrt{2} \times \sqrt{6}$	$2\sqrt{3}$
(b)	$\sqrt{2^2 + 1^2}$	$2.236\,067\,977\,49$
(c)	$\sqrt{\dfrac{144}{121}}$	$1.\overset{.}{0}\overset{.}{9}$

OCR(MEG)

H7 $p = 2 + \sqrt{3}$ $q = 2 - \sqrt{3}$

    (a) (i) Work out $p - q$.

       (ii) State whether $p - q$ is rational or irrational.

    (b) (i) Work out pq.

       (ii) State whether pq is rational or irrational.   Edexcel

H8 Write the recurring decimal $0.21\dot{3}$ as a fraction in its lowest terms.   OCR

Test yourself

T1 (a) Express $0.\dot{4}\dot{5}$ as a fraction. Simplify your answer.

    (b) (i) Simplify $(\sqrt{7} + \sqrt{5})^2$.

       (ii) State whether your answer to (i) is rational or irrational.   OCR

T2 Write $\dfrac{\sqrt{3}}{\sqrt{2}}$ in the form $\dfrac{\sqrt{a}}{b}$, where a and b are whole numbers.   AQA 2003 Specimen

T3 Write $0.0\dot{6}\dot{3}$ as a fraction in its simplest form.   AQA 2003 Specimen

T4 (a) Write $\sqrt{45}$ in the form $a\sqrt{b}$, where a and b are prime numbers.

    (b) Find the value of $(\sqrt{45} - \sqrt{20})^2$.   AQA(SEG) 2000

T5 (a) Express the following in the form $p\sqrt{q}$, where p and q are integers.

$$\frac{4}{\sqrt{2}}$$

    (b) Simplify the following. Give your answer in the form $a + b\sqrt{2}$, where a and b are integers.

$$(1 + \sqrt{2})(3 - \sqrt{2})$$ OCR

T6 (a) Write the number $0.3\dot{2}\dot{5}$ as a fraction.

    (b) Write down an irrational number between 2 and 3.   AQA(SEG) 2000

T7 (a) Write $\sqrt{6} \times \sqrt{3}$ in the form $a\sqrt{b}$, where a and b are prime numbers.

    (b) Simplify fully $5\sqrt{2} + \sqrt{8}$.   AQA(SEG) 2000

T8 Show clearly that $\dfrac{3}{\sqrt{6}} + \dfrac{\sqrt{6}}{3} = \dfrac{5}{\sqrt{6}}$.   OCR

46 *Further trigonometry*

You will revise

◆ using sine, cosine and tangent in right-angled triangles

◆ using Pythagoras' rule

This work will help you

◆ find lengths and angles in any triangle

◆ find the area of any triangle

A *Trigonometry and Pythagoras – revision*

A1 Find the missing lengths (to the nearest 0.1 cm).

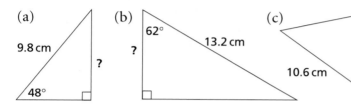

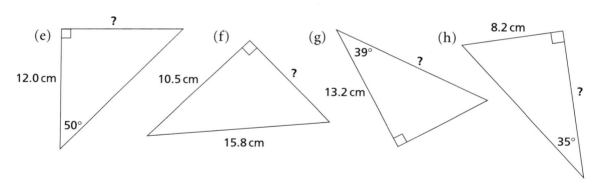

A2 Find the missing angles (to the nearest degree).

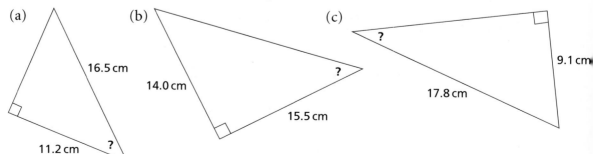

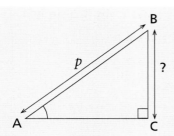

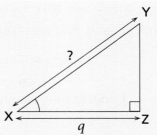

Length BC can be written as an **expression**:

$p \sin A$ (p times the sine of the angle at A)

Length XY can be written as an expression:

$\dfrac{q}{\cos X}$ (q divided by the cosine of the angle at X)

A3 Write an expression for each of the lengths marked x here.

(a)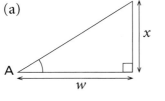

(b)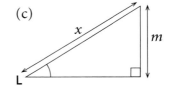

(c)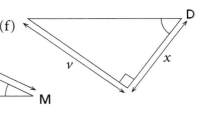

(d)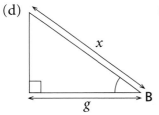

(e)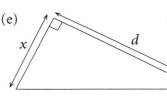

(f)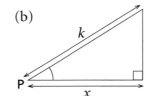

B *The sine rule*

B1 In this triangle, a perpendicular has been drawn from C on to AB. Let the length of the perpendicular be h.

(a) Use $\sin 40°$ to find the value of h to 3 d.p.

(b) Use $\sin 65°$ and your answer to (a) to find BC to 1 d.p.

(c) Check by drawing the triangle accurately from the values given in the diagram, and measuring BC. (Work out angle ACB before you start to draw).

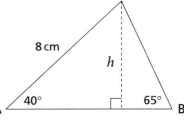

B2 In this triangle, a is the length of the side opposite angle A and b is the length of the side opposite angle B.

(a) Write an expression for h in terms of length b and $\sin A$ (the sine of the angle at A).

(b) Write an expression for h in terms of length a and $\sin B$.

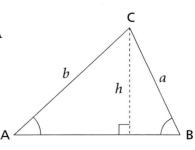

In question B2, you should have found these two expressions for h: $a \sin B$, $b \sin A$

Since they both equal h, they must equal each other: $a \sin B = b \sin A$

Dividing both sides by sin B, we get $a = \dfrac{b \sin A}{\sin B}$

Dividing both sides by sin A, we get $\dfrac{a}{\sin A} = \dfrac{b}{\sin B}$

By drawing a perpendicular from A to side BC, you can prove that $\dfrac{b}{\sin B} = \dfrac{c}{\sin C}$

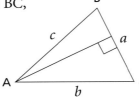

The formula $\dfrac{a}{\sin A} = \dfrac{b}{\sin B} = \dfrac{c}{\sin C}$ is known as the **sine rule**.

Notice how the sides and vertices of a triangle are lettered: side a is opposite vertex A, and so on.

You can use the sine rule to solve some triangle problems.

Example

Find the missing length.

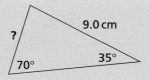

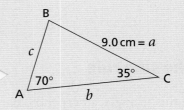

Copy and label the diagram with a opposite A, and so on.

Write the part of the sine rule that you need.

$$\dfrac{a}{\sin A} = \dfrac{c}{\sin C}$$

Substitute the angles and side.

$$\dfrac{9.0}{\sin 70°} = \dfrac{c}{\sin 35°}$$

Make the missing length the subject.

$$c = \dfrac{9.0 \sin 35°}{\sin 70°}$$

Find the answer on a calculator and give it to a suitable degree of accuracy.

$$c = 5.5 \text{ cm (to 1 d.p.)}$$

If letters other than A, B, C are given as vertices of the triangle, you can adapt the sine rule.

For example, $\dfrac{p}{\sin P} = \dfrac{q}{\sin Q} = \dfrac{r}{\sin R}$

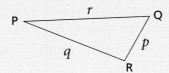

where p is the length of side QR and so on.

B3 Find each missing length to 1 d.p. Check each answer makes sense for the values given in the question.

(a)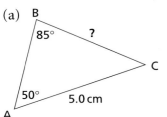

(b)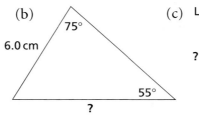

(c)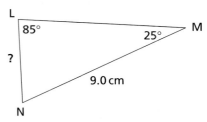

B4

(a) Calculate this angle.

(b) Use the sine rule to find the length of this side.

(c) Use the sine rule to find the length of this side.

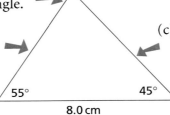

B5 Find the lengths marked with letters.

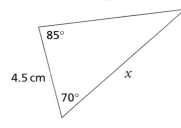

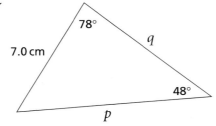

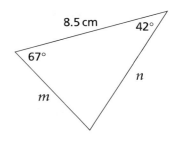

The proof of the sine rule given earlier assumed that all the angles of the triangle were acute.

It is also possible to prove the sine rule when one of the angles is obtuse.

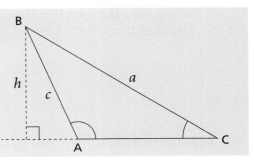

B6 Find the missing lengths here.

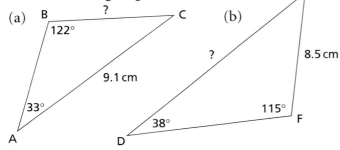

(a)

(b)

(c)

B7 The line MN represents the vertical mast of a radio transmitter.
Points P, Q and N are on level ground.
From P, the angle of elevation of the top of the mast is 28°.
The angle of elevation from Q is 35°.
The distance between P and Q is 50 m.

Find   (a) QM   (b) the height of the mast, MN

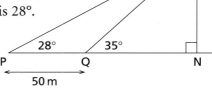

C Finding angles

When you try to find angles with the sine rule,
you need to use sin⁻¹ on your calculator.

But remember there are usually **two** angles
that have a particular sine.

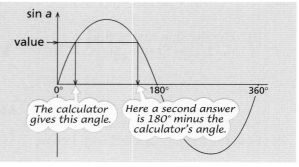

Example

Find the angle at A.

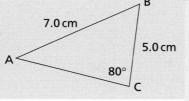

$$\frac{a}{\sin A} = \frac{c}{\sin C}$$

$$\frac{5.0}{\sin A} = \frac{7.0}{\sin 80°}$$

$$\sin A = \frac{5.0 \sin 80°}{7.0}$$

Remember there are **two** angles
between 0° and 180° with this sine.

$\sin A = 0.7034...$ (calculator result)

So A = 45° or 135° to the nearest degree

But if A was 135°, the sum of the angles of
the triangle would be greater than 180°,
because 80 + 135 = 215.

So A = 45°

C1 For each of these...

- Find the sine of
  the missing angle.
- Give two angles
  that have that sine.
- Explain which angle
  is possible in this case.

(a)   (b)

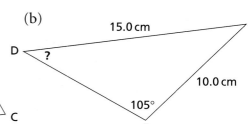

C2 (a) In this triangle…
- Find the sine of the missing angle.
- Give two angles that have that sine.
- What happens when you check to see which angle is possible?

(b) Try drawing the triangle from the information given in the diagram.
Use this to explain what you found in part (a).

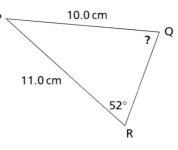

C3 (a) Find the sine of the missing angle here.
What angles have this sine?

(b) Try drawing the triangle from the information given in the diagram.
Use this to explain what you found in part (a).

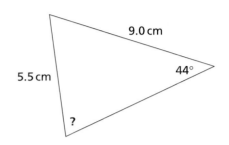

D *The cosine rule*

D1 In this triangle, a perpendicular has been drawn from B on to AC, which it meets at point P.
Let length BP be h.
Let length AP be x.

(a) Use a cosine to work out x to 3 d.p.

(b) Use Pythagoras to find h to 3 d.p.

(c) Use your answer to (a) to find length PC to 3 d.p.

(d) Use Pythagoras with your answers to (b) and (c) to find BC to 1 d.p.

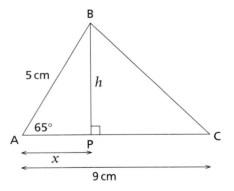

D2 In this triangle, lengths are given as variables.

(a) Write length PC in terms of b and x.

(b) Use Pythagoras and your answer to (a) to express h^2 in terms of a, b and x.

(c) Use Pythagoras to express h^2 in terms of c and x.

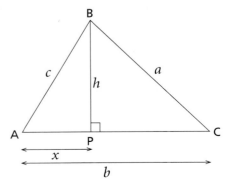

In question D2 you should have found these two expressions for h^2:

$$a^2 - (b - x)^2, \quad c^2 - x^2$$

Since they both equal h^2, they must equal each other.

$$a^2 - (b - x)^2 = c^2 - x^2$$

D3 Make a^2 the subject of the formula.
Expand the brackets and simplify the right-hand side.

D4 Look back at the diagram for D2.
Write an expression for x in terms of c and angle A.

If you replace x in your answer to D3 by the expression you wrote for D4, you should get

$$a^2 = b^2 + c^2 - 2bc \cos A$$

This formula is known as the **cosine rule**.

D5 Substitute the values from question D1 into this formula.
From a^2, work out a.
Does this give you the same answer as in D1(d)?

The cosine rule refers to all three sides of the triangle but only one angle.
So it is useful to have two other versions of it when solving triangle problems:

$$b^2 = a^2 + c^2 - 2ac \cos B$$
$$c^2 = a^2 + b^2 - 2ab \cos C$$

The angle used in the proof above is acute, but the cosine rule also works when the angle it refers to is obtuse.

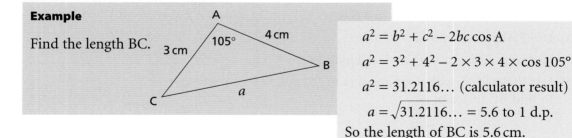

Example

Find the length BC.

$$a^2 = b^2 + c^2 - 2bc \cos A$$
$$a^2 = 3^2 + 4^2 - 2 \times 3 \times 4 \times \cos 105°$$
$$a^2 = 31.2116\ldots \text{ (calculator result)}$$
$$a = \sqrt{31.2116\ldots} = 5.6 \text{ to 1 d.p.}$$
So the length of BC is 5.6 cm.

D6 Work out the missing length in each of these.

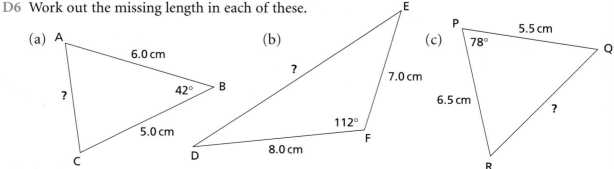

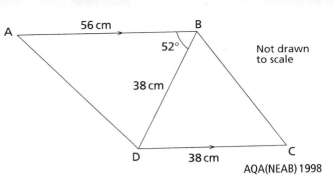

D7 The diagram shows a bicycle frame.
AB is parallel to DC.
DC = DB =38 cm. AB =56 cm.
Angle ABD = 52°.

(a) Calculate length BC.

(b) Calculate length AD.

AQA(NEAB) 1998

D8 P, Q and R are three villages.

Q is 14.3 km due south of P.
R is 18.5 km from P.
The bearing of R from P is 068°.

Calculate the distance from Q to R.

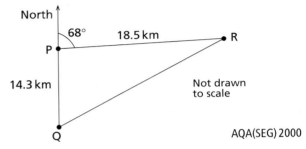

Not drawn to scale

AQA(SEG) 2000

Example

Find the angle at B.

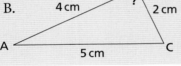

$$b^2 = a^2 + c^2 - 2ac \cos B$$
$$5^2 = 2^2 + 4^2 - 2 \times 4 \times 2 \times \cos B$$
$$16 \cos B = 2^2 + 4^2 - 5^2$$
$$16 \cos B = {}^-5$$

The negative cosine here means the angle is obtuse.

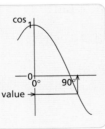

$$\cos B = \frac{{}^-5}{16} = {}^-0.3125$$
$$B = 108° \text{ to the nearest degree}$$

D9 Use the cosine rule to find the missing angles, to the nearest degree.

(a)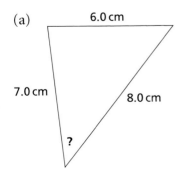

(b)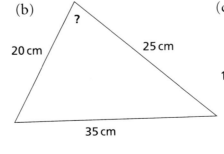

(c)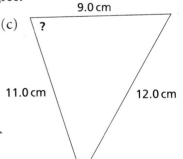

E *The sine formula for the area of a triangle*

E1 Calculate the area of this triangle.

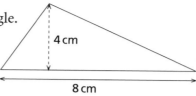

4 cm

8 cm

E2 (a) Use sin 70° to find h to 3 d.p.

(b) Calculate the area of triangle ABC to the nearest cm².

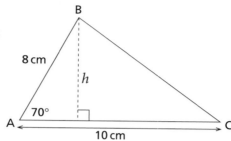

B

8 cm

h

70°

A

10 cm

C

E3 In this triangle, lengths are given as variables.

(a) Write an expression for h in terms of c and angle A.

(b) Write an expression for the area of triangle ABC.

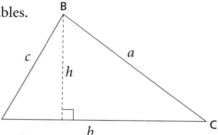

B

c

a

h

A

b

C

Using the standard lettering, the area of any triangle can be given by the formula $\frac{1}{2}ac \sin B$.

This formula can also be written as $\frac{1}{2}ab \sin C$ and $\frac{1}{2}bc \sin A$.

E4 Find the area of each triangle to the nearest cm².

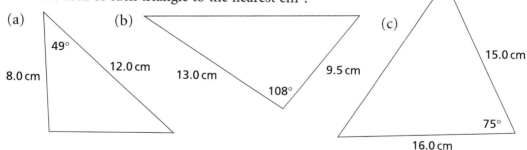

(a)

49°

8.0 cm

12.0 cm

(b)

13.0 cm

108°

9.5 cm

(c)

15.0 cm

75°

16.0 cm

E5 In the triangle XYZ, side XY has length 5 cm, and side YZ has length 6 cm.
Angle XYZ = θ.

(a) Find the area, A, in terms of $\sin \theta$.

(b) What is the maximum value of A as θ takes different values?

(c) Given that $A = 10\,\text{cm}^2$, find the two possible values of θ.

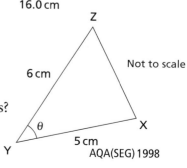

Z

Not to scale

6 cm

θ

Y

5 cm

X

AQA(SEG) 1998

F *Mixed questions*

F1 Each of these triangles represents a problem to be solved.
'G' shows where a length or angle is given.
'?' shows a length or angle that you have to find.

In each case, say whether you should use the sine rule or the cosine rule.
If two different triangles might be drawn from the information given, say so.

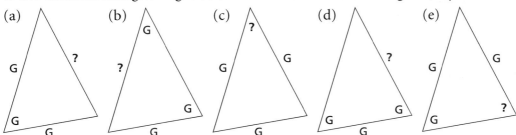

(a)   (b)   (c)   (d)   (e)

F2 (a) Calculate the area of triangle ABD.

(b) Calculate the length of BD.

(c) Calculate the size of angle BCD.

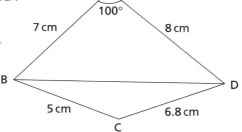

AQA(NEAB) 1997

F3 In triangle ABC the length of AB is 8.3 cm and angle ABC is 20°.
D is a point on BC such that the length of DC is 6.1 cm
and angle ADB is 105°.

Calculate the length of AC.

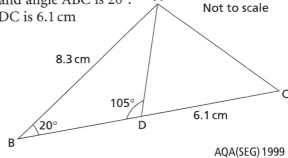

Not to scale

AQA(SEG) 1999

F4 In the triangle ABC, AB = 6 cm, BC = 5 cm and angle BAC = 45°.

There are two possible triangles ABC that can be
constructed with this information.

Calculate the two possible values of the angle BCA.

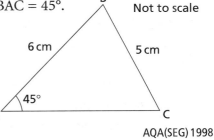

Not to scale

AQA(SEG) 1998

F5 A door wedge is in the shape of a triangular prism.

AB = FE = 8.2 cm,
BC = ED = 4.5 cm and
AF = BE = CD = 3.9 cm.
Angle ABC is 104°.

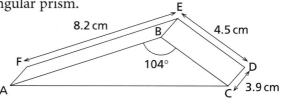

(a) Calculate the length AC.

(b) Find the volume of the door wedge.

AQA(SEG) 1998

F6 The area of this triangle is 32 cm².

(a) Find the obtuse angle at C.

(b) Find the length of AB.

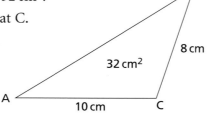

F7 A statue stands on a column.
In the diagram AB represents the statue and
BC represents the column.
Angle ACD = 90°. Angle BDA = 2.8°.
AD = 91.2 m and BD = 88.3 m.
ABC is a vertical straight line.

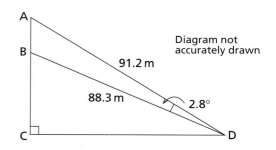

(a) Calculate the height, AB, of the statue.
Give your answer, in metres, correct to
three significant figures.

(b) Calculate the height, BC, of the column.
Give your answer, in metres, correct to three significant figures.

AQA(SEG) 2000

Test yourself

T1 The distance AC is 10.3 m. The distance CD is 16.0 m.
Angle BAC = 47°, angle ACD = 69° and
angle ABC is a right angle.

(a) Calculate the distance BC.

(b) Calculate the distance AD.

(c) Calculate the area of triangle ACD.

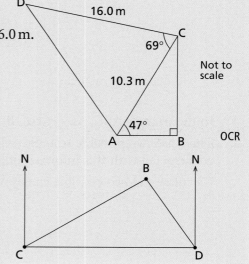

T2 Coastguard station C is 48 km due
west of coastguard station D.

Boat B is 39 km away from C,
on a bearing of 065°.

What is the boat's bearing and distance from D?

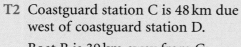

Three dimensions

You will revise how to use Pythagoras, and trigonometric functions and formulas.

This work will help you use

- ◆ Pythagoras and trigonometry to solve three-dimensional problems
- ◆ three-dimensional coordinates

A *Using a grid*

Cut out the grid on sheet G182.
Fold and stick it with the gridlines on the inside.

The grid consists of centimetre squares.

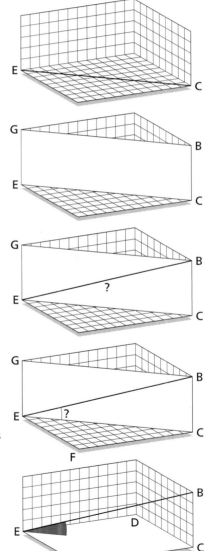

A1 (a) Use Pythagoras to calculate the length of
a straight line from E to C.

Using this result, draw and cut out the rectangle GBCE.
Label its vertices and check that it fits on your grid.
(We can refer to it as the **plane** GBCE.)

(b) Draw the line EB on your rectangle GBCE.
Use Pythagoras to find its length.
Check by measuring.
Keep a record of the calculated length to 1 d.p.

(c) Mark the angle BEC on your rectangle.
Use trigonometry to find this angle.
Check by measuring.

If rectangle EDCF represents horizontal ground, angle BEC is
the **angle of elevation** of B from the point E.

It is also the **angle between the line EB and the plane EDCF.**

A2 Use the same approach as in A1 to find the following, checking by measuring as you go.

    (a) The length and width of the plane ABHI

    (b) The length of line AH

    (c) The angle IAH

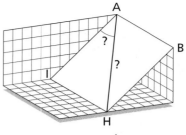

The angle you calculated in (c) is the angle between the line AH and the plane GADE.

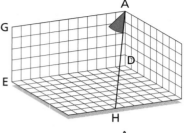

A3 (a) What rectangle (plane) would you use to find the angle GCA?

    (b) Instead of cutting out the rectangle, make a labelled sketch of it.
    Calculate angle GCA, showing your working.

    (c) Copy and complete:

       Angle GCA is the angle between line CG and the plane _____ .

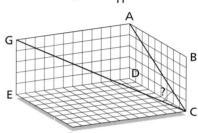

A4 What plane (rectangle) contains lines GH and HA?

A5 What plane contains lines JB and IB?

The following questions refer to your grid.
Try to do them without cutting out a plane.
But always make a labelled sketch of the plane you are using.

A6 Calculate

    (a) the length of HG

    (b) the angle of elevation of G from H

A7 Calculate the angle FHG.

A8 Calculate the angle between EB and the plane GADE.

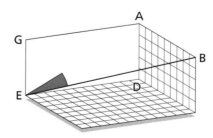

A9 Find (a) the length of HJ (b) the angle FJC (c) the area of triangle FJC

A10 (a) Use Pythagoras to find the lengths of these lines to 3 d.p.

(i) AC (ii) CI (iii) IA

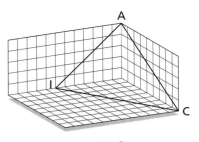

(b) Use the lengths you have found to construct the triangle ACI on plain paper, using compasses.
Write the correct letter inside each vertex.
Cut out the triangle and check that it fits on your grid.

You have met the cosine rule, $a^2 = b^2 + c^2 - 2bc \cos A$.

(c) Use your answers to (a) and the cosine rule to find these angles to the nearest degree.

(i) IAC (ii) ACI (iii) CIA

Check your answers by measuring the angles.

A11 Referring to your grid, sketch triangle GIF.

Work out (a) the lengths of its sides to 3 d.p. (b) its angles

In three-dimensional problems you may sometimes need the sine rule, $\dfrac{a}{\sin A} = \dfrac{b}{\sin B} = \dfrac{c}{\sin C}$, and the formula for the area of a triangle, $\frac{1}{2}ab \sin C$.

B *Without a grid*

Example

The diagram shows a cuboid.

Find the length CE, giving an exact value for your answer.

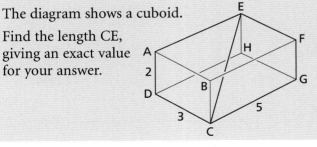

Considering the base CDHG and using Pythagoras,

$CH^2 = 5^2 + 3^2$
$\quad\ = 25 + 9$
$\quad\ = 34$

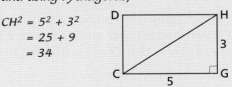

Considering the plane BEHC,

$CE^2 = CH^2 + 2^2$
$\quad\ = 34 + 2^2$
$\quad\ = 38$
$CE \ = \sqrt{38} \text{ units}$

B1 Find the exact value of the length CE above by first finding an expression for DE^2, then considering the plane DEFC.

B2 If, in the example on the previous page, AD was 4 units, DC was 7 units and CG was 10 units, what would be the exact value of CE?

B3 Let the length, breadth and height of the cuboid in the example be l, b and h. Use the approach shown to obtain an expression for the length CE.

B4 The diagram shows a cube with edge length 12 units.

P is $\frac{1}{3}$ of the way along EH.

Q is $\frac{2}{3}$ of the way along BC.

Find the length of PQ, giving an exact value for your answer.

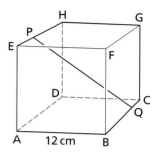

*B5 The diagonal of a cube, shown in the diagram, is 6 units long. What is the edge length of the cube, given as an exact value?

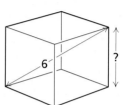

B6 The diagram shows a triangular prism. Calculate the following.

(a) Angle EAD to the nearest degree

(b) EF to 3 d.p.

(c) DF to 3 d.p.

(d) The volume of the prism to 1 d.p.

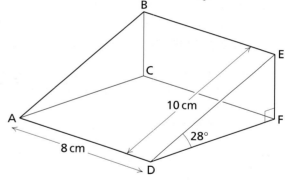

B7 The diagram shows a vertical radio mast, BT, which stands at the corner of a horizontal rectangular plot of ground, ABCD.

The mast is 12 m high.
The angle of elevation of the top of the mast from A is 25°.

(a) Calculate the length of AB.

The length of BC is 15 m.

(b) Calculate the angle of elevation of the top of the radio mast from D.

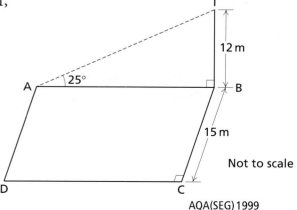

Not to scale

AQA(SEG) 1999

B8 By sketching and labelling a suitable plane in each case, calculate the marked angles, to the nearest degree.

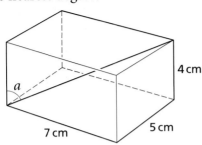

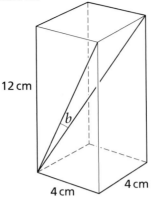

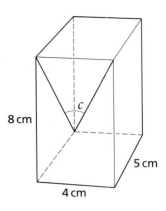

B9 Peter and Queenie are surveying on a horizontal plane PQT, using PQ as a base line.

TM is a vertical television mast of height 70 m.
Angle PTQ = 134°.
The angles of elevation of M from P and Q are 32° and 35° respectively.

(a) Show that the distance from P to Q is approximately 195 m.

(b) Calculate the size of angle TPQ.

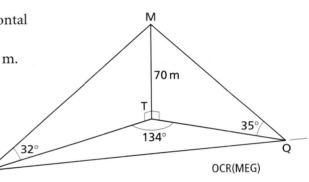

OCR(MEG)

B10 In these cuboids, sketch a copy of the diagram, mark the angle asked for, then calculate the angle.

(a)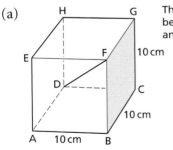

The angle between line FD and plane BFGC

(b)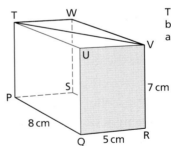

The angle between line TV and plane QRVU

(c)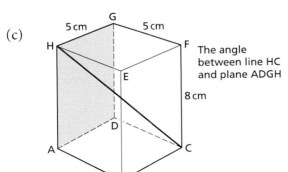

The angle between line HC and plane ADGH

(d)

The angle between line VX and plane UVZY

B11 This inn sign is supported by two wires, AB and AC.

(a) Calculate the length of each of the wires, showing your method.

(b) Calculate the angle that each wire makes with the wall.

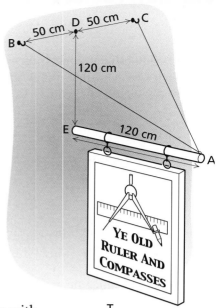

B12 The diagram shows a **right** square-based pyramid (one with its top vertex directly over the centre of the base).

(a) Find the length of PR, a diagonal of the base, to 3 d.p., and hence the length PC.

(b) Draw a sketch of triangle PTR and use it to calculate

   (i) the length of the slant edge PT to 1 d.p.

   (ii) the angle between PT and the base to the nearest degree

M is the midpoint of SR.

(c) Use Pythagoras to find TM, and hence find the area of triangle STR.

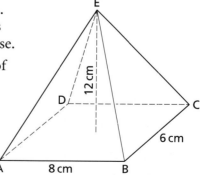

B13 The diagram shows a pyramid ABCDE. The base, ABCD, is a rectangle and E is vertically above the midpoint of the base.

AB = 8 cm, BC = 6 cm and the height of the pyramid is 12 cm.

Calculate the length of CE.

OCR

B14 The diagram represents a right pyramid.

The base is a square of side $2x$ cm.
The length of each of the slant edges is $8\sqrt{3}$ cm.
The height of the pyramid is x cm.

Calculate the value of x.

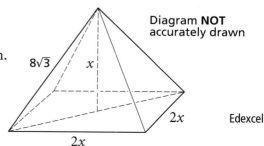

Diagram **NOT** accurately drawn

Edexcel

B15 ABCD is a regular tetrahedron of side 10 cm.

AO is the perpendicular height of the the tetrahedron.
DO meets BC at M, where BM = MC, and MO = $\frac{1}{3}$MD.

The volume of a tetrahedron is
$\frac{1}{3}$ base area × perpendicular height.

Find the volume of the tetrahedron ABCD.

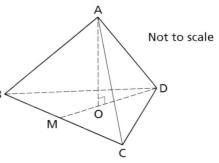

Not to scale

AQA(SEG) 1998

*B16 OABV is a triangular-based pyramid with
vertex V directly above vertex O.

AB = 7 cm, angle VAB = 50° and angle VBA = 64°.
The angle between line AV and plane OAB is 20°.

(a) Calculate the height, OV.

(b) Calculate the angle between BV and the plane OAB.

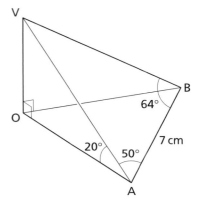

C *Coordinates in three dimensions*

A pair of axes (often labelled *x* and *y*) can give the position of
points on a two-dimensional plane.

To give the position of points in three-dimensional space,
a third axis is needed, called the *z*-axis.
So each point has three coordinates, written in the form (*x*, *y*, *z*).

Here, *x*-, *y*- and *z*-axes have been drawn.
A cuboid 3 units by 4 units by 2 units has been drawn with
the vertex O at the origin (0, 0, 0).

The coordinates of the vertex U are (3, 4, 2).

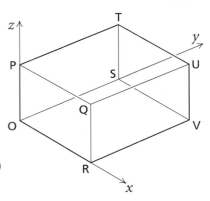

C1 (a) Give the coordinates of P, Q, R, S, T and V.

(b) Say whether each of these points lies on a face of this
cuboid, is inside the cuboid or is outside the cuboid.

(i) (2, 3, 2) (ii) (1, 2, 1) (iii) (2, 3, 3) (iv) (1, 4, 1)

C2 This model is made of cubes of equal size.
The coordinates of A are $(0, 6, 9)$.

Write the coordinates of B, C and D.

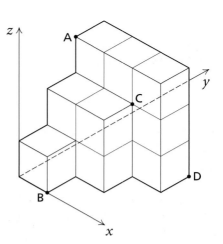

*C3 The three points A, B, C have the coordinates
shown in the diagram.

(a) Calculate the lengths AB, BC, CA, giving
your answers as exact values.

(b) Calculate the angle CAB.

(c) Find the area of the triangle ABC, to 1 d.p.

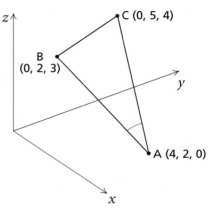

Test yourself

T1 The diagram shows a triangular prism with
the plane PQRS horizontal.
Calculate the following.

(a) The angle of elevation of U from S

(b) The length of PU

(c) The angle between PU
and the plane PQRS

*T2 The diagram shows a right
square-based pyramid.
Calculate its volume.

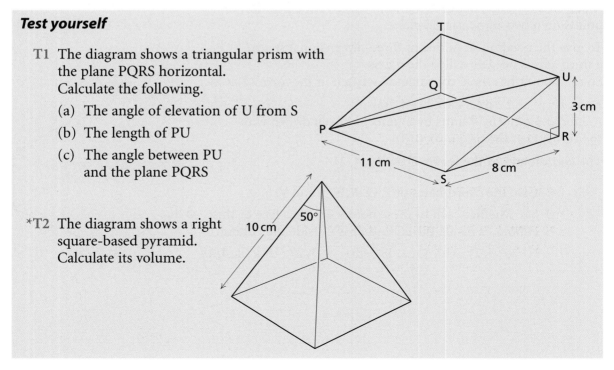

Review 12

☒ 1 (a) Simplify as far as possible the expression $3\sqrt{5} \times 2\sqrt{20}$.

   (b) Write $(\sqrt{3} + \sqrt{6})^2$ in the form $a + b\sqrt{c}$,
      where a, b and c are positive integers and c is as small as possible.

2 P, Q and R are three points on level ground.
Q is 14.5 km from P on a bearing of 027° from P.
R is 19.3 km from P on a bearing of 084° from P.

Calculate, to the nearest 0.1 km, the distance QR.

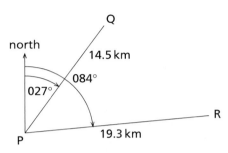

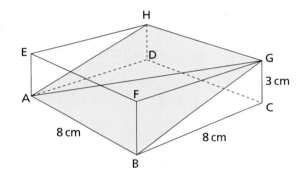

3 The diagram shows a cuboid ABCDEFGH
whose edges are of length 8 cm, 8 cm and 3 cm.

Calculate

   (a) the angle between the plane ABGH and
      the base ABCD

   (b) the length of the diagonal AG

   (c) the angle between AG and the base ABCD

4 (a) The graph of the equation $y = ax^2 + b$ goes through the points $(0, 4)$ and $(4, 6)$.
     Find the values of a and b.

   (b) The graph of the equation $y = \frac{c}{x} + d$ goes through the points $(3, 1\frac{1}{2})$ and $(8, \frac{1}{4})$.
     Find the values of c and d.

5 A and B are two points on level ground.
B is 5.8 km east of A.

The point C is on a bearing of 076° from A
and on a bearing of 033° from B.

   (a) Find the size of angle ACB.

   (b) Calculate the distance AC, to the nearest 0.1 km.

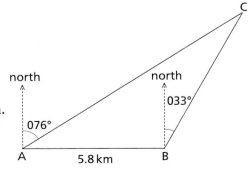

6 In this fairground game, a person spins the arrow twice. The person wins if the total score is 6 or less.

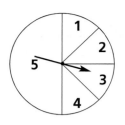

   (a) Calculate the probability of winning.

   (b) People pay 10p each to play. If they win, they get their 10p back and another 10p.

      If 3000 people play during the day, approximately how much profit would you expect the game to make?

7 The weight of a solid rubber ball is proportional to the cube of the diameter. A ball of diameter 7.8 cm weighs 0.96 kg.

   Calculate, to a reasonable degree of accuracy,

   (a) the weight of a ball of diameter 9.4 cm

   (b) the diameter of a ball that weighs 0.56 kg

8 The sides of a rectangle are of length a cm and b cm. The perimeter is P cm and the area is A cm^2. The length of the diagonal of the rectangle is d cm.

   (a) Show that $P^2 = 4d^2 + 8A$.

   (b) Rearrange this formula to make d the subject.

   (c) Use the new formula to find the diagonal of a rectangle whose area is 10 cm^2 and perimeter 16 cm.

9 Find the equation of the line that passes through the point $(1, 1)$ and is perpendicular to the line $2x + 3y = 8$.

10 (a) Find the exact value of $(1 - \frac{9}{25})^{-\frac{3}{2}}$.

   (b) Express $\dfrac{8\sqrt{5}}{\sqrt{2}}$ in the form $a\sqrt{b}$, where a and b are integers, and b is

      as small as possible.

11 Triangle OPQ is mapped on to triangle OQR by a rotation about O followed by an enlargement with centre O.

   (a) What is

      (i) the angle of the rotation

      (ii) the scale factor of the enlargement

   (b) If the same pair of transformations is applied to triangle OQR, what are the coordinates of the images of the points Q and R?

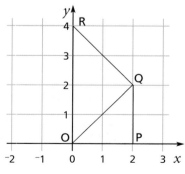

✗ 12 In this right-angled triangle, angle BAC is of size x where $\sin x = \frac{5}{13}$.

   Calculate

   (a) $\cos x$ (b) $\tan x$ (c) BC

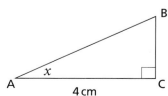
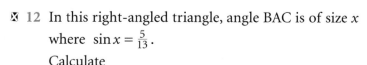

13 A company making electrical components tested a sample of 100 components to see how long they lasted before failing.

The 100 components were switched on at the same time. After 1 day, 2 days, and so on, the number that were still working was noted.

The results of the test are given in this table.

Time in days	0	1	2	3	4	5	6
Number working	100	96	84	68	40	14	0

(a) What percentage of the components that were still working after 4 days failed during the next day?

The length of time that a component lasts before failing is called its 'lifetime'.

(b) How many of the components had lifetimes, L days, in the interval $0 \leq L < 1$?

(c) Copy and complete this frequency table.

(d) Calculate an estimate of the mean lifetime of the components.

Lifetime, L days	Number of components
$0 \leq L < 1$	
$1 \leq L < 2$	
$2 \leq L < 3$	
$3 \leq L < 4$	
$4 \leq L < 5$	
$5 \leq L < 6$	

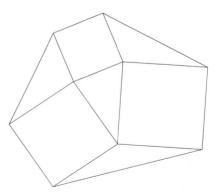

*14 The 15 points shown here are equally spaced round the circle.

(a) Calculate the angles of triangle ABC.

The radius of the circle is 10 cm.

(b) Calculate the area of triangle ABC.

*15 Draw any triangle.
Draw a square on each side of the triangle and join vertices to form three further triangles, as shown here.

There is a relationship between these three triangles and the original triangle.
What is the relationship, and why is it true?

(This result was first discovered by a schoolboy, David Cross, and is called 'Cross's theorem'.)

Challenge

This diagram shows three squares.

- Prove that $x + y = z$.

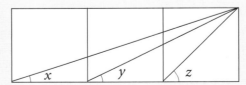

Index